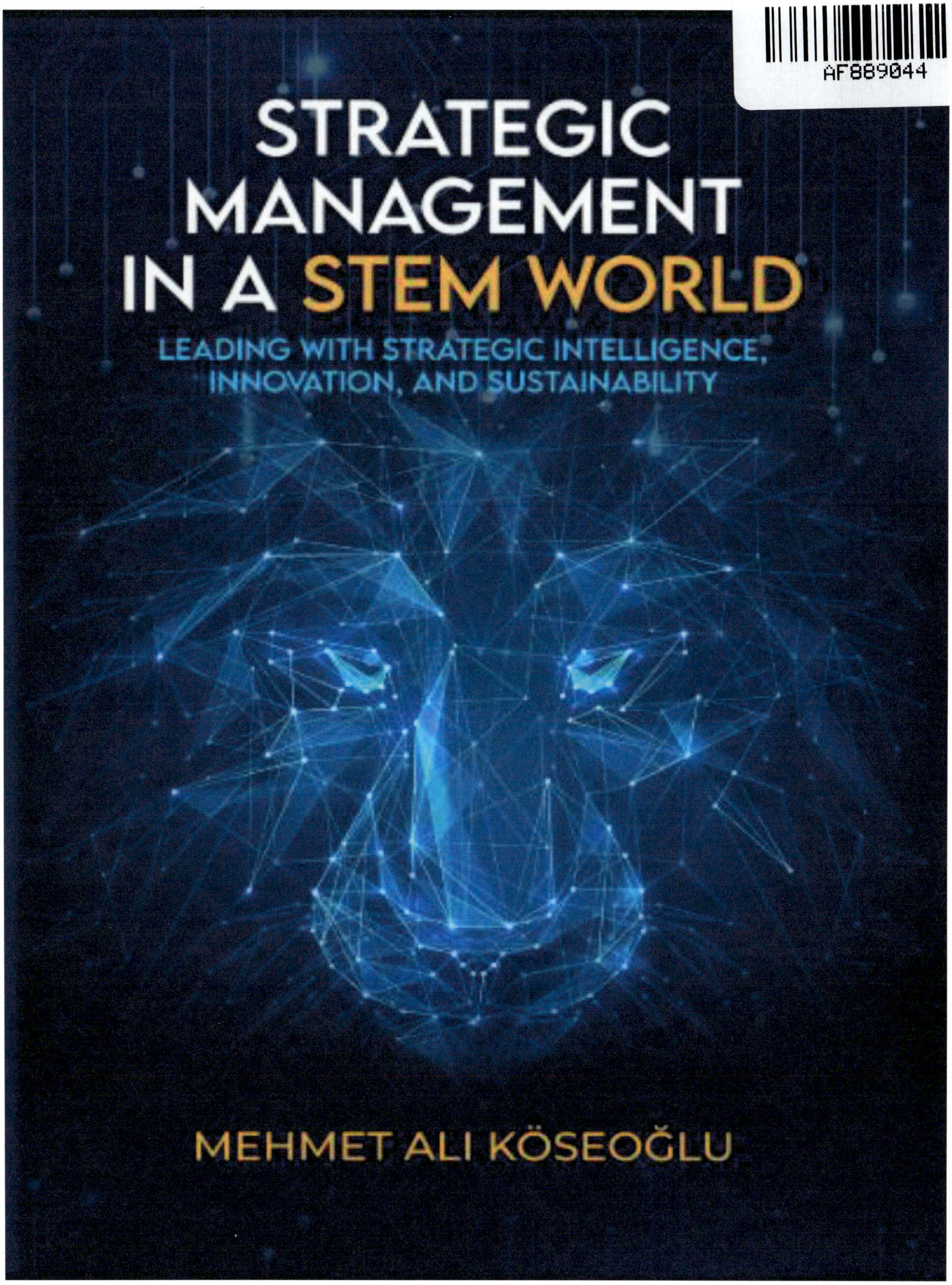
STRATEGIC
MANAGEMENT
IN A STEM WORLD
LEADING WITH STRATEGIC INTELLIGENCE,
INNOVATION, AND SUSTAINABILITY
MEHMET ALI KÖSEOĞLU

ISBN

ebook:	**978-1-966604-35-8**
Paperback:	**978-1-966604-36-5**
Hardcover:	**978-1-966604-37-2**

Preface

Strategy is changing.

No longer confined to static frameworks and paper-based planning, today's strategic management lives at the intersection of data, systems, and dynamic decision-making. The rise of STEM, Science, Technology, Engineering, and Mathematics, has transformed how organizations think, compete, innovate, and lead. This book was born from the urgent need to modernize how we teach and practice strategy in an era shaped by artificial intelligence, predictive analytics, simulation models, and sustainability imperatives.

Strategic Management in a STEM World is not just another textbook; it's a response to a new reality. It integrates timeless strategic principles with modern analytical tools, blending theory with real-world application, and qualitative judgment with quantitative rigor. Whether you are a student preparing for leadership, a manager navigating uncertainty, a scholar seeking new models, or an instructor designing future-ready courses, this book offers a blueprint for understanding and leading in the STEM-empowered economy.

You will still encounter foundational concepts, competitive advantage, industry forces, resource-based views, and strategic positioning, but you'll experience them through a new lens. Here, Porter's frameworks meet machine learning. Value chains become dynamic systems. Strategic thinking is augmented by SHAP values, partial dependence plots, Monte Carlo simulations, and digital twins. Innovation is not just an outcome; it's modeled, forecasted, and optimized.

Each chapter is designed to balance conceptual clarity with analytical depth. You'll find practical cases, modeling examples, and a toolkit of decision intelligence methods that make strategy both learnable and actionable. We bridge the classroom and the boardroom, the algorithm and the agenda, the leadership question and the quantitative answer.

This book is for those who believe that the future of strategy lies not in abandoning tradition, but in reengineering it. In this world shaped by data and complexity, strategic success belongs to those who can think critically, model intelligently, lead ethically, and act decisively.

Welcome to strategy, reimagined for a STEM world.

Mehmet Ali Koseoglu
August 22nd, 2025

DEDICATION

To YOU

INTRODUCTION

When I began my career in strategic management, the world was a different place. I remember sitting in conference rooms, surrounded by flip charts and spreadsheets, debating five-year plans as if the future could be neatly mapped. But as I worked with organizations across industries and continents, I saw the ground shifting beneath our feet. The rise of data, the speed of technological change, and the complexity of global systems demanded a new kind of strategist, one who could blend timeless principles with the tools of science, technology, engineering, and mathematics.

This book is the product of that journey. It is shaped by the lessons I learned from students who challenged my assumptions, from colleagues who pioneered new models, and from organizations that dared to reinvent themselves. My hope is that it will serve as both a guide and an inspiration for you, wherever you are on your strategic path.

Strategic management has long served as the intellectual backbone of organizational competitiveness. Rooted in time-tested concepts like competitive advantage, strategic positioning, and resource-based views, the discipline has historically enabled businesses to chart their course through stable, predictable markets. Decades ago, organizations focused on crafting five-year plans, maximizing operational efficiency, and outmaneuvering well-known rivals within clearly defined industries. Models such as Porter's Five Forces, SWOT, and VRIO became standard fixtures in classrooms and boardrooms alike.

However, the 21st century has introduced unprecedented levels of volatility, uncertainty, complexity, and ambiguity (VUCA). Markets have become hyper-competitive, business cycles have shortened, and industries are now disrupted by digital platforms, AI-powered startups, and global technological shifts. Today's organizations operate in an environment where competitive advantage is fleeting, customer preferences are fluid, and technological innovations can overturn industries in a matter of months.

This emerging complexity demands a new approach to strategy. It calls for leaders and decision-makers to augment traditional strategic frameworks with the predictive power, analytical rigor, and systemic thinking provided by Science, Technology, Engineering, and Mathematics (STEM). The integration of STEM into strategic management is no longer optional; it is an essential competency for modern leaders to navigate data-saturated environments and technology-driven disruption.

What does it mean to practice strategy in a STEM world? It means that the boundaries between business and technology, between leadership and analytics, between vision and simulation, are dissolving. STEM is no longer a set of technical disciplines at the periphery of management. It is the new core.

A vivid example: Ping An's Strategic Reprogramming

Imagine a company in the financial sector, not in Silicon Valley or New York, but in Shenzhen, China. For decades, Ping An was a traditional insurance firm, following the classic playbook of market positioning and operational efficiency. But by 2014, its leaders recognized that the old rules were being rewritten. Operational efficiency was no longer a competitive advantage—it was table stakes. New digital entrants, cross-sector competition, and rising customer expectations signaled the end of the static strategy era.

Ping An's response was not to abandon traditional strategy, but to upgrade it. The company embedded AI, big data, and platform engineering into its strategic core. Insurance became one product in a digital ecosystem that included healthcare, banking, telemedicine, facial recognition, and cloud infrastructure. They didn't just improve operations, they reimagined what an insurance company could be.

What's remarkable is how STEM principles fueled their new strategy: machine learning for risk pricing and customer forecasting, digital twins to simulate systemic risks, and automation to transform the value chain. Strategy became a dynamic model, constantly updated through sensor data, market feedback, and scenario simulations. Ping An didn't abandon Porter; they upgraded him. They didn't discard trade-offs; they made smarter ones, faster. They didn't ignore the Five Forces—they built algorithms to detect when forces were morphing, even before competitors noticed.

This is the kind of transformation that defines strategy in a STEM world, and it is the kind of thinking this book will help you master.

Strategic Management in a STEM World responds to this evolving strategic landscape by presenting a comprehensive framework that merges classical strategy with modern STEM methodologies. The purpose of this book is not to discard the wisdom of traditional strategy but to elevate and evolve it. It equips readers with a balanced understanding of strategic theory and modern analytical tools, enabling them to make decisions that are both insight-driven and action-oriented.

The rationale behind this book is clear: while timeless strategic principles remain relevant, their application must adapt to the realities of AI, big data, systems modeling, and sustainability imperatives. This book provides a structured guide for students, scholars, managers, and executives to integrate these advanced capabilities into their strategic thinking and practice.

The book serves multiple audiences:

- Students preparing for careers in leadership and management;
- Academics seeking to incorporate cutting-edge methodologies into their teaching;
- Practitioners and managers striving for better decision-making in their organizations;
- Executives and strategists navigating complex business environments;
- Policymakers and consultants require a holistic understanding of modern strategy in practice.

The book follows a logical, progressive structure that reflects how strategic competence develops in real-world settings. It begins with foundational concepts to ensure readers understand the timeless underpinnings of strategy before transitioning into advanced, STEM-powered tools and frameworks.

The pedagogical logic unfolds as follows:

- Start with fundamentals, exploring strategy's history and limitations;
- Advance to systems thinking and decision intelligence, introducing data-driven strategy;
- Transition to applied domains such as corporate strategy, innovation, business models, financial strategy, and international strategy;
- Progress into advanced strategic challenges like competitive dynamics, ecosystem participation, and sustainability;
- Conclude with execution, leadership, and future strategic frontiers to ensure the knowledge translates into action and future readiness.

Each chapter maintains a consistent internal structure to maximize learning impact:

- Theoretical explanation of core concepts.
- Practical micro-cases (e.g., Zara, Ping An);
- Wrap-up summaries distilling key lessons;
- Interactive exercises that encourage application.
- Recommended HBR case studies for extended learning.
- Curated strategic films to enhance engagement and visual understanding.

Chapter 1: STEM and the Strategic Paradigm Shift

Establishes the foundation by introducing the pressing need for a paradigm shift in strategic thinking. It explains how traditional strategic management, built on long-term planning and static competitive analysis, is no longer sufficient in a rapidly changing world. The chapter explores the limitations of conventional tools and introduces the reader to the idea of strategic intelligence as a new, essential capability. It frames STEM not as an alternative to strategic thinking, but as a necessary augmentation that enables more accurate, adaptive, and forward-looking decisions.

Chapter 2: Systems Thinking for Strategic Leaders

Focuses on systems thinking, explaining why traditional linear thinking falls short in today's complex, interconnected business environments. Readers are introduced to feedback loops, dynamic interactions, and holistic perspectives that enable managers to understand the bigger picture rather than isolated symptoms. The chapter equips readers to view organizations, industries, and markets as adaptive systems, introducing modeling techniques such as causal loop diagrams and dynamic simulations.

Chapter 3: Strategic Analytics and Decision Intelligence

Introduces the critical analytical capabilities necessary for modern strategic leadership. It explains the evolution from descriptive to predictive and prescriptive analytics, emphasizing why contemporary strategists must understand data flows, statistical models, and AI-supported decision-making processes. The chapter also addresses the challenges and responsibilities that come with data-driven strategy, including algorithmic transparency, bias mitigation, and ethical analytics.

Chapter 4: Innovation as a STEM Process

Rethinks innovation as a structured, systematic, and repeatable process powered by STEM principles. Readers are introduced to frameworks and techniques such as innovation roadmapping, predictive modeling, generative design, and technology foresight. The chapter demonstrates how organizations can use data to anticipate trends, simulate product lifecycles, and strategically invest in R&D.

Chapter 5: Business Models and Digital Value Architecture

Building on the foundations of analytics and innovation, this chapter explores how business models themselves are being transformed by digital technologies, data, and platform thinking. Readers learn how to map, simulate, and redesign business models using tools like the Business Model Canvas, digital overlays, and scenario analysis. The chapter emphasizes the shift from linear value chains to adaptive, ecosystem-based architectures, preparing strategists to thrive in platform economies and digital-first markets.

Chapter 6: Corporate Strategy in a STEM World

This chapter explores how organizations make corporate-level decisions about growth, diversification, vertical integration, and portfolio management using AI-powered forecasting, digital twins, and dynamic simulation models. It equips readers to design agile, data-driven corporate strategies that orchestrate complex multi-business environments, adapt to market disruptions, and maintain systemic coherence. Through case examples, modeling exercises, and practical STEM tools, readers will learn how to navigate corporate expansion in a hyper-dynamic world.

Chapter 7: Financial Strategy Analytics

With business models in place, the focus turns to the financial dimension of strategy. This chapter introduces advanced financial tools such as real-options thinking, Monte Carlo simulations, and predictive credit analytics. Readers discover how to model uncertainty, allocate capital ethically, and design adaptive budgeting systems that respond to volatility. The chapter ensures that financial strategy is not just about control, but about enabling innovation, resilience, and long-term value creation in a STEM-driven world.

Chapter 8: International Strategy through STEM

As organizations expand globally, strategy must account for cultural, regulatory, and technological complexity. This chapter demonstrates how predictive analytics, machine learning, and system modeling can inform market selection, entry modes, and risk management. Readers learn to integrate geopolitical, climate, and ESG data into international strategy and to use tools like Hofstede's cultural dimensions and the GLOBE study for cross-cultural leadership and adaptation.

Chapter 9: Competitive Advantage in a Data-Driven Economy

Competitive advantage is no longer static; it is continuously constructed and eroded in real time. This chapter explores how data, AI, and dynamic capabilities are redefining the sources of strategic differentiation. Readers are introduced to advanced modeling tools—agent-based simulation, digital twins, and system dynamics—to anticipate and respond to evolving competitive landscapes. The chapter emphasizes the importance of strategic renewal loops and the need to embed intelligence, speed, and adaptability into the DNA of the organization.

Chapter 10: Entrepreneurial Ecosystems through the STEM Lens

Strategy today is not just about individual firms, but about the ecosystems they inhabit and help shape. This chapter teaches readers how to map, model, and participate in entrepreneurial ecosystems using systems thinking, agent-based modeling, and network analysis. It covers the design of ecosystem-based strategies, the orchestration of innovation networks, and the measurement of ecosystem vitality and fit. The chapter prepares leaders to move from competition to co-creation, building systemic advantage in complex, multi-actor environments.

Chapter 11: Strategic Sustainability Analytics

Sustainability is now a core strategic imperative. This chapter provides a rigorous, STEM-based approach to ESG (Environmental, Social, and Governance) strategy. Readers learn how to design ESG dashboards, model trade-offs, and use scenario planning and multi-objective optimization to balance profitability with social and environmental impact. The chapter covers lifecycle assessment, circular economy design, and the integration of sustainability into financial and operational strategy.

Chapter 12: Strategy Implementation and Execution with STEM

Strategy succeeds or fails in execution. This chapter shows how to translate strategic intent into action using digital governance, agile frameworks, real-time dashboards, and feedback systems. Readers learn to design execution stacks that connect strategy, technology, and people, and to use behavioral engineering and ethically aligned decision protocols to reinforce strategic behavior. The chapter emphasizes the importance of adaptive, learning-oriented execution systems in a world of constant change.

Chapter 13: Strategic Leadership and STEM Capabilities

Leadership is being redefined by data, systems, and intelligent technologies. This chapter explores the competencies required for strategic leadership in a STEM world, including systems thinking, data literacy, ethical intelligence, and adaptive learning. Readers are introduced to AI-augmented coaching, behavioral analytics, and diagnostic tools for leadership development. The chapter prepares leaders to build and lead organizations that are not only technically proficient but also ethically grounded and future-ready.

Chapter 14: Strategic Futures – Where Strategy Goes Next

The final chapter looks ahead, equipping readers with the tools and mindsets needed for anticipatory, future-oriented strategy. It covers quantum strategy, generative AI, synthetic data, planetary boundaries, and post-human leadership. Readers learn to use foresight tools such as futures wheels, backcasting, and digital sandboxes to design resilient strategies for multiple plausible futures. The chapter concludes with a call to action: to become future architects who can model, simulate, and shape the systems that will define tomorrow's world.

Together, these chapters form a progressive, integrated curriculum, one that starts with foundational strategy and builds, step by step, toward the advanced, STEM-powered capabilities required for leadership in the 21st century. Each chapter reinforces and extends the last, equipping readers to make smarter decisions at the business, corporate, and ecosystem levels. By the end of the book, readers will not only understand strategy in theory but will also possess the data-driven decision frameworks and modeling skills needed to navigate complexity and lead in volatile markets.

This book uniquely combines:

- Practical case studies based on companies like Zara, Ping An, Tesla, and Amazon;
- Interactive exercises and simulations that challenge the reader to apply concepts in real-world scenarios;
- Advanced tools like XGBoost, SHAP, digital twins, and Monte Carlo simulations integrated into strategy;
- Multidisciplinary learning, connecting strategy with sustainability, innovation, leadership, and digital transformation;
- A multi-modal learning experience blending text, cases, exercises, HBR recommendations, and films.

Readers of this book will gain:

- Mastery in blending timeless strategy with modern STEM methodologies;
- Practical tools to enhance decision-making in volatile, data-driven environments;
- Foresight skills are essential for navigating digital disruption and sustainability challenges.
- Leadership capabilities attuned to the technological, ethical, and systemic demands of the 21st century.

A Call to Action: Strategy, Re-Engineered for the Future

In the face of relentless technological change, strategic management must evolve—or risk becoming obsolete. This book was designed with a singular purpose: to equip leaders at every level with the mindset, analytical tools, and ethical compass required to thrive in a world shaped by STEM.

The challenges of the 21st century are unlike any we have faced before. Disruption is constant. Data is abundant, but insight is rare. The boundaries between industries, disciplines, and even nations are dissolving. In this environment, the old playbooks are no longer enough. Leaders must become systems thinkers, data modelers, and innovation architects. They must be able to harness the power of artificial intelligence, predictive analytics, and digital platforms, not as buzzwords, but as core components of their strategic toolkit.

This is not a call to abandon the wisdom of the past. The foundational principles of strategy, clarity of purpose, competitive advantage, and disciplined execution, remain as vital as ever. But their application must be reimagined for a world where uncertainty is the norm and adaptability is the ultimate advantage.

This book invites you to step into that future. It challenges you to:

- **Question assumptions** and embrace complexity, rather than shy away from it.
- **Experiment boldly** with new tools, frameworks, and ways of thinking.
- **Model uncertainty** and simulate outcomes before committing resources.
- **Lead ethically**, recognizing that every data-driven decision has human and societal consequences.
- **Build organizations that learn, adapt, and renew themselves continuously.**

Whether you are a student preparing for your first leadership role, a manager navigating digital transformation, or a senior executive shaping the future of your industry, the journey ahead demands more than knowledge. It demands the discipline to model complexity, the vision to harness innovation, and the courage to make evidence-driven decisions under uncertainty.

The future of strategy will not be written by those who cling to the comfort of the familiar, but by those who are willing to re-engineer their thinking, their organizations, and even themselves. This is your invitation to join that movement, to become not just a strategist, but a builder of futures.

Welcome to strategy, re-engineered for the future.

About the Author

Mehmet Ali Koseoglu, Ph.D., is an internationally recognized scholar, educator, and thought leader in the fields of strategic management, innovation ecosystems, and data-driven decision-making. His work bridges academic theory and real-world application, advancing how organizations think, compete, and evolve in complex, technology-enabled environments.

Dr. Koseoglu currently serves as a professor of strategic management and the Director of the MBA Program at a public university in the United States, where he leads curricular innovation and champions the integration of STEM thinking into business education. He is consistently recognized for his leadership in research and teaching, receiving accolades from students, academic peers, and industry partners alike.

He has published extensively in top-tier journals and is best known for pioneering studies on the intersection of entrepreneurship ecosystems, sustainability, governance, artificial intelligence, and public policy. His research has appeared in journals such as *Business Strategy and the Environment*, *Sustainable Development*, *Journal of Strategy and Management*, *Tourism Economics*, *Management Decision*, *International Journal of Hospitality Management,* and *Journal of Entrepreneurship and Public Policy*, and has contributed to the global academic discourse on strategic agility, innovation metrics, and ESG performance.

In the last three years, he was named among the top 2% of scientists globally by Stanford University, a distinction that highlights both the impact and visibility of his scholarly work across disciplines.

A strategic thinker by nature and systems designer by training, Dr. Koseoglu has advised startups, nonprofit organizations, and academic institutions on strategic planning, curriculum transformation, and innovation design. He is also the founder of MAK Strategy & Research, a platform that translates academic insights into actionable strategy tools for decision-makers.

Through this book, Dr. Koseoglu combines decades of academic expertise with a future-focused vision, offering students and professionals a new blueprint for navigating strategic complexity in the era of AI, data, and sustainability.

TABLE OF CONTENTS

Preface..

DEDICATION ....

INTRODUCTION..

About the Author ....

CHAPTER 1 STEM and the Strategic Paradigm Shift.....................................

1.1. Foundations and Limitations of Traditional Strategic Management.................................. 29

1.1.1 What is Strategy? 30

1.1.2. Operational Effectiveness vs. Strategy 33

1.1.3. Strategic Trade-Offs: Choosing What Not to Do... 35

1.1.4. What Strategy Is Not.. 37

1.1.5. The Value of Clear Strategic Thinking.. 37

1.2. What is Strategic Management? 39

1.2.1. Strategic Management as a Process 40

1.2.2. Rethinking Strategic Management: Limits and the Case for Strategic Thinking 41

1.2.3. Why STEM? Understanding the Strategic Paradigm Shift.. 43

1.3. Strategic Intelligence: The New Core Competency.. 46

1.3.1 What Is Strategic Intelligence? 47

1.3.2 Why Strategic Intelligence Matters Now.. 49

1.3.3 STEM as the Foundation for Strategic Intelligence.. 51

1.3.4 Developing Strategic Intelligence as a Leader.. 53

1.3.5 Strategic Intelligence in Action... 55

1.4. Chapter 1 Wrap-Up: Strategic Intelligence in a STEM World 58

1.4.1. Discussion Questions.. 58

1.4.2. Micro-Case: Zara's Strategic Sprint 59

1.4.3. Interactive Exercise: Strategy Under Uncertainty 60

1.4.4. Recommended HBR Case Studies.. 60

1.4.5. Strategy-in-Film: Learn by Watching 61

CHAPTER 2 Systems Thinking for Strategic Leaders ....

2.1 From Strategy Fundamentals to Systemic Thinking 69

2.1.1 Reimagining the Value Chain: From Linear Flow to Feedback Loop 69

2.1.2 Strategic Fit: From Static Alignment to Dynamic Equilibrium 70

2.1.3 Organizational Alignment: From Structure to Synchronization 70

2.2 What Is Systems Thinking—and Why Strategists Need It 71

2.2.1 Defining Systems Thinking: Beyond Cause and Effect 72

2.2.2 STEM Foundations of Systems Thinking 74

2.2.3 Systems Thinking vs. Analytical Thinking 78

2.3 Modeling Strategic Systems 88

2.3.1 What Is a Strategic System Model? 89

2.3.2 Types of Strategic Models 91

2.3.3 From Mental Models to Dynamic Simulations 94

2.3.4 STEM Tools for System Modeling 103

2.4 Strategic Modeling in Practice: Ethics, Governance, and Cross-Disciplinary Insights 108

2.4.1 Ethical Considerations in Strategic Modeling 108

2.4.2 Governance and Transparency in Model Use 111

2.4.3 Cross-Disciplinary Lessons from System Design 113

2.5. Chapter 2 Wrap-Up: Modeling Strategy as a System 117

2.5.1. Discussion Questions 118

2.5.2. Micro-Case: Zara's Strategy Loop 118

2.5.3. Interactive Exercise: Modeling Organizational Strategy 119

2.5.4. Recommended HBR Case Studies 120

2.5.5. Strategy-in-Film: Learn by Watching 121

CHAPTER 3 Strategic Analytics and Decision Intelligence ....

3.1 From Insight to Intelligence 127

3.1.1 Why Strategic Decisions Fail Without Intelligence 128

3.1.2 Intelligence as a Strategic Capability 129

3.1.3 From Pattern Recognition to Pattern Simulation 130

3.1.4 Strategic Intelligence in Action: The New Decision Blueprint 132

3.2 From Data to Strategy – The Analytics Continuum 133
3.2.1 Descriptive Analytics: Understanding What Has Happened 133
3.2.2 Predictive Analytics: Anticipating What Could Happen 134
3.2.3 Prescriptive Analytics: Recommending What Should Be Done 134
3.2.4 Integrating the Continuum into Strategic Decision-Making 135
3.3 Strategic Analytics Toolkit – Tools That Power Intelligent Strategy 137
3.3.1 XGBoost – The Predictive Engine for Strategy 137
3.3.2 Bayesian Optimization – Intelligent Exploration for Strategic Decisions 139
3.3.3 SHAP – Making Strategy Models Transparent and Trustworthy 141
3.3.4 PDP – Visualizing Strategic Sensitivity 143
3.3.5 Monte Carlo Simulation – Modeling Uncertainty for Strategic Resilience 145
3.3.6 Digital Twins – Simulating Strategy in a Living System 147
3.3.7 Selecting the Right Tool – A Strategic Comparison 149
3.4. Reinventing Classic Strategy with Analytics 151
3.4.1 VRIO + SHAP – Dynamic Value Identification 152
3.4.2 SWOT + Predictive Trend Data: Augmenting Strategic Awareness with Foresight.. 157
3.4.3 Environmental Scanning Powered by ML and NLP 169
3.4.4 Strategic Fit as a Modeling Problem 181
3.5 Ethical Intelligence in Strategic Modeling 193
3.5.1 Bias, Transparency, and Responsibility in Model Use 194
3.5.2 Long-Term Consequences of Automated Decisions 196
3.5.3 Strategic Accountability with Advanced Tools 198
3.6. Chapter 3 Wrap-Up: Strategic Analytics and Decision Intelligence 203
3.6.1. Discussion Questions 204
3.6.2. Micro-Case: Zara's Strategic Sprint 204
3.6.3. Interactive Exercise: Strategy Under Uncertainty — The Analytics Playbook Challenge 205
3.6.4. Recommended HBR Case Studies 206
3.6.5. Strategy-in-Film: Learn by Watching 208
CHAPTER 4 Innovation as a STEM Process ....
4.1. Why Innovation Must Be Engineered 214

4.1.1 Innovation Is No Longer Accidental ... 215
4.1.2 Innovation as a Repeatable Process ... 216
4.1.3 Structured Innovation in a STEM World ... 216
4.1.4 The Strategic Payoff of Engineering Innovation ... 217
4.1.5 Building Innovation Capability in Emerging Leaders ... 217
4.2 Structured Innovation Methods in Strategic Contexts ... 218
4.2.1 TRIZ: The Science of Inventive Problem Solving ... 219
4.2.2 Stage-Gate and Innovation Funnels ... 219
4.2.3 Technology Roadmapping for Strategic Planning ... 220
4.2.4 Integrating Structured Methods with Data and Analytics ... 221
4.3 AI-Driven Creativity and Design ... 223
4.3.1 Generative Design: From Constraints to Concepts ... 223
4.3.2 Machine Creativity and Generative AI ... 224
4.3.3 Personalization Engines and Design Adaptivity ... 225
4.3.4 Strategic Implications of AI-Driven Design ... 225
4.4 Technology Forecasting in Strategy ... 227
4.4.1 Why Forecasting Is Strategic ... 228
4.4.2 Technology Roadmapping ... 228
4.4.3 Emerging Tech Scanning and Weak Signals ... 229
4.4.4 Scenario Planning for Technological Uncertainty ... 230
4.4.5 Forecasting with STEM-Enhanced Tools and the Cost of Failure ... 230
4.5. Chapter 4 Wrap-Up: Engineering Innovation for Strategic Impact ... 232
4.5.1. Discussion Questions ... 233
4.5.2. Micro-Case: Zara's Innovation Forecast ... 233
4.5.3. Interactive Exercise: Structured Innovation Sprint ... 234
4.5.4. Recommended HBR Case Studies ... 235
4.5.5. Strategy-in-Film: Innovation on Screen ... 236
CHAPTER 5 Business Models and Digital Value Architecture ...
5.1 Mapping Business Models in the Digital Era ... 244
5.1.1 From Product Thinking to Business Model Thinking ... 244

5.1.2 Anatomy of a Modern Business Model 245
5.1.3 Digital Overlays: Augmenting the Canvas with Data 245
5.1.4 From Linear Models to Platform Models 246
5.1.5 Strategic Group Mapping with Clustering Algorithms 247
5.2 Business Model Innovation and Analytics 248
5.2.1 Why Business Model Innovation Matters More Than Product Innovation 249
5.2.2 Using Analytics to Identify Business Model Levers 249
5.2.3 From Assumptions to Evidence: Hypothesis-Driven Modeling 250
5.2.4 Simulation and Scenario Analysis for Strategic Resilience 250
5.3 Platform Strategies and Ecosystem Modeling 252
5.3.1 Understanding Platforms as Strategic Architectures 252
5.3.2 Data-Driven Design of Platform Models 252
5.3.3 Ecosystem Strategy: Beyond the Platform Core 253
5.3.4 Simulating Platform Dynamics with STEM Tools 253
5.4 Business Model Resilience and Adaptive Architecture 255
5.4.1 What Makes a Business Model Resilient? 255
5.4.2 Simulation and Adaptive Strategy Design 255
5.4.3 Scenario Analysis and Strategic Flexibility 256
5.4.4 Modular Business Model Architecture 256
5.5. Chapter 5 Wrap-Up: Business Models in a Digital Architecture 258
5.5.1. Discussion Questions 258
5.5.2. Micro-Case: Zara's Modular Reinvention 259
5.5.3. Interactive Simulation: Platform Pivot or Product Focus? 259
5.5.4. Recommended HBR Case Studies for Chapter 5 260
5.5.5. Strategy-in-Film: Learn by Watching 261
CHAPTER 6 Corporate Strategy In a STEM World Navigating Growth Diversification & Ecosystems ....
6.1 Foundations of Corporate Strategy 268
6.1.1 Classic Corporate Strategy: Growth, Stability, Retrenchment 269
6.1.2 Corporate-Level vs. Business-Level Strategy 270

6.1.3 Revisiting Scope: Diversification and Integration 271
6.2 STEM-Powered Growth Strategies 273
6.2.1 Predictive Analytics for Diversification Decisions 274
6.2.2 AI-Augmented M&A Screening and Target Evaluation 276
6.2.3 Simulation Models for Expansion Under Uncertainty 277
6.3 Managing Corporate Portfolios in Complex Markets 278
6.3.1 Portfolio Strategy in a STEM World: BCG Matrix 2.0 279
6.3.2 Dynamic Portfolio Rebalancing with Real-Time Data 281
6.3.3 System Dynamics for Resource Allocation 282
6.4 Vertical Integration, Outsourcing, and Platform Ecosystems 283
6.4.1 Strategic Decision-Making on Vertical Scope Using Digital Twins 285
6.4.2 Ecosystem Orchestration vs. Vertical Integration 286
6.4.3 AI in Make-or-Buy Decisions 289
6.5 Strategic Coherence and Corporate Advantage 290
6.5.1 How STEM Tools Ensure Coherent Corporate Strategy 293
6.5.2 Modeling Synergies and Strategic Fit Across Business Units 295
6.5.3 Detecting Strategic Misalignment Using Analytics 296
6.6 Chapter 6 Wrap-Up: Corporate Strategy Reinvented for a STEM World 297
6.6.1. Discussion Questions 298
6.6.2 Micro-Case: Zara's Data-Driven Corporate Expansion Strategy 299
6.6.3 Interactive Exercise: Designing a Data-Driven Corporate Portfolio 300
6.6.4. Recommended HBR Case Studies for Chapter 6 301
6.6.5 Strategy-in-Film: Learn by Watching 302
CHAPTER 7 Financial Strategy Analytics .....
7.1 Real-Options Thinking in Strategic Finance 308
7.1.1 Moving Beyond Static Valuation 309
7.1.2 Strategic Value Under Uncertainty 309
7.1.3 Categories of Real Options in Strategy 309
7.1.4 STEM Applications of Real Options 310
7.2 Monte Carlo Simulations for Financial Forecasting 312

7.2.1 Fundamentals of Monte Carlo Simulation 312
7.2.2 Strategic Applications in Finance 312
7.2.3 Integration with STEM Capabilities 313
7.3 Ethical and Strategic Allocation of Capital 315
7.3.1 Traditional Capital Allocation: Power and Pitfalls 315
7.3.2 The Strategic Pivot: Capital as a Moral and Adaptive Resource 316
7.3.3 STEM Tools for Ethical Allocation 316
7.3.4 Case Example: Capital Allocation at Ping An 317
7.4 Predictive Credit Scoring and Strategic Lending 318
7.4.1 From Traditional Scoring to Predictive Intelligence 318
7.4.2 Strategic Impacts of Credit Analytics 319
7.4.3 Ethical Risk and Bias in Algorithmic Lending 320
7.5 Strategic Budgeting and Adaptive Finance Systems 321
7.5.1 From Fixed Budgets to Strategic Adaptability 321
7.5.2 STEM Tools Powering Adaptive Finance 322
7.5.3 Strategy-as-Hypothesis: Budgeting for Learning 322
7.6. Chapter 7 Wrap-Up: Financial Intelligence in a STEM-Driven Strategy 323
7.6.1. Discussion Questions 324
7.6.2. Micro-Case: Zara's Financial Foresight 324
7.6.3. Interactive Exercise: Budgeting in an Uncertain World 325
7.6.4. Recommended HBR Case Studies for Chapter 6 326
7.6.5. Strategy-in-Film: Learn by Watching 328
CHAPTER 8 International Strategy Through STEM ....
8.1 STEM-Based Market Selection and Entry Strategy 334
8.1.1 Why Traditional Market Selection Falls Short 335
8.1.2 Building a Data-Rich Market Entry Dashboard 335
8.1.3 Predictive Analytics for Entry Mode Selection 337
8.1.4 Integrating Geopolitical and Climate Risk Modeling 338
8.2 Cultural Modeling with Hofstede and GLOBE Data 339
8.2.1 Hofstede's Cultural Dimensions: Strategic Applications 340

8.2.2 GLOBE Study: Leadership Fit and Strategic Communication 341
8.2.3 Practical Use: Building Culture-Responsive Strategy Models 341
8.2.4 STEM-Enhanced Cultural Intelligence: New Frontiers 342
8.3 Forecasting Political Stability with Machine Learning Models 343
8.3.1 Key Data Sources for Political Stability Modeling 344
8.3.2 Machine Learning Techniques in Political Forecasting 345
8.3.3 Strategic Applications of Political Risk Forecasting 345
8.3.4 From Prediction to Action: Embedding Insights into Strategy 346
8.4 Strategic Intelligence in Global Expansion: AI-Enhanced Market Entry Planning 347
8.4.1 AI-Driven Market Entry Strategy Mapping 347
8.4.2 Optimizing Entry Timing with Predictive Analytics 348
8.4.3 Entry Mode Selection and Strategic Fit 348
8.4.4 System Modeling for Global Strategy Alignment 349
8.5. Chapter 8 Wrap-Up: Strategic Expansion in a Borderless World 349
8.5.1. Discussion Questions 350
8.5.2. Micro-Case: Zara's Algorithmic Leap into Latin America 350
8.5.3. Interactive Exercise: Global Risk Simulation 351
8.5.4. Recommended HBR Case 352
8.5.5. Strategy-in-Film: Learn by Watching 354
CHAPTER 9 Competitive Advantage In a Data Driven Economy ....
9.1 Competing on Data, Not Just Products 360
9.1.1 The Rise of Data as a Strategic Asset 360
9.1.2 Beyond Dashboards: Data as Competitive Infrastructure 361
9.1.3 Data Velocity and the Feedback Advantage 361
9.1.4 Strategic Shifts: From Products to Data-Driven Ecosystems 362
9.2 Dynamic Capabilities and AI-Based Differentiation 364
9.2.1 What Are Dynamic Capabilities? 365
9.2.2 AI as a Differentiation Engine 366
9.2.3 From Resource Possession to Capability Orchestration 366
9.2.4 Embedding Adaptiveness into Strategy 368

9.3 Modeling Evolving Competitive Landscape 369
9.3.1 The Strategy Landscape Is No Longer Static 370
9.3.2 Competitive Intelligence as Simulation, Not Surveillance 372
9.3.3 Tools for Modeling Competition in a Data-Driven World 373
9.4 Strategic Renewal and the Erosion Curve 375
9.4.1 Competitive Advantage Is Now Perishable 375
9.4.2 Forecasting the Advantage Erosion Curve 376
9.4.3 Data-Driven Strategic Renewal Loops 378
9.5. Chapter 9 Wrap-Up: Competing on Intelligence, Not Just Products 380
8.5.1. Discussion Questions 381
9.5.2. Micro-Case: Zara's Data-Driven Renewal Loop 381
9.5.3. Interactive Exercise: The Advantage Erosion Lab 382
9.5.4. Recommended HBR Case Studies 382
9.5.5. Strategy-in-Film: Learn by Watching 385
CHAPTER 10 Entrepreneurial Ecosystems Through The Stem Lens ....
10.1 Mapping the Architecture of Entrepreneurial Ecosystems 391
10.1.1 Core Components of the Entrepreneurial Ecosystem 392
10.1.2 Mapping the Interdependent Ecosystem 394
10.1.3 Institutional Quality and Ecosystem Leadership 397
10.2 Measuring Ecosystem Vitality and Strategic Fit 398
10.2.1 Beyond Vanity Metrics: What Really Matters 398
10.2.2 Strategic Fit: Aligning Ecosystem Context with Organizational Goals 399
10.3 Strategic Participation in Entrepreneurial Ecosystems 400
10.3.1 From Presence to Participation: Strategic Roles 400
10.3.2 Building Ecosystem Capital: Relationships, Trust, and Reputation 401
10.3.3 Timing and Strategic Entry: First Movers, Fast Followers, and Catalysts 401
10.4 Orchestrating Ecosystem Innovation and Collaboration 402
10.4.1 Platforms, APIs, and Innovation Marketplaces 402
10.4.2 Managing Intellectual Property in Open Systems 403
10.4.3 Balancing Openness and Control: The Governance Challenge 403

10.5 Strategic Modeling of Ecosystems as Dynamic Systems 404
10.5.1 Why Static Analysis Fails in Ecosystems 405
10.5.2 Agent-Based Modeling and System Dynamics.. 405
10.5.3 Scenario Engines and Digital Twins for Ecosystems..................................... 406
10.6 Designing Ecosystem-Based Strategy for Sustainable Growth 407
10.6.1 Firms as Ecosystem Architects 407
10.6.2 From Competitive Advantage to Systemic Advantage 408
10.6.3 Co-Creation, Governance, and Resilience 409
10.7. Chapter 10 Wrap-Up: Ecosystem Thinking as Strategic Advantage 410
10.7.1. Discussion Questions 411
10.7.2. Micro Case: Zara as an Ecosystem Architect 411
10.7.3. Interactive Exercise: Designing a Resilient Ecosystem Strategy 412
10.7.4. Recommended HBR Case Studies 413
10.7.5. Strategy-in-Film: Learn by Watching 414
CHAPTER **11** Strategic Sustainability Analytics..
11.1 ESG Strategy in a STEM World 421
11.1.1 From Greenwashing to Strategic Sustainability... 421
11.1.2 Triple Bottom Line and ESG as Strategic Data 422
11.1.3 CSR vs. CSV: Beyond Compliance 422
11.1.4 Stakeholder Theory Reimagined with STEM Tools 423
11.2 ESG Dashboards and Impact Analytics 424
11.2.1 Anatomy of an ESG Dashboard: From Metrics to Meaning.............................. 425
11.2.2 Scoring Models: Environmental, Social, and Governance Dimensions 425
11.2.3 Visualizing Impact: Heat Maps, Time Series, and Risk Zones 426
11.2.4 Auditability, Bias, and the Limits of ESG Quantification.................................. 426
11.3 Modeling ESG Trade-Offs in Strategic Planning.. 428
11.3.1 Sustainability Is Full of Trade-Offs—So Model Them...................................... 428
11.3.2 Multi-Objective Optimization: Carbon vs. Cost vs. Inclusion............................. 428
11.3.3 Scenario Modeling for ESG Futures 429
11.3.4 Decision Intelligence and Stakeholder Simulation 429

11.4 ESG Strategy in Climate, Circular Economy, and Social Impact 431
11.4.1 Climate Analytics: Emissions, Offsets, and Decarbonization Paths 432
11.4.2 Circular Strategy: Designing for Regeneration 432
11.4.3 Just Transition: From Job Risk to Social Equity 433
11.4.4 Governance Foresight: Board Design and ESG Accountability 433
11.5. Chapter 11 Wrap-Up: ESG Strategy in a STEM World 434
11.5.1. Discussion Questions 434
11.5.2. Micro-Case: Zara's ESG Sprint 435
11.5.3. Interactive Exercise: ESG Trade-Off Simulation 436
11.5.4. Recommended HBR Case Studies for Chapter 10 436
11.5.5. Strategy-in-Film: Learn by Watching 438
CHAPTER 12 Strategy Implementation and Execution with STEM ....
12.1 Translating Strategy into Actionable Systems 443
12.1.1 From Strategy Maps to Execution KPIs 444
12.1.2 KPI Automation and Real-Time Monitoring 444
12.1.3 Aligning KPIs Across Functions and Levels 445
12.1.4 Designing KPIs with STEM Intelligence 446
12.2 Strategic Alignment through Digital Governance 447
12.2.1 Aligning Across Functions and Levels 448
12.2.2 Digital Governance: From Control to Coordination 449
12.2.3 Dashboards and Role-Based Visibility 450
12.2.4 Ethically Aligned Decision Protocols 452
12.3 Agile Execution and Continuous Adaptation 453
12.3.1 Agile Frameworks for Strategic Projects 454
12.3.2 Analytics-Driven Sprint Management 455
12.3.3 Agile Tools and Platform Integration 456
12.3.4 Learning Loops and Adaptive Strategy Cycles 458
12.4 Managing Strategic Change in Tech-Enabled Firms 459
12.4.1 Diagnosing Change Readiness with Data 460
12.4.2 STEM Tools for Change Management 461

12.5 Designing Execution Systems in STEM Organizations 463
12.5.1 Building the Strategic Execution Stack 464
12.5.2 Connecting Strategy, Technology, and People 465
12.5.3 Feedback Systems and Behavioral Engineering 467
12.5.4 Platforms That Reinforce Strategic Behavior 468
12.6. Chapter 12 Wrap-Up: Strategy Execution in a STEM World 470
12.6.1. Discussion Questions 471
12.6.2. Micro-Case: Zara's Execution Engine 471
12.6.3. Interactive Exercise: Execution by Design – The Strategic Systems Challenge 472
12.6.4. Recommended HBR Case Studies 473
12.6.5. Strategy-in-Film: Learn by Watching 475
CHAPTER **13** ....
Strategic Leadership and STEM Capabilities ....
13.1 STEM-Enabled Leadership Models 480
13.1.1 From Command to Collaboration: Redefining Leadership Norms 481
13.1.2 Cognitive Biases and the STEM-Aware Leader 482
13.1.3 Feedback Loops and Adaptive Leadership 483
13.1.4 STEM Skills for Strategic Influence 485
13.2 Intelligent Diagnostics for Leadership Development 486
13.2.1 Leadership Pipeline 2.0 486
13.2.2 Behavioral Analytics for Strategic Insight 488
13.2.3 Predictive Modeling of Leadership Potential 489
13.2.4 Ethical Boundaries in Leadership Data 490
13.3 AI-Augmented Leadership Coaching 491
13.3.1 Micro-Coaching Models and Feedback Loops 492
13.3.2 AI–Coach Symbiosis 493
13.3.3 Real-Time Coaching via Digital Twins 494
13.3.4 Interpretable AI in Personalized Coaching 495
13.4 Strategic Integration in Practice: Leading Across Systems and Disciplines 496
13.4.1 Bounded Rationality and Decision Traps 497

13.4.2 Algorithm-Supported Leadership 498
13.4.3 STEM-Enhanced Leadership Diagnostics 499
13.5 Building the AI-Augmented Coaching Ecosystem 501
13.5.1 Behavior Tracking and Feedback Loops 502
13.5.2 AI Nudges and Simulation Tools 502
13.5.3 Human-AI Coaching Collaboration 503
13.5.4 Ethical Governance in Coaching Systems 503
13.6. Chapter 13 Wrap-Up: Strategic Leadership and STEM Capabilities 504
13.6.1. Discussion Questions 505
13.6.2. Micro-Case: Zara's Decision Integrity Sprint 505
13.6.3. Interactive Exercise: Leadership Calibration Challenge 506
13.6.4. Recommended HBR Cases 507
13.6.5. Strategy-in-Film: Learn by Watching 508
CHAPTER 14 Strategic Futures – Where Strategy Goes Next ....
14.1 Quantum Strategy and the Future of Strategic Logic 514
14.1.1 Quantum Advantage as Strategic Advantage 515
14.1.2 Rewriting Optimization: From Linear Models to Quantum Superposition 516
14.1.3 Strategic Applications: Logistics, Finance, and Cybersecurity 518
14.2 Generative Intelligence and Synthetic Strategy 519
14.2.1 Synthetic Data as Strategic Infrastructure 520
14.2.2 Generative AI as a Strategic Co-Pilot 520
14.2.3 Simulating the Unseen: From Scenario Planning to Imagination Engineering 520
14.3 Planetary Strategy and the Climate-Strategy Nexus 521
14.3.1 Strategic Decision-Making Within Planetary Boundaries 522
14.3.2 Circular Systems and Regenerative Value Loops 522
14.3.3 AI Governance, Ethics, and Climate Intersections 523
14.4 Strategic Leadership in a Post-Human Era 524
14.4.1 Human-Machine Strategic Collaboration 524
14.4.2 Trust, Explainability, and Algorithmic Integrity 524
14.4.3 Augmented Cognition and Bio-Enhanced Strategy 525

14.5 Strategic Intelligence Tools for the Futures Architect 526
14.5.1 Futures Wheel, Backcasting, and Wind Tunneling 526
14.5.2 Simulation Engines and Decision Sandboxes 527
14.5.3 Building an Anticipatory Culture 527
14.6. Chapter 14 Wrap-Up: Strategic Futures – Where Strategy Goes Next 528
14.6.1. Discussion Questions 529
14.6.2. Strategy Micro-Case: Ping An and the Predictive Planet 530
14.6.3. Interactive Exercise: Designing a Strategy for the Post-Human Boardroom 531
14.6.4. Recommended HBR Case Studies 532
14.6.5. Strategy-in-Film: Learn by Watching 533

CHAPTER 1

STEM and the Strategic Paradigm Shift

Strategy without analytics is intuition without insight
innovation without systems is creativity without direction
sustainability without data is intention without impact

by MAK

Learning Objectives

After completing this chapter, readers will be able to:

The strategist must become the octopus— fluid, intelligent, and always sensing the system.

- ❖ Define strategic management and explain how it has evolved from traditional approaches to modern, STEM-integrated models.
- ❖ Differentiate between operational effectiveness and strategy and understand why this distinction is critical in a technology-driven world.
- ❖ Apply Porter's definition of strategy to contemporary challenges and environments shaped by data, speed, and innovation.
- ❖ Analyze the role of strategic trade-offs and why they remain essential, even in algorithmically optimized settings.
- ❖ Evaluate the limitations of classic tools like Porter's Five Forces and understand how STEM-based enhancements (e.g., dynamic modeling, simulations) can improve them.
- ❖ Compare traditional vs. STEM-enhanced strategic thinking, using conceptual contrasts introduced in this chapter.
- ❖ Recognize why a paradigm shift is underway, and how this textbook equips future strategists with the tools and mindsets needed to compete in a digitally complex, STEM-centric world.

Strategic management is at a crossroads.

For decades, strategy was guided by frameworks like SWOT, Porter's Five Forces, VRIO, and the value chain. These tools helped organizations analyze industries, position themselves competitively, and plan their future. But the environment they were designed for, slow, relatively stable, and information-constrained--is gone. Today's world is fast, complex, digitalized, and interconnected. Static frameworks, intuition-led decision making, and quarterly plans no longer suffice. In their place, strategy must evolve, becoming smarter, faster, and more adaptive.

That evolution begins with STEM: Science, Technology, Engineering, and Mathematics.

In a world shaped by artificial intelligence, real-time data, complex systems, global risk, and sustainability pressures, strategy is no longer just about vision and differentiation. It's about modeling uncertainty, simulating outcomes, forecasting disruption, and designing intelligent responses. STEM doesn't replace strategic thinking; it transforms it. It adds precision to insight, structure to creativity, and adaptability to ambition.

Consider companies like Ping An, Moderna, Tesla, or Amazon. They didn't just innovate, they redefined strategic management itself. Their strategies are powered by algorithms, sensors, cloud platforms, digital twins, and dynamic feedback systems. They don't guess. They simulate. They don't react. They anticipate. These organizations show us that in a STEM world, strategy is no longer a once-a-year planning ritual. It is a continuous, real-time, feedback-driven process, part experiment, part execution, all intelligence.

This chapter marks the beginning of a new way of thinking. You'll learn why traditional strategy models are necessary but insufficient. You'll discover how operational effectiveness differs from strategy, and why clear trade-offs are still critical. And you'll see how modern tools- from machine learning to system dynamics- can enhance, not replace, the core of strategic thought.

Above all, you'll be invited to reimagine your role, not just as a student or manager, but as a STEM-enabled strategist. One who thinks in feedback loops, builds models, designs for uncertainty, and leads transformation.

The future of strategy isn't just a theory. It's a toolkit. And it starts here.

Opening Case: Ping An's Strategic Reprogramming

Imagine a company in the financial sector; not Silicon Valley, not New York, but in **Shenzhen, China**. Now imagine that this company, originally a traditional insurance firm, becomes one of the most **AI-powered, cloud-integrated, and platform-based strategic machines** on the planet.

That company is **Ping An**.

For decades, strategy was about choosing a position in the market, establishing trade-offs, and carving a defensible space. Ping An followed that playbook, at first. But by 2014, their leadership recognized that the old rules were being rewritten. Operational efficiency was no longer a competitive advantage; it was table stakes. Worse, the Five Forces were shifting in unpredictable ways: new digital entrants, cross-sector competition, and rising customer expectations all signaled the end of the static strategy era. Ping An's response wasn't to abandon traditional strategy; it was to **upgrade it**.

The company embedded **AI, big data, and platform engineering** into its strategic core. Insurance became one product in a digital ecosystem that also included healthcare, banking, telemedicine, facial recognition, and cloud infrastructure. They didn't just improve operations, they reimagined the strategic logic of what an insurance company could be. What's remarkable is how **STEM principles fueled their new strategy**:

- ❖ They used machine learning not only for pricing risk, but for forecasting customer churn and identifying emerging needs.
- ❖ They deployed digital twins to simulate systemic risks across their portfolio in real time.
- ❖ They redefined the value chain, from underwriting to claims, by automating 95% of service interactions.
- ❖ They treated strategy not as a plan, but as a dynamic model, constantly updated through sensor data, market feedback, and scenario simulations.

Ping An didn't abandon Porter. They upgraded him.

They didn't discard trade-offs. They made smarter ones, faster.

They didn't ignore the Five Forces. They built algorithms to detect when forces were morphing, even before competitors noticed.

Play the Strategist

- ❖ Imagine you were the Chief Strategy Officer at Ping An. What decisions would you have made differently?
- ❖ How would you convince traditional managers to embrace AI, digital twins, and platform models?

1.1. Foundations and Limitations of Traditional Strategic Management

In order to understand how STEM transforms strategic thinking, we must first revisit the foundations. For decades, strategic management has been built on powerful and elegant frameworks that offered clarity in a world of relative stability. However, in today's digitally complex and high-velocity environment, these models, while still relevant, often fall short. This section explores the foundational concepts of strategy and highlights where they must evolve to remain effective in a STEM-powered age.

Traditional strategic thinking emphasized long-term planning, deliberate positioning, and optimal resource allocation. These approaches assumed relatively predictable environments and slower cycles of change. Organizations were expected to analyze their surroundings, commit to a strategic direction, and execute consistently over time. Foundational frameworks such as **SWOT analysis**, **Porter's Five Forces**, **the value chain**, and the **VRIO model** (Value, Rarity, Imitability, Organization) provided structured methods for understanding industry dynamics and achieving sustainable advantage.

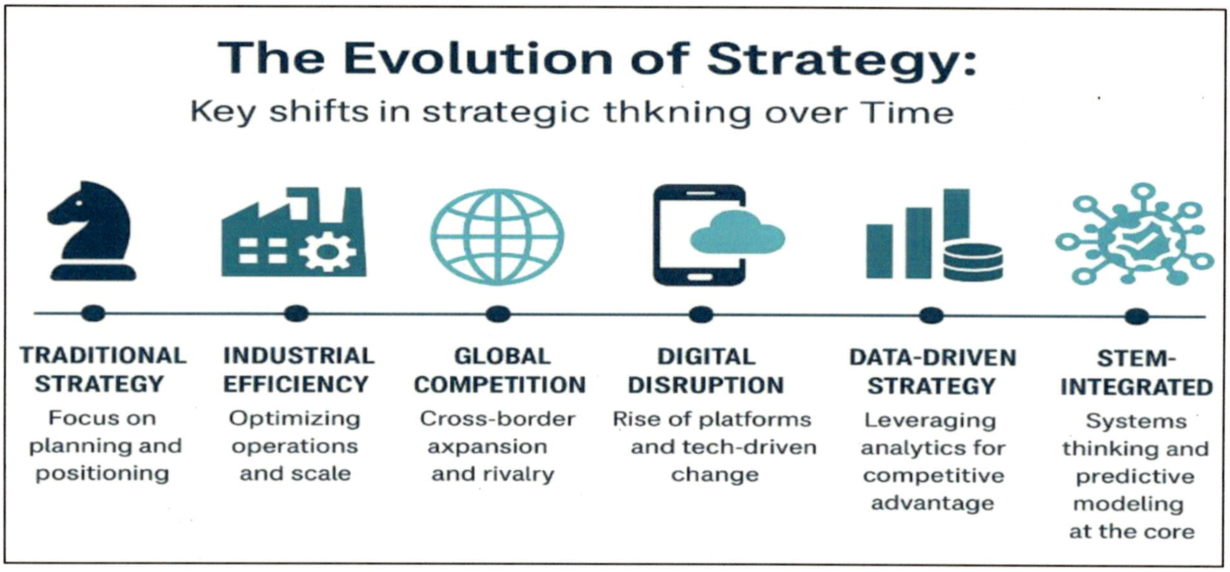

The evolution of strategy has unfolded through several distinct phases, each marked by different challenges and capabilities. The timeline below highlights this transformation, illustrating the shift from traditional models toward data-driven and STEM-integrated strategy approaches.

Today, however, change is constant and nonlinear. Disruption no longer arrives in generational cycles, but in rapid bursts driven by technological innovation, global complexity, and platform-based competition. Strategy can no longer rely on static planning alone. Emerging competitors may appear with no warning, and entire industries can be reshaped by a single platform, algorithm, or AI breakthrough. While the core principles of differentiation and value creation still apply, their application now demands more than conceptual insight. It requires a new kind of strategist, one who can combine classical frameworks with predictive modeling, systems thinking, and data-enabled decision-making in real time.

1.1.1 What is Strategy?

At its essence, strategy is about making informed choices under uncertainty. It involves determining what an organization will do, and equally, what it will not do, to create and sustain value in a competitive environment. Strategy aligns internal resources and capabilities with external opportunities and threats, guiding decisions across the organization in pursuit of long-term advantage.

The word *strategy* derives from the ancient Greek *strategos*, meaning "the art of the general." In its earliest uses, strategy referred to military leadership, battlefield tactics, and the orchestration of campaigns. These martial origins remain visible in today's business lexicon: terms such as *competitive advantage*, *positioning*, and *tactics* echo that legacy. Yet modern strategy transcends its military roots. In contemporary organizations, strategy encompasses how firms compete, innovate, grow, adapt, and collaborate, often across multiple industries and geographies.

Over the past fifty years, strategy scholars have proposed a wide range of definitions, each reflecting different assumptions about how firms operate and succeed. Michael Porter famously defined strategy as "the creation of a unique and valuable position, involving a different set of activities." According to Porter, an effective strategy involves clear trade-offs and a coherent activity system that differentiates the firm from rivals. In contrast, Henry Mintzberg emphasized that strategy is not always planned in advance; instead, it often emerges as a "pattern in a stream of decisions." This emergent strategy evolves as managers adapt to unexpected challenges, technologies, and market feedback.

These and other perspectives were synthesized into the widely cited *5 Ps of Strategy*, introduced by Mintzberg:

- **Plan**: a deliberate course of action or roadmap
- **Pattern**: a consistent behavior observed over time
- **Position**: a firm's place in the competitive landscape
- **Perspective**: the collective mindset or ideology of the organization
- **Ploy**: a specific tactical move to outmaneuver rivals

Together, these views highlight the multifaceted nature of strategy. It is not merely about long-term planning or market positioning; it is also about culture, adaptability, experimentation, and context. Strategy is both intentional and emergent, both analytical and behavioral, both structured and fluid.

More recently, the *strategy-as-practice* perspective has added a critical dimension. Spearheaded by scholars such as Richard Whittington, this approach shifts the focus from what strategy is to what strategists do. It investigates the daily actions, routines, tools, and conversations through which strategy is enacted inside organizations. Strategy may emerge not from executive retreats or 100-page reports, but through data dashboards, agile sprints, Slack threads, or informal decision rituals. This approach emphasizes three core elements:

- **Practitioners**: the individuals involved in making and shaping strategy
- **Practices**: the tools, routines, and artifacts they use
- **Praxis**: the flow and rhythm of actual strategic activity

This is especially relevant in today's digital and STEM-intensive economy. Strategy is no longer confined to annual planning cycles or boardroom presentations. It is enacted continuously and iteratively, often in real time. Strategic decisions now unfold through algorithms, simulations, interfaces, and automated alerts, not just memos or slides. As such, strategic thinking today requires not only insight but also technological fluency, data interpretation skills, and systems thinking.

The evolution of strategy reflects shifts in the business environment. In the 1950s and 60s, large firms such as General Electric and consultancies like Boston Consulting Group developed formal approaches to managing complexity and scale. By the 1980s and 1990s, strategy focused heavily on competitive positioning, emphasizing industry structure, resource advantages, and long-term planning. This era produced enduring tools such as SWOT analysis, Porter's Five Forces, and the Value Chain framework.

Yet in the 21st century, those stable environments have given way to rapid, disruptive change. Globalization, digital platforms, generative AI, and algorithmic competition have redefined the nature of strategic advantage. Firms can no longer count on static positioning; they must constantly reinvent. Competitive edge is now earned through resilience, adaptability, and real-time learning. Strategy must move beyond prediction to embrace probabilistic thinking, scenario modeling, and simulation. It must transition from linear planning to dynamic orchestration. Table 1.1. summarizes the evolution of strategy perspectives.

Table 1.1. Evolving Perspectives on Strategy

Perspective	Description	Key Contributor	Example in Practice
Plan	Strategy as a deliberate, documented course of action	Michael Porter	A 5-year expansion roadmap into emerging markets
Pattern	Strategy as recurring behavior that reveals direction	Henry Mintzberg	Reinvesting in AI infrastructure annually
Position	Strategy as the firm's stance relative to competitors	Ansoff, Porter	Competing as a budget vs. luxury airline
Perspective	Strategy as the organizational worldview or ideology	Hamel & Prahalad	Prioritizing sustainability over short-term gain
Ploy	Strategy as a calculated maneuver to gain advantage	Mintzberg	Underpricing to block a new entrant
Practice	Strategy as the micro-level routines and tools used daily	Whittington	Using dashboards and agile methods for decision-making

Strategy in the Age of Systems, Models, and Machines

The perspectives above remain valuable, but the strategist's world has changed. In a STEM-powered environment, execution is no longer delayed or static. Strategy is not just devised; it is performed through systems, models, and software.

> What Strategy Means to You
>
> Think of an organization, startup, or team you admire.
>
> * Which of the 5 Ps best describes their strategy?
> * Could they benefit from a more STEM-enabled approach?

Strategists today don't just craft roadmaps; they build models, run simulations, interpret dashboards, and react to dynamic data in real time. These shifts have redefined the tools, timelines, and roles involved in strategy-making. Hence, Table 1.2. demonstrates the difference between traditional and STEM-driven thinking,

Table 1.2. Traditional vs. STEM-Driven Thinking

Element	Traditional Strategy	STEM-Driven Strategy
Time Horizon	Long-term planning (3–5 years)	Real-time adjustment and forecasting
Decision Style	Top-down, hierarchical	Distributed, model-informed, algorithmic
Tools Used	SWOT, Porter's Five Forces, Value Chain	Digital Twins, Monte Carlo Simulations, Machine Learning
Data Usage	Historical, static reports	Dynamic, real-time, predictive modeling
Execution	Strategic plans, quarterly goals	Dashboards, agile iterations, simulations
Practitioner Role	Executives, planners	Cross-functional teams, data scientists, and digital strategists

In a STEM-powered strategy world, models, systems, and simulations are not just enhancements; they are core to decision-making. This book embraces that shift. We honor traditional strategic thinking, but we also extend it into a STEM-integrated framework for the complexity of modern business. In the next section, we'll build on this foundation by introducing **strategic management**: the discipline that operationalizes strategy, coordinates its execution, and measures its outcomes across the organization.

1.1.2. Operational Effectiveness vs. Strategy

Not everything that improves performance is strategic. One of the most enduring contributions to strategic thought comes from Michael Porter's distinction between **operational effectiveness** and **strategy**. While both are essential to organizational success, they are not the same. Operational effectiveness means performing similar activities better than rivals, faster, cheaper, with higher quality, or fewer errors. Strategy, by contrast, means performing **different activities** or performing similar activities **in different ways** to create a unique and sustainable position.

Operational effectiveness is about excellence in execution. It includes practices such as total quality management (TQM), lean operations, agile workflows, Six Sigma, and advanced productivity tools. These approaches improve efficiency, reduce waste, and increase speed. However, they are not substitutes for strategy. Why? Because best practices are quickly imitated. When competitors adopt the same technologies or management techniques, the advantage erodes. Competing on operational effectiveness alone often leads to a race to the bottom, lower costs, lower margins, and minimal differentiation.

Strategy, in contrast, is about making deliberate choices that involve **trade-offs**. A company cannot be all things to all people. Choosing what **not** to do is as important as deciding what to pursue. Strategy defines the unique value a firm offers, the customers it serves, and the configuration of activities that deliver that value. Whereas operational effectiveness can be benchmarked and copied, a well-crafted strategy creates defensible uniqueness.

In today's STEM-driven world, the line between operational effectiveness and strategy can become blurred. Table 1.3. presents the differences between these two terms. Technologies such as robotic process automation, AI-powered optimization, and predictive maintenance deliver extraordinary gains in effectiveness, but they are widely accessible. The key strategic question becomes: **how are these tools integrated differently** to reinforce a firm's unique position? For example, two airlines may use AI for pricing, but only one might use it to design a personalized loyalty ecosystem tied to traveler behavior, giving it a strategic edge.

Table 1.3. Operational Effectiveness vs. Strategy

Dimension	Operational Effectiveness	Strategy
Focus	Doing things better	Doing different things
Objective	Efficiency, speed, quality	Unique positioning
Time Horizon	Short-to-medium term gains	Long-term differentiation
Replicability	Easily copied by competitors	Difficult to imitate
Examples	Lean manufacturing, automation, agile workflows	IKEA's self-service model, Tesla's direct sales model
STEM Integration	Optimization algorithms, process automation	Unique AI-enabled experiences, system design

Effectiveness vs. Innovation

In STEM-intensive environments, operational gains are increasingly powered by algorithms and systems. But strategic advantage requires configuring these technologies into a differentiated value proposition.

- A logistics firm using IoT to monitor fleet performance is improving operations.
- A rival that integrates IoT, customer-facing apps, and sustainability reporting into a seamless user experience has a strategy.

STEM tools enhance both sides, but it's the integration and purpose that define whether it's operational or strategic. Are You Efficient or Strategic?

Think about an organization you're familiar with (your workplace, a startup, a university unit, etc.).

- What STEM-enabled processes have they adopted to improve efficiency?
- Are these enhancements widely available to competitors?
- How could the organization configure these capabilities into a unique value proposition?

1.1.3. Strategic Trade-Offs: Choosing What Not to Do

Strategy is as much about what an organization chooses **not** to do as it is about what it chooses to pursue. This principle of **strategic trade-offs** lies at the heart of sustainable competitive advantage as seen in Table 1.4. In an era where data is abundant and possibilities are endless, the discipline of focus becomes more, not less, important. Strategic trade-offs help organizations clarify identity, sharpen positioning, and avoid the trap of trying to be everything to everyone.

A trade-off arises when an organization must **sacrifice one strategic benefit to gain another**. For example, offering customized products may require sacrificing scale efficiencies. Competing on low cost often means foregoing the deep customer relationships needed for a premium experience. Choosing a narrow customer segment may limit market size, but allows for more tailored solutions.

Michael Porter famously argued that trade-offs are essential to strategy because they create **sustainability**. Without trade-offs, competitors can imitate a strategy by simply adding new features or capabilities. But when strategic choices are mutually exclusive, when doing one thing well means **not** doing something else, imitation becomes harder. Southwest Airlines, for instance, achieves low costs by avoiding hub-and-spoke models, skipping meal service, and flying a single aircraft type. These choices make it hard for full-service airlines to copy Southwest's low fares without undermining their own business models.

In today's fast-paced and STEM-powered environment, the temptation to "do it all" is even stronger. Data-driven platforms encourage expansion into adjacent markets, real-time dashboards reveal new performance gaps, and AI tools promise both personalization and scale. But **strategy remains a game of focus**. Sustainable advantage comes from deep, coherent systems, not scattered capabilities. Firms must choose which metrics to prioritize, which technologies to embed deeply, and which user needs to ignore.

Table 1.4. Classic Strategic Trade-Offs in Business

Strategic Focus	Trade-Offs Accepted	Example Organization
Cost Leadership	Forego product variety and customization	Ryanair: no-frills, low-price model
Differentiation	Higher operational costs	Apple: premium experience, high margins
Niche Market	Smaller total addressable market	Warby Parker: direct-to-consumer eyewear
Customer Intimacy	Slower scalability	Zappos: intensive customer service
Speed and Agility	Less standardization	Spotify: rapid feature testing and rollout
Platform Ecosystem	Loss of direct control	Airbnb: hosts create value, not the company

Algorithmic Trade-Offs and Design Constraints

In a STEM-intensive strategy context, trade-offs often take place **within systems, code, and models**. For instance:

- A machine learning model trained for accuracy may sacrifice explainability.
- A digital twin that optimizes logistics for carbon efficiency might reduce speed.
- An AI-driven pricing system might maximize revenue but erode customer trust if perceived as unfair.

Strategists must now manage **algorithmic trade-offs,** deciding between transparency vs. performance, personalization vs. privacy, or short-term gains vs. long-term resilience. Smart strategy isn't about avoiding trade-offs; it's about making them intentional, visible, and in alignment with the firm's identity.

Facing Trade-Offs in Practice

Reflect on a project, product, or organization you've encountered.

- What trade-offs were made, explicitly or implicitly?
- Were these choices communicated clearly to stakeholders?
- Did the firm sacrifice something meaningful to gain a strategic edge?

1.1.4. What Strategy Is Not

While it is essential to understand what strategy is, it is equally important to understand what it is not. Misconceptions about strategy often lead organizations down paths of confusion, inefficiency, or superficial success. Without a clear distinction, many leaders mistakenly equate operational efficiency, tactical maneuvers, or ambitious goals with strategy. This section clarifies these critical differences.

Strategy is not operational effectiveness. Doing existing tasks better, faster, cheaper, or more efficiently is undoubtedly important, but it is not strategy. Operational effectiveness refers to executing similar activities better than rivals. These improvements are easily replicable and thus do not lead to sustainable advantage. As Michael Porter emphasized, "operational effectiveness is necessary, but not sufficient" for strategy.

Strategy is not a goal or aspiration. Stating that an organization wants to "be the market leader" or "increase revenue by 20%" does not constitute a strategy. These are outcomes, not pathways. A real strategy defines *how* the organization will achieve such goals uniquely and sustainably.

Strategy is not simply planning. Detailed budgets, annual plans, or project timelines are useful tools, but they are not strategy. A plan details what will happen and when, but it may lack the analytical backbone and competitive logic that defines true strategy.

Strategy is not copying competitors. Benchmarking best practices and imitating successful firms may bring short-term gains, but they fail to produce distinctive positioning. Strategy is inherently about making *unique* choices and doing things differently, not just better.

In a technology-saturated environment, the temptation to equate advanced tools (like CRM systems, AI platforms, or digital dashboards) with strategic advantage is common. But merely adopting technology without a coherent competitive logic is not strategy, it's an operational upgrade. Strategy arises when these technologies are deployed in ways that create differentiated value.

1.1.5. The Value of Clear Strategic Thinking

Strategic clarity is the foundation of competitive advantage. When leaders, teams, and organizations understand the distinct nature of strategy, they can focus efforts, prioritize effectively, and execute with purpose.

Clear strategy provides direction. It acts as a north star in turbulent environments. Rather than chasing every new trend or reacting impulsively to market changes, organizations with clear strategy can assess what fits and what doesn't. This focus is especially vital in fast-changing, STEM-intensive industries.

Strategic thinking fosters differentiation. In a world of commoditized offerings and hypercompetition, differentiation is not optional; it is essential. Strategy helps define what makes an organization distinct and why customers, partners, and investors should care.

Strategic clarity supports disciplined execution. Without it, execution becomes fragmented. Teams may pull in different directions, wasting resources and time. But when strategy is clearly understood, day-to-day decisions, from budgeting to hiring to innovation, can align seamlessly with the firm's strategic intent.

Consider a retail company using AI to forecast demand. If it merely applies AI to reduce inventory costs, it's practicing operational effectiveness. But if it integrates AI to redesign its supply chain and offer real-time personalization, it moves into strategic territory, creating unique customer value that's hard to copy. Table 1.5. highlights misconceptions about strategy.

Table 1.5. Clarifying Strategy

Misconception	Reality
Operational improvements = Strategy	Operational gains are replicable; strategy creates unique positioning.
Goal statements = Strategy	Strategy defines the *how*, not just the *what.*
Planning documents = Strategy	Plans are tools; strategy is the logic guiding those tools.
Copying winners = Strategy	True strategy emerges from unique insights and decisions.

Think of a recent initiative in your organization, workplace, or a company you follow.

- Was it a true strategy, or more of an operational move, goal, or plan?
- What would have made it more strategic in nature?

So far, we have examined the foundational dimensions of strategy, its definitions, historical evolution, misconceptions, and the importance of clarity in strategic thinking. We have seen that strategy is not a singular act or a static document, but a dynamic discipline that involves making deliberate choices, creating unique value, and navigating uncertainty. It is both a mindset and a practice, rooted in centuries of theory, yet increasingly shaped by digital tools, real-time data, and computational modeling.

The shift from traditional strategic thinking to a STEM-powered approach does not abandon classical frameworks; rather, it extends them. Concepts like positioning, planning, and perspective now operate within environments defined by systems thinking, algorithmic insight, and automated decision-making. Still, the essence of strategy endures: crafting coherent choices, staying focused, and aligning actions with long-term objectives in a rapidly evolving world.

With this foundation in place, we now turn to the discipline that operationalizes strategy: **strategic management**. If strategy defines the *what* and the *why*, then strategic management delivers the *how*. It orchestrates the design, execution, and continuous refinement of strategy across the organization. In the next section, we will explore how strategic management emerged, how it functions in modern enterprises, and how it is transforming in an era driven by STEM capabilities.

1.2. What is Strategic Management?

Strategic management is the discipline that transforms strategy from abstract intention into coordinated, purposeful action. While strategy defines a vision of competitive advantage, what an organization aspires to achieve and how it seeks to differentiate, strategic management is the process by which that vision is executed, monitored, and adapted. It connects analysis to execution, plans to learning, and long-term goals to short-term decisions. In doing so, it enables organizations to sustain coherence across changing environments and competing priorities.

The origins of strategic management trace back to the post-war corporate expansion era, when firms began to formalize long-range planning and resource coordination. However, strategic management has evolved far beyond planning offices and organizational charts. It now encompasses a continuous cycle of strategic analysis, formulation, implementation, and performance evaluation. Modern strategic management operates at the intersection of leadership, operations, innovation, and change, requiring not only foresight but also agility, learning capability, and execution discipline.

In today's STEM-enabled world, strategic management is undergoing a profound transformation. Decision-making is increasingly supported by predictive models, optimization algorithms, and real-time performance dashboards. Implementation is coordinated through digital tools, agile workflows, and integrated platforms. Adaptation is driven by data feedback loops and simulations, not just quarterly reports or management reviews. Strategic management is no longer just the "managerial part" of strategy; it is a technologically augmented system for sensing, deciding, and acting in dynamic conditions.

1.2.1. Strategic Management as a Process

Strategic management is the discipline that turns strategy from an abstract concept into a coordinated, living practice. While strategy defines where an organization wants to go and why, strategic management explains how to get there and how to adjust course along the way. It provides the systems, tools, and processes that connect long-term vision with day-to-day execution. In this sense, strategic management is the engine room of strategy: translating purpose into performance, insight into impact, and design into delivery.

Traditionally, strategic management has been viewed as a linear sequence of steps: assess the environment, formulate a strategy, implement it, and then evaluate results. While these stages still matter, they are no longer sufficient in isolation. In a world of rapid feedback loops, real-time data, and AI-assisted decision-making, the process must be continuous, adaptive, and systems-based. Strategic management is no longer a static planning function; it is a dynamic operating capability.

At its core, strategic management consists of four interdependent stages, forming a circular and iterative process:

1. **Environmental Analysis**

   Organizations begin by analyzing both the internal environment (resources, capabilities, organizational culture) and external environment (market trends, competitors, technologies, regulations). Traditional tools such as SWOT analysis, PESTEL, and industry structure models remain useful—but are now augmented by data analytics, dashboards, and predictive scanning technologies. For example, firms may use AI-based trend detection tools to identify emerging risks before they materialize.

2. **Strategy Formulation**

   Informed by environmental insights, leaders craft a coherent strategy. This involves defining a mission and vision, setting long-term goals, identifying competitive positioning, and determining the means to achieve advantage. Strategic formulation includes trade-offs, scenario planning, and value creation logic. In modern organizations, formulation often includes simulations, machine learning models, and probabilistic forecasting. For example, a renewable energy company might use climate models to evaluate future regulatory scenarios before deciding on market entry.

3. **Strategy Implementation**

   Strategy must be embedded into structures, systems, and behaviors. Implementation is where strategy often fails, not because it is flawed, but because it is not translated effectively into operations. This step requires aligning resources, managing change, allocating budgets, building culture, and developing metrics. Agile methodologies, digital project management tools, and cloud-based execution platforms now allow real-time coordination across teams and geographies.

4. **Performance Evaluation and Feedback**

   Strategic management is incomplete without measurement. Organizations track key performance indicators (KPIs), financial outcomes, innovation rates, customer metrics, and more. In STEM-enhanced environments, this includes automated dashboards, A/B testing, algorithmic alerts, and performance heatmaps. These tools feed insights back into the system, allowing for learning, reconfiguration, and strategic renewal.

1.2.2. Rethinking Strategic Management: Limits and the Case for Strategic Thinking

Strategic management, in its traditional form, has long served as a foundational discipline for planning and executing firm-level strategies. However, the challenges of today's STEM-driven economy expose critical limitations in conventional strategic management approaches, especially when divorced from deeper strategic thinking. The problem isn't that traditional tools lack utility; it's that they were designed for a world that moved slower, assumed stability, and didn't revolve around data.

Let's examine where these legacy models break down and why the future of strategic management must blend execution discipline with dynamic thinking, real-time data, and systems-level awareness.

Static Thinking in a Dynamic World

Traditional tools such as SWOT, BCG Matrix, or PESTLE capture environmental snapshots. They are grounded in the assumption of environmental stability. However, in modern business, strategic variables, technologies, markets, customer preferences, and regulations change continuously. Strategy must be fluid, not static. Legacy tools like SWOT assume the environment stays still long enough for decisions to matter. But the pace of change has outgrown the tools themselves.

Consider Peloton's miscalculated post-pandemic strategy. The company failed to anticipate post-COVID demand shifts and inflation effects, leading to excess inventory and layoffs. Similarly, Zoom and Robinhood overrelied on early success patterns, ignoring rapidly evolving user needs and competitive ecosystems.

Over-Reliance on Managerial Intuition

Legacy strategic planning often placed intuition at the center. Experience and gut feel were critical when data was scarce. But in today's complex and data-rich environments, relying on instinct without empirical validation creates strategic blind spots. Without data to anchor decision-making, strategic planning becomes opinion-driven rather than evidence-based.

Think Blockbuster, Kodak, or Yahoo; each relied heavily on past formulas and executive confidence rather than adaptive modeling or user data. In contrast, firms like Netflix and Amazon have embedded algorithmic foresight into the core of their strategic models.

Linear Models in a Nonlinear World

Classical frameworks are rooted in linear assumptions, such as "increase market share to increase profits." But today's businesses operate in nonlinear ecosystems, where outcomes arise from feedback loops, tipping points, and compounding effects. Traditional models rarely capture the complexity and feedback inherent in real business ecosystems.

Tesla's strategy is not about price or cost leadership; it's about vertically integrated innovation. Meta scaled through network effects. Uber adapts supply and demand through dynamic pricing. These models defy linear cause-and-effect predictions and thrive on dynamic complexity.

Blind Spots in Technology and Data

Pre-digital models were never meant to process real-time data or run simulations. Today, static strategy cycles struggle to keep up. Firms like Zara, UPS, and TikTok exemplify data-powered agility, using analytics not just for insights but for real-time decisions. Old frameworks weren't built for big data. Without analytics, they can't anticipate; they can only explain.

STEM-powered strategic thinking means incorporating machine learning, scenario modeling, and simulation into every phase of strategy, from planning to execution to adaptation.

Peripheral Treatment of DEI, Sustainability, and Innovation

Traditional models viewed sustainability, innovation, DEI, and ethics as secondary. But modern markets demand they be central to value creation. Today's most resilient firms, Tesla, Patagonia, Microsoft, BlackRock, and Unilever, make purpose, inclusion, and innovation the core of their strategies. In the 21st century, sustainability and innovation are not side topics; they are the strategy. Table 1.6 provides failure and success examples.

Table 1.6. Center or Sideline? Strategic Lessons in DEI, Sustainability, and Innovation

Firm	Static Thinking Failure	Strategic Thinking Lesson
Zoom	Assumed post-COVID dominance	Failed to anticipate platform convergence and integration shifts
Kodak	Invented digital camera but didn't pivot	Let intuition override behavioral trends and innovation potential
TikTok	Succeeded by embedding real-time feedback	Continuous user data analysis and algorithmic adaptation
Tesla	Redefined product cycles with integration	Systems strategy, not product planning

Modern strategic management must embed **thinking within doing**. STEM tools, digital twins, predictive models, and optimization engines allow strategy to evolve as reality evolves. Strategic management can no longer rely on annual plans or static playbooks. It must become a **thinking system**, drawing insight from live data, testing scenarios through simulation, and recalibrating in real time. That's the essence of STEM-based strategic intelligence.

1.2.3. Why STEM? Understanding the Strategic Paradigm Shift

STEM doesn't replace strategic judgment, it strengthens it with evidence, adaptability, and systems-level thinking.

We are living in an era where the very nature of strategic thinking is being redefined. It is no longer sufficient to analyze markets, set goals, and execute fixed plans. In today's volatile, data-rich, and interconnected environment, strategic leadership demands precision, adaptability, and foresight, capabilities more closely aligned with STEM than with traditional business intuition. This section explores the rationale behind the shift toward STEM-based strategy and why it is becoming essential for modern leaders.

At its core, a STEM-based approach integrates empirical thinking, algorithmic modeling, and digital simulation into the heart of strategic decision-making. Whereas legacy frameworks often rely on one-time analysis or consensus-based intuition, STEM encourages a continuous loop of testing, validating, and adapting. Strategic decisions, like engineering designs, must now be stress-tested before implementation. Just as a bridge must be modeled for load-bearing forces, strategic plans must be modeled for market volatility, supply chain disruptions, and customer behavior changes.

The power of STEM in strategy lies in its ability to handle complexity and uncertainty. Decision-makers now face challenges that defy linear cause-and-effect reasoning: climate risk, cybersecurity threats, demographic shifts, and platform-based competition. These challenges require nonlinear tools, such as machine learning, Monte Carlo simulation, geospatial analysis, or systems dynamics modeling, that reflect the feedback-rich, interconnected reality of modern business.

Moderna's strategic advantage during the COVID-19 crisis came not just from its scientific breakthroughs but from its strategic use of STEM infrastructure. Using cloud-based models and AI to simulate thousands of mRNA combinations, the company shortened its research and development cycle from years to months. Moderna did not rely on long-range static planning. It modeled, iterated, and adapted its strategy in real time.

Similarly, firms like Google, Netflix, and Amazon have transformed strategy into a continuous learning loop. Google constantly A/B tests new features. Netflix relies on user data to shape content creation. Amazon adjusts its supply chain with predictive algorithms. These companies treat strategy not as a set of quarterly goals, but as an evolving system informed by live data, experimental feedback, and analytical insight.

The Manager of Tomorrow: From Intuition to Intelligence

For today's students and future managers, this shift isn't just a theoretical evolution; it's a career imperative. Employers increasingly seek professionals who blend strategic thinking with technical fluency. Being "strategic" now means being able to interpret data patterns, run simulations, build scenario models, and assess risks through quantitative reasoning. This isn't about turning managers and MBAs into engineers; it's about preparing them to lead in a world where data fluency, digital platforms, and systems intelligence are foundational to competitive advantage.

STEM-based business education responds to this shift. Across universities worldwide, STEM-designated MBA and management programs are growing rapidly, reflecting the need to connect management science with technology, analytics, and sustainability. These programs are not a trend; they are a transformation of what it means to lead.

This is not about replacing business acumen with technical jargon, it's about blending rigor and creativity to navigate uncertainty.

From Strategy Retreats to Strategic Systems

Organizations embracing STEM thinking are replacing opinion-driven debates with evidence-based dialogue. They prioritize real-time experimentation over annual strategy reviews. Their leaders think like scientists: formulate hypotheses, test variables, evaluate outcomes, and iterate rapidly. The entire strategic process becomes a living system, constantly sensing, learning, and evolving.

Table 1.7 below compares each stage of the traditional strategic management process with its STEM-enhanced counterpart. It illustrates how data analytics, machine learning, simulation tools, and digital collaboration platforms reshape how strategic planning is executed in real time.

Table 1.7. Strategic Management Process — Traditional vs. STEM-Driven Practice

Stage	Traditional Approach	STEM-Driven Enhancement	Example in Practice
Environmental Analysis	SWOT, PESTEL, Industry Reports	AI-driven trend scanning, sentiment analysis, and dashboards	Using NLP tools to scan global media for emerging market signals
Strategy Formulation	Mission statements, Porter's models	Simulations, scenario modeling, optimization algorithms	Simulating multiple futures for electric vehicle adoption
Strategy Implementation	Org charts, budget allocation, and training	Agile teams, cloud collaboration, OKRs, digital execution tools	Executing product rollout using Jira, Slack, and OKR alignment
Performance Evaluation	KPIs, Balanced Scorecard, quarterly reviews	Real-time dashboards, predictive analytics, and machine learning alerts	Using anomaly detection to identify deviations from sales targets

In today's world, strategic management operates like an intelligent control system, constantly sensing, responding, and recalibrating. STEM integration turns the process from periodic and reactive to continuous and predictive. Advanced analytics allow strategists to detect weak signals, anticipate disruptions, and stress-test strategies under multiple assumptions.

For instance, digital twins can model business ecosystems in real time; Monte Carlo simulations help assess financial and operational risks; and machine learning identifies hidden patterns in customer behavior or supply chain volatility. The strategist's dashboard becomes the nerve center for sensing, aligning, and acting. This shift marks a fundamental transformation, from **static cycles** to **dynamic loops**. The table below contrasts traditional strategy models with STEM-evolved approaches that embed technology at the core of decision-making:

Table 1.8. From Static Cycles to Dynamic Loops

Element	Traditional View	STEM-Evolved View
Nature of Process	Periodic, sequential	Continuous, iterative
Feedback	End-of-cycle reviews	Real-time alerts and learning loops
Decision Inputs	Historical data, expert judgment	Predictive models, live data streams
Role of Technology	Supporting documentation	Embedded in sensing, planning, and action
Execution	Hierarchical rollout	Agile, tech-enabled, cross-functional

1.3. Strategic Intelligence: The New Core Competency

In today's hyper-connected, data-saturated world, strategy is no longer just about choosing a direction and committing to it. The most successful organizations don't simply *plan* the future; they continuously *sense*, *learn*, and *adapt* to it. At the heart of this adaptive capacity lies **strategic intelligence**: the dynamic capability to gather, integrate, and act upon insights in a fast-changing environment.

Strategic intelligence is more than market research or competitive analysis. It fuses traditional strategic thinking with advanced tools of data science, behavioral insight, and systems modeling. Where classical strategy asked, "Where are we now, and where do we want to go?" strategic intelligence asks, "What's changing, what might happen next, and how can we adapt in time?"

This competency is especially critical in an era defined by volatility (e.g., geopolitical risks, climate change), exponential technologies (e.g., AI, IoT), and rapidly shifting consumer preferences. Relying on rear-view mirrors is no longer enough. Leaders must cultivate forward-facing dashboards, strategic intelligence systems, that continuously analyze data, interpret patterns, generate scenarios, and trigger timely action.

Strategic intelligence transforms the strategist into an architect of foresight. It enhances situational awareness, improves decision quality, and reduces response time. But achieving it requires the integration of several components: organizational culture, analytical infrastructure, cross-functional collaboration, and digital fluency. These must all be aligned to support real-time sensing, hypothesis-driven experimentation, and rapid feedback.

In a STEM-enabled context, strategic intelligence becomes the connective tissue between data and decision. It turns complexity into clarity and equips leaders to navigate not just the known, but the unfolding unknown.

Reflect on a project or organization you've studied or worked with:

- Which stage of the strategic management process was most visible?
- Was the process linear or iterative?
- How did the use (or absence) of technology impact the success of implementation?

1.3.1 What Is Strategic Intelligence?

Strategic intelligence is the capability to anticipate, interpret, and respond to complex environments in ways that create sustained organizational advantage. It represents the fusion of analytical acuity, technological fluency, and strategic foresight. More than just data gathering or reporting, strategic intelligence is an integrated process of generating actionable insights from multiple, often ambiguous, signals.

In traditional strategy, intelligence was often equated with static market research, competitor analysis, or performance reviews. These approaches, while valuable, tend to deliver snapshots in time. In contrast, strategic intelligence operates as a **real-time sensing system**, much like the neural network of an organism, continuously scanning internal and external environments, detecting weak signals, interpreting patterns, and generating predictive insights.

Strategic intelligence includes the following key components:

- **Environmental Scanning**: Monitoring trends, disruptions, regulatory shifts, and technological advances across industries, geographies, and stakeholders.
- **Data Interpretation and Pattern Recognition**: Identifying non-obvious connections, emerging risks, and opportunities by integrating quantitative and qualitative, structured and unstructured data.
- **Scenario Modeling and Forecasting**: Using Monte Carlo simulations, AI-driven tools, and sensitivity testing to explore futures and stress-test decisions.
- **Insight Generation and Decision Support**: Converting data into clear, actionable insights through dashboards, visual analytics, and collaborative decision platforms.
- **Organizational Learning**: Embedding feedback loops to refine sensing and strategic response based on past decisions and real-world outcomes.

Strategic Intelligence vs. Business and Competitive Intelligence

While the terms are often used interchangeably, they represent different layers of organizational insight. The table below highlights how **business intelligence**, **competitive intelligence**, and **strategic intelligence** differ in scope, tools, and strategic value.

Table 1.9. Comparing Business, Competitive, and Strategic Intelligence

Aspect	Business Intelligence	Competitive Intelligence	Strategic Intelligence
Primary Focus	Internal performance and operations	Monitoring competitors and market position	Anticipating broad external change and shaping long-term direction
Typical Users	Analysts, functional managers	Marketing, product, or strategy teams	Executives, board members, and strategic leaders
Time Horizon	Short- to medium-term	Medium-term	Long-term and foresight-based
Data Sources	Internal systems (CRM, ERP, sales reports)	Public competitor data, industry reports, and customer trends	Macro-trends, geopolitical shifts, tech disruption, stakeholder sentiment
Key Tools	Dashboards, KPIs, performance reports	SWOT on rivals, benchmarking, war-gaming	Scenario modeling, Monte Carlo, trend mapping, AI-driven scanning
Analytical Depth	Descriptive and diagnostic	Competitive profiling and tactical analysis	Predictive, systemic, exploratory
Strategic Value	Improves decision-making at the operational level	Enhances positioning and market response	Enables agility, vision, and resilience in uncertain environments

1.3.2 Why Strategic Intelligence Matters Now

Strategic decisions today must be made in increasingly unpredictable and high-stakes environments shaped by digital disruption, climate volatility, geopolitical uncertainty, and shifting societal expectations. In such complexity, traditional strategy tools, such as static planning cycles or executive intuition, are no longer sufficient on their own. Strategic intelligence has emerged as a necessary competency, enabling leaders to move from hindsight to foresight, from reaction to anticipation.

In a data-rich, algorithmically influenced world, the rules of competitive advantage are changing. Success no longer depends solely on who can plan best in the past, but on who can sense earliest, simulate fastest, and adapt continuously. Strategic intelligence replaces static assumptions with iterative learning, active sensing, and data-grounded insight. It turns strategy into a living process, resilient, responsive, and future-facing.

Table 1.10 compares reactive and intelligence-driven strategies across ten core dimensions and illustrates each with real-world examples.

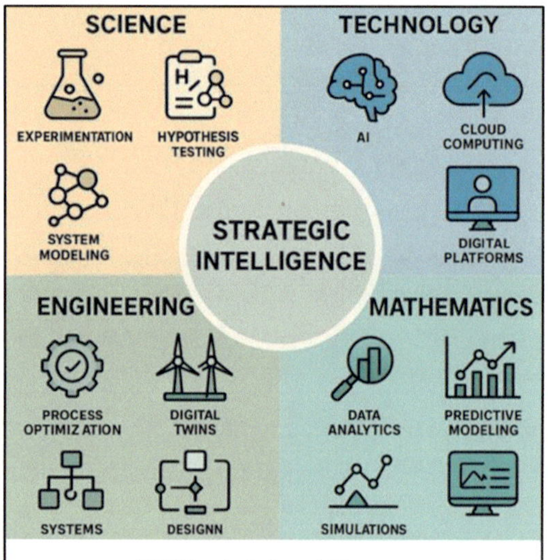

STEM Domains in Strategic Management

Strategic intelligence turns uncertainty into an advantage. Consider how **Google** constantly adjusts its strategy using data from millions of search queries and user tests, enabling rapid shifts in product design. In contrast, **Yahoo**, once a leader, failed to adapt with the same fluidity, overwhelmed by competing visions and delayed reactions.

Similarly, **Starbucks** uses geospatial intelligence to optimize store locations and product launches, combining customer data with environmental trends. This contrasts with brands that still make major expansion decisions based on historical sales performance alone.

These examples show that intelligence-driven strategy is not reserved for tech giants; it's a replicable mindset and skillset that any organization can cultivate. Whether responding to a supply chain shock, navigating a reputational crisis, or launching a new product in an uncertain market, strategic intelligence provides the foundation for informed, agile, and resilient action.

Table 1.10. Reactive vs. Intelligence-Driven Strategy

Dimension	Reactive Strategy	Intelligence-Driven Strategy	Real-World Example
Decision Timing	Reacts after disruption strikes	Acts on early signals before disruption unfolds	*Kodak* responded to digital cameras too late, whereas *Apple* preemptively pivoted into smartphones, disrupting itself.
Data Usage	Relies on backward-looking reports and lagging indicators	Uses predictive analytics, sentiment analysis, and real-time feedback	*Netflix* predicts viewing trends using real-time data; *Blockbuster* relied on past rental models.
Planning Style	Annual, rigid planning processes	Dynamic, rolling, scenario-driven planning	*Amazon* iterates its strategy quarterly and adapts based on customer behavior patterns.
Risk Management	Responds during crisis, often improvisational	Model risks ahead of time using simulations and decision trees	*Tesla* uses scenario modeling for supply chain risk; *GM* suffered during chip shortages.
Tools & Methods	Traditional dashboards, scorecards, and financial reports	Monte Carlo simulation, machine learning, geospatial modeling, digital twins	*Pfizer* used AI to simulate molecule interactions; *older pharma firms* relied on trial-heavy cycles.
Organizational Behavior	Siloed responses by function or department	Cross-functional collaboration, real-time information sharing	*Microsoft Teams* enables live alignment; *traditional firms* often misalign between strategy and execution.
Competitive Positioning	Market follower—reactive to trends set by others	Market shaper—introduces new categories and models	*Tesla* defined the EV market; *Nokia* failed to lead the smartphone shift.
Leadership Orientation	Relies on authority, experience, and top-down decision-making	Combines expert insight with data-driven decision support	*Moderna's* AI-supported R&D empowered scientists to make strategic calls; *Merck* had more centralized decisions.
Learning Approach	Post-failure reviews and internal audits	Embedded feedback loops for continuous learning	*Toyota* implements "kaizen" continuous learning; *others* perform quarterly post-mortems.
Strategic Mindset	"Wait and see"—assumes the future will resemble the past	"Sense, simulate, and act"—treats the future as uncertain, dynamic, and designable	*Unilever* uses foresight tools for sustainability; *older FMCGs* waited for regulatory pressure.

1.3.3 STEM as the Foundation for Strategic Intelligence

The rise of strategic intelligence marks a profound shift in the competencies required of today's strategic leaders. At the heart of this transformation lies STEM, not as a set of technical skills, but as a way of thinking, reasoning, and acting in complex environments. The fusion of STEM disciplines with strategic decision-making enables leaders to not only understand the forces shaping their industries but to anticipate them, simulate their impacts, and engineer intelligent responses.

Strategic intelligence is no longer sustained by intuition, experience, or retrospective analysis alone. It thrives on the ability to detect patterns in vast data streams, test multiple scenarios under conditions of uncertainty, and optimize decisions using algorithmic reasoning. These capabilities are inherently STEM-driven.

This is not about turning managers into software engineers or data scientists. Rather, it is about enabling them to think like scientists, curious, hypothesis-driven, and evidence-based; to act like engineers, systematic, model-oriented, and solution-focused; and to lead like systems designers. Navigating complexity, interdependencies, and emergent behavior.

STEM as a Cognitive Operating System for Strategists

Consider the strategic challenges facing a global firm: predicting customer preferences, responding to supply chain disruptions, launching products in volatile markets, or adapting to sustainability pressures. Each of these requires more than a well-crafted mission statement. It demands tools and mindsets that STEM equips us with:

> Think of a real or hypothetical organization navigating a major disruption (e.g., pandemic, digital shift, geopolitical risk).
>
> - Would it benefit more from a reactive or intelligence-driven approach?
> - How could predictive tools or foresight methods improve its strategic response?
> - What role should you play as a strategic thinker in embedding intelligence?

- **Data Analysis and Visualization:** Leaders today must be able to interpret high-dimensional data, detect anomalies, and extract insight from dashboards, geospatial maps, and analytics platforms.
- **Predictive Modeling and Forecasting:** From revenue projections to climate-related risk modeling, strategies now depend on the ability to simulate possible futures using time-series analysis, regression, and machine learning.
- **Systems Thinking and Simulation:** Strategic leaders must model dynamic systems, markets, ecosystems, operations, as networks of feedback loops, delays, and compounding effects. Tools such as system dynamics, agent-based modeling, or digital twins provide the scaffolding for such simulations.

- **Algorithmic Reasoning:** Decision trees, optimization algorithms, and AI models enhance decision-making by exploring vast combinations of choices and constraints faster than human deliberation alone can manage.

These STEM capabilities are not abstract; they are already reshaping how strategy is practiced in real-world settings. Table 1.11 illustrates how each capability translates into a strategic intelligence function, along with concrete corporate examples.

Table 1.11. STEM Applications in Strategic Intelligence

STEM Capability	Strategic Intelligence Application	Example in Practice
Data Science	Identify trends, segment markets, optimize pricing	*Spotify* analyzes listening patterns to personalize user content
Systems Engineering	Model supply chains, simulate stress scenarios	*Unilever* uses digital twins to simulate logistics disruptions
Mathematical Modeling	Forecast demand, test sensitivity of assumptions	*Airbnb* uses Monte Carlo simulations for pricing and occupancy
Artificial Intelligence	Predict churn, detect fraud, forecast performance	*Ant Financial* uses AI to assess credit risk in real-time
Visualization & Interfaces	Support executive dashboards, scenario mapping	*GE* uses interactive dashboards for operations & decision support

From Linear Plans to Live Models

Classic strategic planning relied on historical data, expert consensus, and top-down decision-making. In contrast, the STEM-powered strategist uses live models, continuously refined by real-time data and machine learning feedback. Instead of asking "what happened last year?", they ask "what could happen tomorrow, and how can we model it today?"

Take **Moderna's** mRNA vaccine development as an example. Rather than follow the traditional multi-year R&D pathway, Moderna integrated AI and cloud simulations to rapidly test thousands of genetic sequences. The company's leadership did not just authorize R&D, they were engaged in strategic modeling of supply chain capacity, regulatory responses, and distribution logistics, all underpinned by STEM frameworks.

In another case, **Maersk**, the global shipping giant, uses real-time satellite data and predictive weather modeling to optimize routes, reduce fuel costs, and avoid disruption. These decisions are not left to static planning, they are shaped by a STEM-enabled command center that supports continuous navigation and strategy alignment.

In the end, strategic intelligence is not a department, it's a capability embedded across the organization, and **STEM is its cognitive infrastructure**. As this textbook unfolds, you will see how each function of the strategy process, analysis, formulation, implementation, and adaptation, can be reimagined through STEM. Not to replace strategic judgment, but to **enhance it with clarity, rigor, and dynamic responsiveness**.

Quick Reflection

- Which STEM skills, data science, simulation, modeling, do you think will be most critical in your career?
- How can you begin to shift your own strategic thinking from intuition-based to evidence-based?
- What tools or platforms could help you apply this mindset in a real-world business setting?

1.3.4 Developing Strategic Intelligence as a Leader

Strategic intelligence is not a fixed trait, it is a cultivated capability. It grows through intentional exposure to complexity, analytical rigor, and the development of a learning-oriented mindset. In a world defined by volatility, uncertainty, complexity, and ambiguity (VUCA), traditional leadership styles, relying on charisma, intuition, or past experience, are no longer enough. What distinguishes effective leaders today is their ability to interpret weak signals, navigate ambiguity, simulate scenarios, and align action with foresight.

This textbook is designed to help you build those skills. Through the chapters ahead, you will move:

- From reactive decision-making to predictive insight
- From static reports to real-time data loops
- From narrow expertise to system-wide awareness

But strategic intelligence is more than a collection of tools, it reflects a transformation in leadership identity. Strategic leaders must become rigorous learners, systems thinkers, and ethical decision-makers who act on evidence while anticipating long-term impacts. Four critical capacities drive this transformation:

1. **Intellectual Curiosity** – A drive to question assumptions, explore emerging trends, and investigate beyond the obvious. Curious leaders resist confirmation bias and actively seek out conflicting data to sharpen insight.

2. **Systems Awareness** – The ability to see organizations as dynamic systems with feedback loops and interdependencies. This allows leaders to anticipate ripple effects and unintended consequences.
3. **Ethical Foresight** – The capacity to consider not only what is strategically advantageous, but also what is socially responsible and sustainable.
4. **Computational Thinking** – A way of breaking complex problems into structured components. This includes abstraction, modeling, and algorithmic logic, even for leaders who do not write code.

These leadership attributes are not abstract, they guide real-world decision-making. To illustrate the difference strategic intelligence makes, consider the comparison in Table 1.12:

Table 1.12. Reactive vs. Intelligence-Driven Strategy: Real-World Applications

Strategic Characteristic	Reactive Strategy	Intelligence-Driven Strategy	Case Example
Trigger	Crisis or lagging performance	Ongoing sensing and anticipation	*Target* adjusted prices only after losing share to Amazon vs. *Amazon* using predictive pricing models daily
Decision Basis	Historical data, gut instinct	Real-time data, simulations, forecasting	*Kodak* relied on film legacy; *Canon* modeled digital adoption early
Timing	Periodic (quarterly, annually)	Continuous, real-time	*Nokia* stuck to fixed plans; *Apple* adapted rapidly to smartphone demand
Execution Mode	Top-down, rigid plans	Agile, data-informed, adaptive teams	*Blockbuster* held to legacy plans; *Netflix* constantly A/B tests content
Technology Role	Support tool	Core to strategy formulation and adjustment	*Unilever* uses digital twins for logistics; *Traditional FMCG firms* lag
Risk Management	Reactive after loss	Modeled and stress-tested in advance	*Tesla* models raw material risks with AI; others responded after shortages

These cases reflect the core value of developing strategic intelligence: resilience, responsiveness, and advantage in a fast-changing world. Becoming an intelligence-driven leader does not require perfection, it requires practice. That means engaging with dashboards and simulation tools, participating in foresight exercises, learning from interdisciplinary collaboration, and refining your ability to communicate insights with clarity and impact. It also means confronting ethical dilemmas and recognizing that every data-driven decision has a human and societal consequence.

Strategic intelligence is the ability to translate complex, dynamic inputs into actionable foresight. It is what allows leaders not just to understand the present but to shape the future. By cultivating this mindset now, you are preparing not only to lead in your organization, but to transform it.

1.3.5 Strategic Intelligence in Action

Strategic intelligence is not confined to corporate boardrooms or academic models, it is embedded in how leaders across industries interpret signals, navigate uncertainty, and act with foresight. Its impact can be seen in the worlds of commerce, public policy, healthcare, technology, entertainment, hospitality, and beyond. The unifying thread is the capacity to sense complexity, simulate future conditions, and respond with precision.

Consider the case of **Netflix**. What once began as a DVD rental service has evolved into a global content powerhouse through data-driven storytelling. Netflix doesn't just analyze ratings; it mines viewing behavior, regional trends, and user sentiment to shape what it produces and how it releases it. Strategic intelligence guides decisions from script approval to global marketing campaigns, allowing the company to anticipate preferences and reduce creative risk.

In **public health**, the **South Korean government's** strategic intelligence framework was key to its world-leading response to the COVID-19 pandemic. While many countries lagged behind in detection and containment, South Korea employed digital dashboards, real-time testing data, and location tracking to identify outbreaks before they escalated. Instead of reactive policy, their strategy was anticipatory, built on a platform of integrated data streams, simulations, and rapid decision cycles.

In **shipping and logistics**, **Maersk**, the global maritime leader, uses real-time satellite and weather data to make strategic routing decisions. Rather than depend on seasonal forecasts or fixed schedules, Maersk continuously adjusts its operations based on predictive analytics—reducing delays, lowering fuel consumption, and increasing reliability in an unpredictable world.

In **healthcare innovation**, **Moderna's** strategic intelligence wasn't just about scientific capability, it was the infrastructure to simulate thousands of mRNA combinations in silico using cloud computing and AI. Its leaders coordinated regulatory, supply chain, and R&D strategies based on predictive insights, not static assumptions. This intelligence-enabled approach allowed Moderna to move faster and smarter than traditional pharmaceutical players.

In **entertainment**, we see the power of strategic intelligence in action. **Taylor Swift's** record-breaking concert tours are not just driven by artistry, they're powered by social sentiment analysis, geographic fan data, and real-time simulations of ticket demand. Her team doesn't guess, they model. Strategic foresight allows them to craft marketing campaigns, merchandise releases, and media engagements that are aligned to dynamic audience feedback loops.

In the **hospitality industry**, **Marriott International** has integrated advanced analytics into its pricing, staffing, and customer engagement strategies. By using real-time booking data, weather patterns, local event forecasts, and travel trends, Marriott dynamically adjusts room rates, predicts occupancy, and personalizes guest experiences, all informed by machine learning and simulation models.

In **finance**, **Goldman Sachs** employs algorithmic trading, scenario stress tests, and real-time sentiment analysis to manage investment strategies and portfolio risks. Strategic intelligence helps the firm navigate volatility, regulatory shifts, and macroeconomic trends with precision and speed.

In **education**, **Arizona State University (ASU)** exemplifies intelligence-driven transformation. By using student performance analytics, predictive enrollment modeling, and AI-supported advising systems, ASU improves student outcomes, enhances retention, and drives institutional planning.

Each of these examples (Table 1.13) highlights how strategic intelligence enables decision-makers to move from instinct to insight, from fixed planning to adaptive foresight, and from reaction to anticipation. It is not limited to one sector or profession, it is the new operating system for modern leadership.

Table 1.13. Strategic Intelligence Across Sectors

Sector	Organization / Individual	Strategic Intelligence in Action
Technology	Google	Uses continuous A/B testing and real-time analytics to refine user experience and guide product development.
Healthcare	Moderna	Applied AI and cloud modeling to simulate mRNA combinations and coordinate strategy across R&D, regulation, and logistics.
Shipping	Maersk	Leverages satellite data and weather forecasts to optimize maritime routes and reduce disruption risks.
Public Sector	South Korean Government	Integrated live dashboards and location tracking to forecast pandemic spread and implement agile public health policy.
Entertainment	Netflix	Uses machine learning and regional viewing data to shape content development and release strategies globally.
Music Industry	Taylor Swift	Applies fan analytics, predictive modeling, and geo-data to plan concerts and merchandise strategies with pinpoint timing.
Hospitality	Marriott International	Dynamically adjusts pricing and staffing based on real-time booking data, events, and demand forecasts.
Finance	Goldman Sachs	Uses algorithmic trading, sentiment analysis, and stress testing to manage portfolio strategy under uncertainty.
Education	Arizona State University	Uses predictive analytics to improve student outcomes and guide institutional strategy and advising.

1.4. Chapter 1 Wrap-Up: Strategic Intelligence in a STEM World

As we close this opening chapter, it's clear that strategic management is no longer confined to abstract planning or intuition. In today's dynamic landscape, shaped by data, disruption, and systemic uncertainty, leaders must develop **strategic intelligence**: the ability to interpret complexity, sense opportunity, and respond with precision.

Throughout this chapter, we examined the transformation of strategy from a linear, business-only discipline into a **STEM-enabled process** of continuous learning, simulation, and insight generation. We defined strategic intelligence as the core competency of future leaders and explored how disciplines like **data science, engineering, mathematics**, and **systems thinking** are fundamentally reshaping how strategy is conceived and executed.

From the evolution of the strategic management process to the case examples of firms like **Moderna**, **GE**, and **Maersk**, one message stands out: modern strategy is not only about where to compete and how to win, it's about how to sense, simulate, and adapt in real time. Strategic decisions can no longer rely solely on static models or retrospective analysis; they must be **proactive, evidence-based, and iterative**.

We also looked at how strategic intelligence is applied across industries, including hospitality, government, tech, and even non-profits, and why **computational thinking**, **ethical foresight**, and **systems awareness** are vital attributes for the next generation of strategic thinkers. The rise of STEM-based strategy doesn't mean replacing business acumen; it means augmenting it with clarity, speed, and foresight.

As you move through the rest of this textbook, you'll explore how each phase of the strategy process, analysis, formulation, implementation, and adaptation, can be **reimagined through a STEM lens**, giving you the tools to navigate ambiguity and lead intelligently in a rapidly evolving world.

1.4.1. Discussion Questions

1. How does strategic intelligence differ from traditional business intelligence?
2. What STEM skills are most relevant for strategic leadership in your field?
3. Why is systems thinking essential in today's business environment?
4. How can leaders balance ethical foresight with speed and competitiveness?
5. What are the risks of relying solely on intuition for strategy?
6. Which tool introduced in this chapter do you see yourself using first?
7. How does scenario planning enhance strategic decision-making?
8. In what ways can strategic intelligence be applied in non-profits or government?
9. What did the Moderna or Maersk examples teach you about real-time strategy?
10. How does visualization help communicate strategic insights to stakeholders?

1.4.2. Micro-Case: Zara's Strategic Sprint

Zara is a rising strategic leader at a fast-growing fintech firm known for leveraging advanced analytics to reshape financial decision-making. Recognized for her systems thinking and data fluency, she is tapped by the executive team to lead a high-stakes initiative: launching a new **AI-powered ESG (Environmental, Social, and Governance) scoring platform,** in just 60 days.

The platform is designed to help investors evaluate companies' ESG performance using real-time data and machine learning. It promises dynamic scoring models, policy integration, and investor-friendly visualizations. The ambition is to launch simultaneously across North America, Europe, and Asia.

But Zara faces a tangled web of challenges. Global ESG standards remain fragmented and in flux—what qualifies as sustainability in one country may be politically controversial in another. Early versions of the AI model exhibit **biases**, including overemphasis on governance indicators in Western firms and underestimation of social metrics in developing economies. Meanwhile, some stakeholders, particularly in conservative markets, are skeptical of ESG scoring in general, seeing it as opaque or ideologically charged.

Her team is interdisciplinary and sharp, but they speak in different technical languages. The data scientists want training data. The legal advisors want guardrails. The climate experts want rigor. Product developers want a roadmap.

Zara knows this is more than a product launch. It's a live strategy sprint, compressed in time, high in uncertainty, and charged with ethical, reputational, and financial implications.

She begins by applying **PESTLE analysis** to scan regulatory and social factors across target regions. Using **system modeling**, she maps interdependencies between data sources, scoring algorithms, and user responses. She scopes the MVP (minimum viable product) to emphasize transparency and stress-tests scoring outcomes under different assumptions.

Zara also proposes a **multi-level KPI structure**: technical accuracy, user adoption, regional ESG alignment, and public sentiment. Feedback loops are embedded into the platform's dashboard, allowing real-time updates, alerts for controversial scores, and dashboards customizable by region.

To address algorithmic bias, Zara brings in an external ethics advisor and launches a parallel testing system for diverse scenarios. She leads cross-functional sprints with shared visual dashboards and insists on weekly retrospectives to adapt quickly.

By Day 45, the team has not only built a functioning prototype, they have also co-created a governance model for AI scoring. The platform will launch on Day 60 in beta form, but with clear transparency protocols and a dynamic update loop.

Zara's sprint illustrates what **strategic intelligence** looks like in action: blending foresight, modeling, stakeholder sensitivity, and fast-cycle execution. Her success lies not in solving every problem up front, but in designing a system that **learns, adapts, and earns trust over time**.

Discussion Questions

1. What strategic tools from Chapter 1 would you use to prioritize Zara's challenges?
2. How could systems thinking improve the platform's long-term adaptability?
3. What ethical considerations should shape Zara's decision-making process?
4. How would you balance precision and speed in a project like this?

If you were leading the team, how would you ensure cross-functional alignment and fast learning?

1.4.3. Interactive Exercise: Strategy Under Uncertainty

Scenario:
You're advising a mid-sized renewable energy company deciding whether to expand into **Morocco**, **Chile**, or **Vietnam**. You have access to:

- Regional climate risk models
- Local policy dashboards (e.g., energy subsidies, ESG reporting mandates)
- Consumer adoption trends from social listening tools

Your Task:

Using any two tools/frameworks from this chapter, such as **PESTLE**, **system modeling**, or **predictive analytics,** develop a one-page strategic recommendation. Your response should:

- Identify key uncertainties and describe how to model them
- Provide a data-driven rationale for the recommended country
- Propose at least two post-launch evaluation metrics

This simulation reflects how modern strategy blends structured frameworks with real-time digital tools.

1.4.4. Recommended HBR Case Studies

❖ Core Innovation Capital: Investing in Fintech for Good

- **Overview**: This case examines how Core Innovation Capital, a venture capital firm, evaluates investments in fintech startups that aim to deliver both financial returns and positive social impact. It delves into the challenges of assessing ESG factors and the strategic decision-making processes involved in aligning investments with ethical considerations.

- **Citation**: Kluender, Ray, Natalia Rigol, Benjamin Roth, and Nicole Tempest Keller. "Core Innovation Capital: Investing in Fintech for Good." Harvard Business School Case 825-064, December 2024.

❖ **PortageBay and ESG Analytics**

- **Overview**: This case examines how PortageBay, a sustainable analytics platform, leverages AI to aggregate and analyze ESG data. It addresses challenges such as the lack of standardized ESG data, mitigating greenwashing risks, and balancing financial returns with sustainability goals.
- **Citation**: Gandhi, Vikram S., and Radhika Kak. "PortageBay and ESG Analytics." Harvard Business School Case 324-065, January 2024. (Revised May 2024.)

❖ **Generation Investment Management: Sustainable Investing in a Warming World**

- **Overview**: This case explores how Generation Investment Management evaluates companies for investment, focusing on sustainability and ESG factors. It delves into the firm's decision-making processes and the integration of ESG considerations into investment strategies.
- **Citation**: Gandhi, Vikram S., Michael Norris, and David Allen. "Generation Investment Management: Sustainable Investing in a Warming World." Harvard Business School Case 324-043, January 2024. (Revised May 2024.)

❖ **CPP Investments—The Road to Zero**

- **Overview**: This case focuses on how CPP Investments, one of the world's largest pension funds, commits to achieving net-zero greenhouse gas emissions across its portfolio by 2050. It examines the strategic considerations and challenges involved in aligning investment practices with long-term sustainability goals.
- **Citation**: Gandhi, Vikram, and Radhika Kak. "CPP Investments—The Road to Zero." Harvard Business School Case 324-054, February 2024. (Revised April 2024.)

1.4.5. Strategy-in-Film: Learn by Watching

Film offers a powerful lens to observe how strategy evolves in response to disruption and complexity. The selections below illustrate organizations navigating changing environments, technological shifts, and competitive pressures. As you explore the paradigm shift discussed in this chapter, use these narratives to reflect on how strategy must adapt, and what happens when it doesn't.

Movie / Documentary	Strategic Theme Highlighted
Moneyball	Data-driven decision-making, challenging tradition
The Social Network	Startup strategy, intellectual property, platform growth
The Big Short	Systemic risk, modeling, ethical foresight
The Imitation Game	Algorithmic reasoning, strategic urgency during crisis
Her	AI integration and ethical foresight in product strategy
Margin Call	Strategic response under uncertainty
Iron Man	STEM-powered innovation and decision-making
Hidden Figures	Data, diversity, and strategic breakthroughs in high-stakes environments
Don't Look Up	Systems denial, strategic communication in complex crises
Ex Machina	Strategic risk and ethics of AI development
Hotel Rwanda	Humanitarian strategy and crisis decision-making in extreme uncertainty

References

Ansoff, H. I. (1957). Strategies for diversification. *Harvard business review*, *35*(5), 113-124.

Ansoff, H. I. (1977). The changing shape of the strategic problem. *Journal of General Management*, *4*(4), 42-58.

Ansoff, H. I. (1991). Critique of Henry Mintzberg's 'The design school: reconsidering the basic premises of strategic management'. *Strategic management journal*, *12*(6), 449-461.

Ayoub, K., & Payne, K. (2016). Strategy in the age of artificial intelligence. *Journal of strategic studies*, *39*(5-6), 793-819.

Bazile, J., Côté, A. M., Toumi, S., & Su, Z. (2025). Strategic intelligence as a resilience capability of global supply chains: Proposal of a conceptual framework based on a systematic literature review. *Journal of Global Operations and Strategic Sourcing.*

Bonn, I. (2001). Developing strategic thinking as a core competency. *Management decision*, *39*(1), 63-71.

Calof, J. (2017). Canadian competitive intelligence practices–a study of practicing strategic and competitive intelligence professionals Canadian members. *foresight*, *19*(6), 577-589.

Calof, J. L., & Wright, S. (2008). Competitive intelligence: A practitioner, academic and inter-disciplinary perspective. *European Journal of marketing*, *42*(7/8), 717-730.

Cheng, Y. C., & So, W. W. M. (2020). Managing STEM learning: A typology and four models of integration. *International Journal of Educational Management*, *34*(6), 1063-1078.

Collis, D. J., & Rukstad, M. G. (2008). Can you say what your strategy is?. *Harvard business review*, *86*(4), 82-90.

Craig, C. A., Sayers, E. L. P., Gilbertz, S., & Karabas, I. (2022). The development and evaluation of interdisciplinary STEM, sustainability, and management curriculum. *The International Journal of Management Education*, *20*(2), 100652.

Davies, W. (2000). Understanding strategy. *Strategy & Leadership*, *28*(5), 25-30.

Dess, G. G., McNamara, G., Eisner, A. B., & Sauerwald, S. (2023). *Strategic management: Text and cases* (11th ed.). McGraw Hill. ISBN: 9781264124329

Deveci, İ., & Seikkula-Leino, J. (2023). The link between entrepreneurship and STEM education. In *Enhancing entrepreneurial mindsets through STEM education* (pp. 3-23). Cham: Springer International Publishing.

Durand, R., Grant, R. M., & Madsen, T. L. (2017). The expanding domain of strategic management research and the quest for integration. *Strategic Management Journal*, *38*(1), 4-16.

Eden, C., & Ackermann, F. (1998). *Making strategy: The journey of strategic management.* Sage.

Evered, R. (1983). So what is strategy?. *Long range planning*, *16*(3), 57-72.

Felin, T., & Powell, T. C. (2016). Designing organizations for dynamic capabilities. *California management review*, *58*(4), 78-96.

Ghasemaghaei, M. (2019). Does data analytics use improve firm decision making quality? The role of knowledge sharing and data analytics competency. *Decision Support Systems*, *120*, 14-24.

Gitelman, L. D., Kozhevnikov, M. V., & Chebotareva, G. S. (2021). *Strategic intelligence of an organization amid uncertainty* (pp. 294-305). WIT Press: Ashurst, Southampton, UK.

Hacıoğlu, Y., & Gülhan, F. (2021). The effects of STEM education on the students' critical thinking skills and STEM perceptions. *Journal of Education in Science Environment and health*, *7*(2), 139-155.

Hambrick, D. C., & Fredrickson, J. W. (2001). Are you sure you have a strategy?. *Academy of Management Perspectives*, *15*(4), 48-59.

Hamel, G., & Prahalad, C. K. (1996). *Competing for the Future.* Harvard Business Press.

Herrmann, P. (2005). Evolution of strategic management: The need for new dominant designs. *International Journal of management reviews*, *7*(2), 111-130.

Hitt, M. A., Ireland, R. D., & Hoskisson, R. E. (2023). *Strategic management: Concepts and cases: Competitiveness and globalization* (14th ed.). Cengage Learning.

Iansiti, M., & Lakhani, K. R. (2020). *Competing in the age of AI: Strategy and leadership when algorithms and networks run the world.* Harvard Business Press.

Jamali, S. M., Ale Ebrahim, N., & Jamali, F. (2023). The role of STEM Education in improving the quality of education: a bibliometric study. *International Journal of Technology and Design Education*, *33*(3), 819-840.

Jourdan, Z., Rainer, R. K., & Marshall, T. E. (2008). Business intelligence: An analysis of the literature. *Information systems management*, *25*(2), 121-131.

Keding, C. (2021). Understanding the interplay of artificial intelligence and strategic management: four decades of research in review. *Management Review Quarterly*, *71*(1), 91-134.

Ketchen, D. and Short, J. (2020). *Mastering Strategic Management.* (2.1). FlatWorld.

Kidman, G., Gesthuizen, R., Tan, H., & Dielenberg, H. (2023). An entrepreneurial STEM teaching framework: integrating business and STEM education. In *Enhancing*

Entrepreneurial Mindsets Through STEM Education (pp. 93-113). Cham: Springer International Publishing.

Köseoglu, M. A., Ross, G., & Okumus, F. (2016). Competitive intelligence practices in hotels. *International Journal of Hospitality Management, 53*, 161-172.

Köseoglu, M. A., Morvillo, A., Altin, M., De Martino, M., & Okumus, F. (2020). Competitive intelligence in hospitality and tourism: a perspective article. *Tourism Review*, *75*(1), 239-242.

Köseoglu, M. A., Mehraliyev, F., Altin, M., & Okumus, F. (2021). Competitor intelligence and analysis (CIA) model and online reviews: integrating big data text mining with network analysis for strategic analysis. *Tourism Review*, *76*(3), 529-552.

Kühl, S. (2024). Strategy is structure. A systems theory-based definition of strategy. *Journal of Organizational Sociology*, *1*(3), 251-286.

Levine, S. S., Bernard, M., & Nagel, R. (2017). Strategic intelligence: The cognitive capability to anticipate competitor behavior. *Strategic Management Journal*, *38*(12), 2390-2423.

Liebowitz, J. (2006). *Strategic intelligence: business intelligence, competitive intelligence, and knowledge management.* Auerbach Publications.

Lynch, R. (2018). *Strategic management.* Pearson UK.

Mintzberg, H. (1987). The strategy concept I: Five Ps for strategy. *California management review*, *30*(1), 11-24.

Montgomery, D. B., & Weinberg, C. B. (1998). Toward strategic intelligence systems. *Marketing Management*, *6*(4), 44.

Moon, B. J. (2013). Antecedents and outcomes of strategic thinking. *Journal of business research*, *66*(10), 1698-1708.

Mudau, T. N., Cohen, J., & Papageorgiou, E. (2024). Determinants and consequences of routine and advanced use of business intelligence (BI) systems by management accountants. *Information & Management*, *61*(1), 103888.

Nag, R., Hambrick, D. C., & Chen, M. J. (2007). What is strategic management, really? Inductive derivation of a consensus definition of the field. *Strategic management journal*, *28*(9), 935-955.

Nickols, F. (2016). Strategy, strategic management, strategic planning and strategic thinking. *Management Journal*, *1*(1), 4-7.

Niu, Y., Ying, L., Yang, J., Bao, M., & Sivaparthipan, C. B. (2021). Organizational business intelligence and decision making using big data analytics. *Information Processing & Management*, *58*(6), 102725.

Nuntamanop, P., Kauranen, I., & Igel, B. (2013). A new model of strategic thinking competency. *Journal of Strategy and Management*, *6*(3), 242-264.

Parnell, J. A. (2020). *Strategic management: Theory and practice* (6th ed.). Academic Media Solutions.

Porter, M. E. (1991). Towards a dynamic theory of strategy. *Strategic management journal*, *12*(S2), 95-117.

Porter, M. E. (1985). *Competitive advantage: Creating and sustaining superior performance.* Free Press.

Porter, M. E. (1996). What is strategy? *Harvard Business Review*, 74(6), 61–78.

Ronda-Pupo, G. A., & Guerras-Martin, L. Á. (2012). Dynamics of the evolution of the strategy concept 1962–2008: a co-word analysis. *Strategic management journal*, *33*(2), 162-188.

Rouach, D., & Santi, P. (2001). Competitive intelligence adds value:: Five intelligence attitudes. *European management journal, 19*(5), 552-559.
Rothaermel, F. T. (2019). *Strategic management.* McGraw-Hill.
Saloner, G., Shepard, A., & Podolny, J. (2005). *Strategic management.* John Wiley & Sons.
Santa Soriano, A., & Valdes, R. M. T. (2021). Engaging universe 4.0: The case for forming a public relations-strategic intelligence hybrid. *Public Relations Review, 47*(2), 102035.
Schoemaker, P. J., Heaton, S., & Teece, D. (2018). Innovation, dynamic capabilities, and leadership. *California management review, 61*(1), 15-42.
Shapira, I. (2020). Strategic intelligence as an art and a science: creating and using conceptual frameworks. *Intelligence and National Security, 35*(2), 283-299.
Sheu, M. (2025). Why underserved business students show a lack of interest in developing sought-after STEM skills. *Industry and Higher Education, 39*(1), 61-73.
Shrestha, Y. R., Ben-Menahem, S. M., & Von Krogh, G. (2019). Organizational decision-making structures in the age of artificial intelligence. *California management review, 61*(4), 66-83.
Singer, J. G. (2008). What strategy is not. *MIT Sloan Management Review, 49*(2), 96.
Skok, J. E. (1989). Toward a definition of strategic management for the public sector. *The American Review of Public Administration, 19*(2), 133-147.
Spender, J. C. (2014). *Business strategy: Managing uncertainty, opportunity, and enterprise.* Oxford University Press.
Steen, E. V. D. (2017). A formal theory of strategy. *Management Science, 63*(8), 2616-2636.
Steptoe-Warren, G., Howat, D., & Hume, I. (2011). Strategic thinking and decision making: literature review. *Journal of strategy and management, 4*(3), 238-250.
Teece, D., Peteraf, M., & Leih, S. (2016). Dynamic capabilities and organizational agility: Risk, uncertainty, and strategy in the innovation economy. *California management review, 58*(4), 13-35.
Teece, D. J., Pisano, G., & Shuen, A. (1997). Dynamic capabilities and strategic management. *Strategic management journal, 18*(7), 509-533.
Voulvoulis, N., Giakoumis, T., Hunt, C., Kioupi, V., Petrou, N., Souliotis, I., & Vaghela, C. J. G. E. C. (2022). Systems thinking as a paradigm shift for sustainability transformation. *Global Environmental Change, 75*, 102544.
Welch, J. (2021). Success beyond STEM: an analysis of educational background of the Fortune 50 CEOs. *Higher Education, Skills and Work-Based Learning, 11*(2), 557-575.
Whittington, R. (2006). Completing the practice turn in strategy research. *Organization studies, 27*(5), 613-634.
Zaza, S., Abston, K., Arik, M., Geho, P., & Sanchez, V. (2020). What CEOs have to say: Insights on the STEM workforce. *American Business Review, 23*(1), 9.
Zeeshan, K., Watanabe, C., & Neittaanmäki, P. (2021, October). Problem-solving skill development through STEM learning approaches. In *2021 IEEE Frontiers in Education Conference (FIE)* (pp. 1-8). IEEE.

CHAPTER 2

Systems Thinking for Strategic Leaders

"If you only see the parts you miss the strategy"

by MAK

Learning Objectives:

By the end of this chapter, readers will be able to:

- Reframe traditional strategy concepts, such as strategic fit, value chain analysis, and organizational alignment, through a systems thinking lens to understand interdependencies, dynamic feedback, and non-linear effects.
- Define systems thinking and distinguish it from linear approaches to strategic planning, explaining why it is essential for navigating complexity, uncertainty, and interconnected challenges in the modern business environment.
- Analyze feedback loops (reinforcing and balancing) and delays within strategic systems using tools like causal loop diagrams and systems archetypes to anticipate systemic behavior over time.

A single bee knows only its task, but the hive thrives through connection, feedback, and collective intelligence.

- Apply systems thinking principles to identify leverage points in strategic contexts and model potential unintended consequences, ethical, operational, and social, before they emerge.
- Utilize STEM-based tools, including digital twins, system dynamics software, agent-based modeling, and predictive analytics, to simulate, test, and evaluate strategic scenarios across disciplines and industries.
- Evaluate real-world strategy challenges using adaptive systems approaches, demonstrating how resilient, ethically informed strategies outperform static plans in volatile or uncertain conditions.
- Develop modeling mindsets by translating strategy from static frameworks into evolving systems of interaction, supporting evidence-based decisions, ethical foresight, governance alignment, and cross-functional leadership.

In today's volatile, interconnected, and data-rich world, strategy can no longer be approached as a static sequence of isolated decisions. Competitive advantage no longer comes just from choosing the right industry position or optimizing internal operations; it comes from understanding how everything connects. Systems thinking is not a buzzword; it is a strategic necessity. It enables leaders to map complexity, diagnose problems beneath the surface, and design interventions that ripple through organizations with insight and precision.

This chapter introduces systems thinking as a foundational capability for 21st-century strategic leaders. Where traditional models like the value chain and SWOT offer snapshots, systems thinking reveals motion. It shifts the strategist's lens from parts to patterns, from events to structures, and from symptoms to root causes. The strategist of today must be able to see interdependencies, model feedback loops, anticipate unintended consequences, and act with clarity in environments of constant change.

To understand this shift, we begin with a company that didn't just evolve its strategy; it reprogrammed it: Ping An. Originally a traditional insurance company in China, Ping An has become a global model for how systems thinking, powered by data, AI, and platform integration, can transform strategy into a living, adaptive system. As we'll explore, Ping An's real advantage was not its technology, but its mindset; it saw the system when others saw silos.

In the face of climate disruption, AI breakthroughs, geopolitical instability, and rapidly shifting stakeholder expectations, strategic leaders must abandon linear thinking and embrace complexity as a design space. This chapter equips you with the cognitive tools and modeling approaches to do just that. Whether you're leading a startup, managing a global supply chain, or advising on sustainability initiatives, the ability to think systemically will allow you to identify leverage points, shape organizational behavior, and create resilient strategies that scale intelligently. This isn't just about seeing the big picture; it's about understanding how the picture moves.

Most importantly, you'll begin to think like a systems strategist, one who doesn't just respond to change but shapes the system in which strategy unfolds. This shift is not just technical; it is intellectual and cultural. It requires curiosity, humility, and a willingness to zoom out before acting. Systems strategists ask better questions, listen for patterns, and design solutions that reflect the interconnected realities of modern business. In this chapter, you'll take your first steps toward that mindset, learning how to visualize complexity, intervene with foresight, and lead strategy as a dynamic, evolving system.

Opening Case: Ping An's Strategic Reprogramming

Ping An began in 1988 as a conventional insurance company. For years, it followed a familiar strategic playbook: optimize underwriting, grow policyholder base, reduce claims fraud. Its value chain was linear, its operations siloed, and its digital investments modest. But by the early 2010s, Ping An's leadership recognized something most incumbents did not: strategy in a digital world couldn't be treated as a quarterly plan; it needed to function as a system.

The company initiated a radical transformation. It infused AI into underwriting, embedded digital twins to simulate systemic risk, and restructured its organization around a platform ecosystem that spanned finance, health, and smart cities. Customer interactions were no longer touchpoints—they were data nodes feeding predictive models. Cross-functional teams monitored real-time dashboards, simulations guided resource allocation, and feedback loops drove continuous adaptation. Every department, from marketing to claims processing, became part of a living feedback system.

This systemic integration wasn't just operational; it was strategic. Ping An didn't chase growth at any cost. It mapped interdependencies between technology, regulation, customer trust, and service delivery, recognizing that improving one node without understanding its ripple effects could destabilize the whole. As a result, Ping An's strategic model became not a fixed hierarchy but a learning organism, always sensing, adapting, and recalibrating.

> Sustainable advantage comes not from isolated excellence, but from orchestrated adaptation.
> Ping An's transformation shows that strategic success in a STEM world requires more than digitization, it demands systems thinking. It's not about optimizing parts; it's about redesigning the whole.

The outcome? By 2023, Ping An had become one of the most valuable insurance and financial services platforms globally, not by scaling linearly, but by evolving systematically. It outpaced rivals not because of a single breakthrough, but because it learned to think, decide, and lead as a system.

Play the Strategist

* What feedback loops (reinforcing or balancing) fueled Ping An's transformation?
* How did Ping An use real-time data and simulation to build strategic resilience?
* In what ways did the company move from traditional planning to systemic design?
* Could a firm in your industry apply similar systems thinking principles?
* If you were leading a legacy firm, how would you initiate a shift toward systems strategy?

2.1 From Strategy Fundamentals to Systemic Thinking

Strategic thinking didn't begin with systems modeling or digital simulations; it began with structure, segmentation, and clear frameworks. Tools like Porter's value chain, strategic fit, and organizational alignment helped generations of strategists define advantage, streamline operations, and guide execution. But as the pace of change accelerates and the complexity of business environments deepens, these classical models, while still valuable, now risk oversimplifying reality. Today's strategic challenges emerge not from single sources but from dynamic interactions: shifting customer preferences, cross-sector convergence, regulatory flux, and digital feedback loops that amplify small moves into system-wide effects. To lead effectively in this context, strategists must evolve from analyzing parts to understanding wholes.

This section reframes the building blocks of traditional strategy through the lens of systems thinking. It doesn't discard the fundamentals; it upgrades them. You'll learn how the value chain becomes a feedback system, how strategic fit must adapt over time, and how alignment emerges not from design alone, but from sensing, learning, and recalibrating. In doing so, you'll build the conceptual bridge from the tools of the past to the dynamic models of the future.

2.1.1 Reimagining the Value Chain: From Linear Flow to Feedback Loop

Porter's value chain remains one of the most iconic tools in strategic management. Segmenting business operations into primary and support activities allowed firms to analyze where value was created, lost, or enhanced. In the traditional view, the value chain was sequential, procurement led to production, which led to distribution, then marketing, and finally service. Each stage was optimized as a separate function.

But today, this linearity no longer holds. Modern organizations operate more like adaptive networks than assembly lines. Decisions made in one function immediately influence others, often in real time. Customer feedback influences design; logistics data adjusts pricing; marketing sentiment shifts supplier choices. Value is not simply "added"; it is co-produced, measured, and iterated continuously across departments.

Consider **Zara**. In a conventional value chain, the retail function would come at the end of the process. But for Zara, real-time retail data is the starting point. Store managers transmit purchasing trends back to designers and manufacturers daily. This data-driven loop compresses Zara's design-to-shelf cycle to as little as two weeks. Its competitive advantage comes not from any one activity, but from how fast and intelligently its system loops adapt to the market.

Systems thinking reframes the value chain as a dynamic loop, where each node is both a producer and a sensor of strategic intelligence. When mapped this way, firms can uncover reinforcing cycles (e.g., customer insight → product innovation → loyalty → insight) or identify friction points that slow system responsiveness.

2.1.2 Strategic Fit: From Static Alignment to Dynamic Equilibrium

Strategic fit has long been defined as the degree to which a firm's internal capabilities align with external opportunities. When this alignment is strong, firms are positioned to create and capture value. Traditional strategy focused on identifying and locking in this fit, often through market analysis, resource evaluation, and competitive positioning. The assumption: once fit is achieved, it can be sustained through consistent execution.

But in a world of rapid digital disruption, changing stakeholder expectations, and emergent business models, fit is not stable; it is continuously shifting. What fits today may falter tomorrow. Systems thinking reframes **fit as a dynamic equilibrium,** an ongoing state of balance between a firm's evolving capabilities and the changing environment it inhabits.

Netflix offers a vivid illustration. Its first strategic fit came from aligning mail-order DVDs with customer convenience. Later, it evolved to streaming, matching content delivery with broadband expansion. Today, its fit depends on matching hyper-personalized content creation with shifting global viewing patterns, made possible through AI and real-time behavioral analytics. Each phase of Netflix's evolution involved sensing systemic shifts, reconfiguring internal processes, and testing new models, an adaptive dance rather than a fixed formula.

In systems terms, strategic fit is maintained through **continuous feedback** between internal learning and external signals. Firms that model this loop, through simulations, scenario planning, or digital twins, can proactively adjust rather than react after fit erodes.

2.1.3 Organizational Alignment: From Structure to Synchronization

Traditional strategic thinking treated organizational alignment as a structural problem. Ensure that incentives, reporting lines, culture, and strategy are "aligned," and execution will follow. The logic was top-down, often framed as fit between strategy and structure (e.g., Chandler's "structure follows strategy").

But as organizations become flatter, more decentralized, and digitally connected, alignment is no longer about structure alone; it's about synchronization. In a STEM-enabled firm, cross-functional teams operate semi-autonomously, digital tools coordinate in real time, and decisions emerge from shared data environments, not chain-of-command hierarchies.

Ping An, for example, does not manage its sprawling platform of insurance, healthcare, and fintech services through rigid command structures. Instead, it builds shared data systems, synchronized APIs, and feedback-rich dashboards that allow each unit to adjust locally while maintaining strategic coherence globally. Strategic alignment becomes a systems outcome, not a compliance task.

Systems thinking helps organizations model alignment not just as a design principle, but as a **living system of coordination**. Strategic leaders build this through shared platforms, common language, transparency of data, and mutual feedback loops between units, teams, and decision layers. This shift reframes foundational strategy concepts, like value chains, strategic fit, and organizational alignment, not as static structures, but as dynamic, feedback-driven systems. Table 2.1. Reframing the Fundamentals illustrates this contrast between traditional and systems thinking views:

Table 2.1. Reframing the Fundamentals

Concept	Traditional View	Systems Thinking View
Value Chain	A sequence of activities	A feedback-driven loop of interdependent functions
Strategic Fit	Static alignment of resources and environment	Dynamic recalibration in response to ongoing change
Organizational Alignment	Structural consistency	Adaptive coordination through continuous feedback

In a traditional model, Zara's value chain would be mapped linearly, from design to production to retail. But in reality, Zara operates more like a neural network. Sales data becomes the input signal. Design responds almost instantly. Production lines flex based on predictive demand. Logistics routes are optimized by weather and traffic analytics. Every node senses and adjusts. This is systems strategy in action: one where cause and effect are not linear but recursive, and where competitive advantage comes not from stability, but from synchronized responsiveness.

Are You Seeing Systems?

* Think about a company or team you work with.
  Do their operations resemble a value chain or a value system?
* What loops (positive or negative) influence strategic decisions in that organization?
* How would introducing feedback mechanisms (dashboards, simulations, cross-functional data sharing) change the way decisions are made?

2.2 What Is Systems Thinking—and Why Strategists Need It

Strategy no longer unfolds in isolation. In today's global, data-driven economy, business challenges are rarely the result of a single cause, and their solutions are never linear. Market volatility, geopolitical shifts, climate disruptions, platform dynamics, and stakeholder expectations constantly collide in unpredictable ways. In such a context, traditional planning fails not because it lacks rigor, but because it oversimplifies complexity.

Systems thinking offers a different lens. Instead of asking, "What's the next step?", it asks, "How does this system behave, and why?" It helps leaders see relationships instead of parts, patterns instead of snapshots, and leverage points instead of surface fixes. Where conventional strategy may optimize the pieces, systems thinking reimagines the whole.

At its core, systems thinking is a discipline of mapping feedback, tracing causality, anticipating unintended effects, and navigating complexity with clarity. For strategic leaders, this shift is transformational. It moves the strategist from tactician to systems designer, from planning in straight lines to strategizing in loops.

2.2.1 Defining Systems Thinking: Beyond Cause and Effect

Systems thinking is not just a technique; it is a shift in how we understand reality. Traditional strategy tends to rely on linear logic: if X happens, then Y follows. This mindset is useful for short-term tactical decisions, where variables are few and outcomes are predictable. But in strategic environments, where change is multidimensional, decisions interact with each other, and results unfold over time, such linear reasoning falls short.

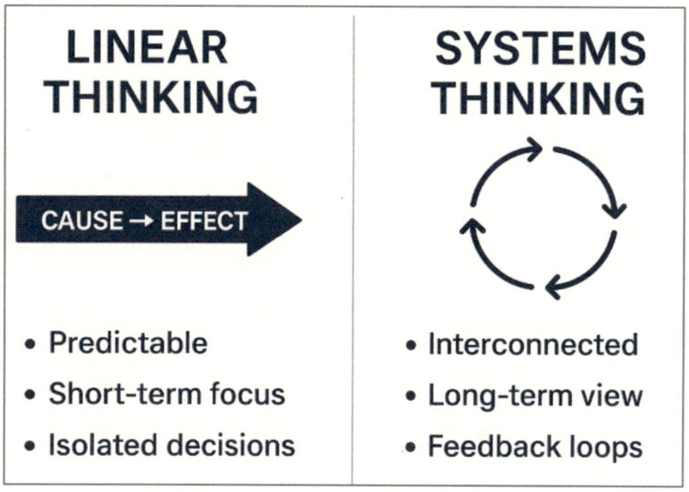

Systems thinking begins where linear analysis ends. It views organizations and industries as **dynamic systems,** networks of interdependent elements, where **structure drives behavior**. Rather than looking for singular causes or isolated metrics, it examines how components influence one another through **feedback loops**, **delays**, and **nonlinear relationships**.

For example, consider a firm experiencing a drop in employee productivity. A linear diagnosis might focus on underperformance or lack of training. A systems thinker, however, would explore deeper patterns: Is the dip linked to rising project overload? Has high turnover created a loop of knowledge loss and rework? Are misaligned incentives encouraging the wrong behaviors? The root cause may lie not in individual effort, but in the **systemic design** of workflows, feedback channels, or leadership models.

Systems thinking also challenges the notion of **isolated solutions**. A price discount may increase short-term sales, but could also set off a reinforcing loop of discount dependency and brand erosion. Hiring more sales staff may seem like a fix, but if onboarding is weak, the long-term result could be burnout and higher churn. In systems, **today's solution can become tomorrow's problem**.

Table 2.2 is more than a conceptual contrast; it is a strategic checklist. It invites leaders to examine whether their current decision-making culture encourages clarity or complexity blindness. Do they optimize functions or system performance? Do they rush to fix problems or pause to map feedback?

Table 2.2. Linear Thinking vs. Systems Thinking in Strategic Contexts

Dimension	Linear Thinking	Systems Thinking
Causality	One cause → one effect	Multiple causes, circular feedback
Focus	Isolated problems or departments	Interconnections across the system
Time Horizon	Short-term actions and results	Long-term effects and delayed outcomes
Response Pattern	Reactive: fix symptoms	Proactive: address root structure
Data Use	Track KPIs and variances	Analyze patterns, loops, and flows
Mental Model	Straight lines	Loops, webs, delays, thresholds
Strategic Risk	Oversimplification, unintended consequences	Complexity awareness, scenario robustness
Tool Examples	Dashboards, benchmarking	Causal loop diagrams, system dynamics simulations

Amazon's Prime membership is more than a pricing model; it's a system of interlocking feedback loops. As more users join Prime, shipping costs drop due to economies of scale. Faster delivery increases customer satisfaction, which boosts platform engagement and spending. Higher sales fund more warehouses and logistics technology, which again improves delivery. These loops reinforce each other, making Prime sticky and Amazon strategically dominant. A linear lens might see Prime as just a loyalty program; systems thinking reveals it as a **self-reinforcing engine** of growth, cost advantage, and customer lock-in.

In healthcare, delayed diagnosis often leads to a systemic backlog that overwhelms capacity. For example, postponing routine screenings due to staffing shortages may seem harmless in the short term. But delayed screenings lead to more late-stage diagnoses, which require more intensive treatment and longer recovery. These patients occupy beds and specialists for extended periods, reducing availability for new cases, worsening the original shortage. The loop continues unless interventions are made at the structural level. A systems view helps hospitals **shift resources upstream**, redesign intake flows, or model capacity under varying demand scenarios.

By training yourself and your teams to "see systems," you gain a diagnostic advantage. You begin to anticipate resistance before change initiatives stall. You recognize hidden reinforcing loops that drive performance or dysfunction. Most importantly, you start making **strategic moves that shape system behavior**, not just patch local pain points.

2.2.2 STEM Foundations of Systems Thinking

Systems thinking is inherently interdisciplinary, but it thrives most where scientific reasoning, technological modeling, engineering design, and mathematical logic intersect. In other words, it thrives in the **STEM domain**. STEM disciplines are uniquely equipped to handle the complexity, interdependence, and feedback-rich nature of systems because they are built on the same principles that define complex strategic environments: precision, iteration, simulation, and feedback.

Where business strategists may traditionally think in terms of plans and goals, STEM thinkers are more likely to think in terms of **constraints, structures, and behavior over time**. They are comfortable asking not just *what* will happen, but *how the system behaves under different conditions*. By integrating STEM-based tools into strategy work, today's leaders can diagnose systemic issues, simulate change, and intervene with confidence, especially in fast-moving or uncertain markets.

2.2.2.1 Scientific Logic: The Foundation of Causal Reasoning

Science is built upon a structured approach to inquiry: observe, hypothesize, test, and refine. Systems thinking adopts this same discipline, applying it not to nature but to organizations and markets. It replaces guesswork with **causal rigor,** modeling how one action influences another across time and space.

Strategists working in turbulent environments must think like scientists. They form working hypotheses about the behavior of their systems, such as why employee turnover is rising, why innovation has slowed, or why customers are defecting, and then test these theories through experimentation, A/B testing, or pilot programs. These are not isolated analytics projects; they are iterations within a system framework.

* **Hypothesis-driven decision-making**: Leaders frame problems as testable questions rather than assumptions. For example, "Is employee churn caused by compensation or career stagnation?" becomes a testable hypothesis using comparative data, focus groups, or time-series analysis.
* **Experimental testing**: Like scientists in a lab, system-minded strategists isolate variables and test interventions in low-risk environments, such as launching pricing

pilots in small markets or testing new workflows in a single business unit before scaling up.
* **Model validation and refinement**: Feedback from each test is used to improve the organization's mental model, making the strategy more adaptive and evidence-based over time.

Google's Project Oxygen evaluated 10 years of data to form hypotheses about management effectiveness. The process wasn't one study; it was a rolling system of hypothesis-testing that led to a strategic shift in leadership development.

2.2.2.2 Technology and Simulation: Seeing Strategy Before It Happens

Technology doesn't just support strategy; it **makes systems visible**. With the rise of cloud computing, IoT sensors, and real-time data collection, organizations now have the infrastructure to simulate decisions before executing them. Systems thinkers use this capability to experiment safely, reduce risk, and forecast long-term effects.

* **Digital twins**: These virtual models replicate real-world entities, from machines and supply chains to customer experiences, allowing leaders to test ideas, simulate changes, and understand cause-and-effect relationships without real-world disruption. For instance, a warehouse's digital twin can simulate how changes in layout affect speed, labor, and costs.
* **System dynamics tools**: Software platforms like Vensim, Stella, and Insight Maker enable strategists to build and manipulate complex feedback systems. Leaders can visualize how variables such as marketing spend, customer sentiment, and competitor moves interact over time, revealing hidden loops and bottlenecks.
* **Agent-based modeling**: Unlike aggregate models, agent-based modeling simulates the behaviors of individual entities, employees, customers, and competitors, and observes how collective outcomes emerge. It's especially useful for strategy in ecosystems or decentralized environments where centralized control is weak.

A national retail chain used system dynamics to forecast how promotions would impact not only short-term sales but long-term store traffic, brand perception, and competitor reactions, leading to smarter pricing strategies.

2.2.2.3 Engineering Design: Structuring Resilience and Feedback

Engineering isn't just about building things; it's about designing them to **perform reliably under constraints and change**. In business strategy, the same principle applies. Engineering-trained leaders often bring with them a deep respect for structure, failure points, and feedback design, traits essential for systems thinking.

* **Feedback loops as control mechanisms**: Engineers design systems that self-correct, like cruise control in cars. Similarly, strategic feedback loops (e.g., performance reviews,

customer NPS tracking) help businesses stabilize operations and adapt to change. Leaders must know where these loops exist and whether they're reinforcing good or bad outcomes.

* **Redundancy and buffering**: Engineers build backup systems (like dual brakes or redundant power supplies) to ensure resilience. Strategists apply this through flexible staffing models, diversified suppliers, or distributed decision-making, all designed to keep the system functioning when one part fails.
* **Designing for failure and recovery**: Smart engineers assume things will go wrong. They build for recoverability, not perfection. Strategists must adopt the same mindset, designing business continuity plans, stress-tested financial models, or scalable infrastructures that grow without collapse.

Netflix's "Chaos Monkey" intentionally disrupts services in its cloud environment to test resilience. Strategically, this mindset translates into preparing not only for expected demand but for failure scenarios—ensuring systems survive pressure and evolve.

2.2.2.4 Mathematics and Statistics: Modeling Nonlinear Dynamics

At the heart of every complex system lies **nonlinearity,** where inputs and outputs do not scale proportionally. Mathematics and statistics give strategists the power to understand exponential change, delayed effects, and threshold dynamics.

* **Bayesian inference and adaptive learning**: Instead of static decisions, strategists use Bayesian methods to continuously update models as new data arrives. For example, if customer churn increases after a new policy, Bayesian logic helps adjust beliefs about its effectiveness, guiding course corrections.
* **Regression and time-series analysis**: These techniques model how variables change over time and how past behaviors influence future trends. Strategic planners use them to forecast sales, staffing needs, supply shocks, or policy outcomes under different conditions.
* **Network analysis**: A growing area of strategic insight, network analysis examines how nodes (people, products, partners) connect and how influence spreads through those networks. It's useful for understanding social media dynamics, partnership ecosystems, or viral marketing.

A fintech startup used network graphs to map user referrals. By identifying super-connectors and mapping trust flows, they redesigned their referral program to maximize systemic growth, not just individual sign-ups.

Digital Twins and the Future of Strategic Simulation

Imagine running your entire organization, supply chain, HR system, customer experience, and environmental footprint, not just in real life, but as a **living digital replica**. That's the promise of **digital twins**.

Born in engineering and manufacturing, digital twins have now entered the strategic realm. Leading companies are creating digital replicas of their businesses to simulate growth strategies, stress-test crisis responses, and optimize transformation plans before taking action.

* A **retail giant** simulates regional store performance based on foot traffic, weather, staffing, and promotions, before making investment decisions. This allows granular, data-driven strategic moves at local and regional levels.
* A **healthcare system** uses a hospital-wide digital twin to test how shifting patient loads, staffing shortages, or policy changes affect outcomes. It helps anticipate system bottlenecks and redesign workflows in advance.
* An **energy company** models wind farm efficiency and grid impacts through a digital twin, aligning operations with sustainability goals. This improves resource planning and supports regulatory compliance.

Digital twins exemplify the power of STEM-enabled systems thinking, blending real-time data, predictive modeling, feedback visualization, and scenario planning to **test strategic futures before committing real-world resources**.

Engineering a Strategic Ecosystem with STEM Thinking

China's **Ping An Insurance Group**, once a traditional financial services firm, has transformed into one of the world's most technologically integrated ecosystems. With over 220 million retail customers and hundreds of millions more in its health, finance, and smart city platforms, Ping An exemplifies how **STEM-powered systems thinking** can redefine strategic boundaries.

Facing intensifying competition in banking and insurance, Ping An needed to expand beyond its core offerings while preserving system-wide value and efficiency. Leadership recognized that its future would depend not on selling more products, but on **orchestrating an intelligent, interconnected system,** where services anticipate needs, reinforce each other, and learn over time.

Systems Thinking in Action

1. Scientific Logic

Ping An's strategic approach starts with **hypothesis-driven experimentation**. When building its health platform "Ping An Good Doctor," leadership hypothesized that combining AI diagnostics with telemedicine and pharmacy delivery could create a self-reinforcing loop of engagement and trust. The system was tested across population segments and refined continuously based on user behavior and feedback loops.

2. Technology and Simulation

Through its Smart City initiatives, Ping An built **digital twins of urban systems,** modeling traffic, waste, energy use, and public health in real time. These simulations allow city managers to anticipate risks (e.g., flu outbreaks, congestion) and optimize interventions. Strategically, Ping An gains embedded access to municipal ecosystems, enhancing data access and service integration.

3. Engineering Design

The company engineered **modular business units,** such as insurance, healthcare, wealth management, and AI labs, that function independently but share a unified data infrastructure. This mirrors engineering principles of **modularity**, **interoperability**, and **fault tolerance**: if one business underperforms, others compensate. Feedback loops between units (e.g., health data influencing insurance pricing) create continuous value creation.

4. Mathematical Modeling

Ping An's AI engines use **Bayesian inference**, **machine learning**, and **risk modeling** to dynamically underwrite policies, price financial products, and detect fraud. For example, its facial recognition algorithms integrate network behavior, transaction patterns, and demographic data to assess credit risk, blending **statistical reasoning** with system-wide visibility.

Strategic Payoff

Ping An’s transformation illustrates how **system-level design** and **STEM principles** coalesce into competitive advantage. Rather than compete on product margins alone, it has created a self-sustaining ecosystem where:

* Customers rarely leave because each service enhances the next.
* Data flows continuously to improve product development and operations.
* Strategic simulations (digital twins) guide policy, pricing, and growth investments.

Ping An doesn’t just *react* to change; it uses systems thinking to shape the environment in which it competes. Ping An’s journey underscores a key lesson: when STEM is integrated into strategic thinking, not as a toolset, but as a mindset, organizations unlock systems-level foresight, feedback learning, and long-term resilience.

2.2.3 Systems Thinking vs. Analytical Thinking

At first glance, **analytical thinking** and **systems thinking** may seem interchangeable. Both seek to make sense of complexity, use data to inform decisions, and aim for better outcomes.

But they operate on **very different mental models**. Analytical thinking is best suited for breaking problems into parts and optimizing each element. Systems thinking, in contrast, seeks to understand how the parts interact, often in unexpected, nonlinear, or delayed ways.

Strategists must know **when to analyze and when to synthesize**. In stable or well-structured environments, analytical tools offer clarity. But in dynamic, uncertain, or interdependent systems, such as global supply chains, innovation ecosystems, or climate policy, **systems thinking becomes essential**. This section explores the distinction and integration of both mindsets.

2.2.3.1 Analytical Thinking: Reduction, Precision, and Control

Analytical thinking is the dominant logic taught in traditional business education. It works by **disassembling complex problems into manageable parts**, studying them independently, and then optimizing each element to improve the whole. This mode of thinking is especially effective in **structured, repeatable environments** where relationships are stable and linear.

The strength of analytical thinking lies in its ability to deliver **clarity**, **precision**, and **efficiency**. It helps managers make decisions based on hard data, isolate causes, reduce ambiguity, and create performance benchmarks. However, its power is also its limitation, because it assumes that **systems behave like machines**: predictable, decomposable, and controllable.

Core Traits of Analytical Thinking:

* **Linear Causality**

  Analytical logic often assumes that **inputs produce proportional outputs,** that if you increase marketing spend by 20%, sales should rise in a predictable manner. This logic simplifies complexity and supports goal setting, forecasting, and performance reviews. It is useful for diagnostic tasks (e.g., "Why did profits decline?") and for narrow optimization (e.g., "Which channel delivers the best ROI?").

* **Problem Decomposition**

  One of the central tools in analytical thinking is the ability to **break down problems into smaller, independently solvable parts**. A supply chain problem, for example, can be segmented into procurement, production, inventory, and delivery, each assigned to specialists for improvement. This allows for focused expertise, but can also blind strategists to interdependencies between parts.

* **Optimization and Control Orientation**

  Analytical frameworks seek the **best answers** under given constraints, whether it's minimizing cost, maximizing efficiency, or boosting KPIs. Tools like Six Sigma, linear programming, or scenario analysis are grounded in control logic: change the variable, measure the result, repeat for optimal output.

* **Snapshot Analysis and Short-Term Focus**

  Because analytical tools often rely on static datasets or time-bound indicators, they provide **snapshots of performance**, rather than motion pictures of system evolution. This makes them great for quarterly reviews, pricing decisions, or variance analysis, but less effective for understanding **long-term transformation or emergent risks**.

A logistics firm seeks to reduce last-mile delivery costs. Using analytical thinking, it studies fuel prices, delivery routes, and package density, identifying route optimization as the highest leverage point. It reconfigures delivery zones and achieves a 12% cost reduction. Success is clear, measurable, and linear. But what if the underlying problem is **not route structure** but **a feedback loop between delivery delays, customer complaints, and driver burnout**? That's where analytical thinking hits its boundary.

In complex systems, **optimizing the parts does not necessarily optimize the whole**. In fact, over-optimizing one part can make the overall system worse. For example, slashing supplier costs may look great analytically, until delivery delays erode customer satisfaction and damage brand trust. Analytical thinking often fails to account for **indirect effects, delays, or system-wide feedback**.

Analytical thinking is best suited for:

* Well-defined problems with clear boundaries
* Short-term decisions with quantifiable variables
* Environments with high data availability and stable cause-and-effect relationships
* Tasks requiring comparative analysis or resource optimization

But when environments are **uncertain, adaptive, or non-linear**, strategy needs something more: **a systemic lens** that captures feedback, emergence, and dynamic change. That's where we now turn, to systems thinking.

2.2.3.2 Systems Thinking: Holism, Feedback, and Dynamics

Where analytical thinking focuses on the parts**, systems thinking focuses on the** relationships **among the parts. It is a holistic mindset that recognizes that organizations, markets, and industries are** dynamic systems**, not static machines. In**

these systems, actions produce effects that are often delayed**,** indirect**, and** amplified or dampened **through feedback loops.**

Systems thinking invites strategists to look beyond surface-level symptoms to uncover the underlying structure that drives behavior over time**. It doesn't replace analytical thinking, but it shifts the frame from solving isolated problems to** navigating complex, interconnected challenges**.**

Core Traits of Systems Thinking:

* **Nonlinear Causality**

  In complex systems, cause and effect are rarely proportional or immediate. A small policy change can have outsized consequences, or no impact at all until a tipping point is reached. For example, lowering product prices may initially boost sales, but could also trigger a price war, undercut profitability, or devalue the brand. Systems thinkers explore **how variables interact over time**, often through reinforcing or balancing feedback loops.

* **Feedback Loops and Delays**

  Strategic actions rarely operate in a vacuum. Systems thinking helps identify **reinforcing loops** (where change begets more change, like customer referrals or viral marketing) and **balancing loops** (where systems self-regulate, like inventory controls or hiring freezes). These loops often involve **time delays**, which makes their impact harder to detect with traditional tools. Misreading them can lead to overreaction or policy resistance.

* **Pattern Recognition and Behavior Over Time**

  Rather than reacting to individual events, systems thinkers look for **patterns and trends**. What is recurring? What's accelerating or decelerating? This approach trains strategists to step back from operational noise and ask: "What's the system doing, and why is it doing that?" The goal is not control, but **leverage,** to find the points in the system where a small, well-placed action can produce outsized change.

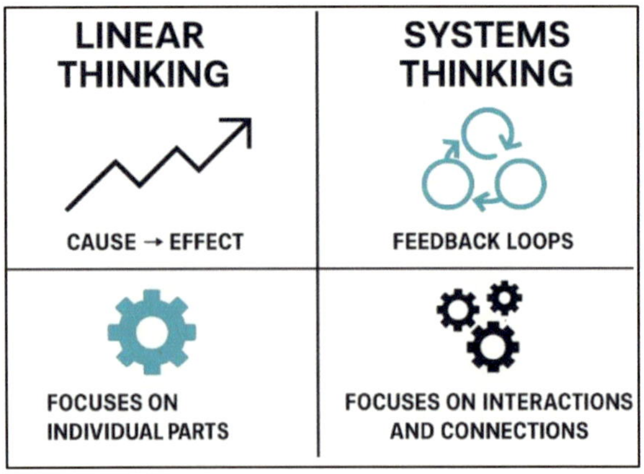

* **Holism and Context**

  Systems thinking demands that we see the **whole before optimizing the parts**. This includes understanding social, technical, economic, and ecological subsystems as intertwined. For instance, in sustainability strategy, a systems lens connects sourcing practices with supply chain risk, brand reputation, investor expectations, and regulatory pressures, revealing ripple effects beyond the factory floor.

In the early 2000s, Boeing focused analytically on cutting production costs by outsourcing more of its aircraft components, including parts of the 787 Dreamliner. The decision looked efficient on spreadsheets, but the system-level effect was **a loss of integration and control**, leading to delays, quality issues, and customer dissatisfaction. A systems thinker would have mapped out the feedback loops, how coordination complexity, supplier knowledge gaps, and timeline delays would amplify each other, and likely designed a more resilient production ecosystem.

Systems thinking can feel overwhelming, especially for those trained to seek definitive answers. It requires **comfort with ambiguity**, the ability to **zoom in and out**, and the discipline to model complex interdependencies without oversimplifying. If used carelessly, it may also lead to **paralysis by analysis**, where every action seems too interconnected to manage.

Systems thinking is especially valuable when:

* Problems are persistent, recurring, or resistant to previous solutions
* Multiple stakeholders or variables are interacting dynamically
* Consequences are delayed, indirect, or distributed across boundaries
* Strategy requires transformation rather than optimization

Where analytical thinkers ask "What's the problem and how do we fix it?", systems thinkers ask "What is the system doing, and what structure is causing that behavior?" The two modes are not in conflict; they are **complementary lenses**. Mastery in modern strategy requires the ability to use both, knowing when to **analyze the parts** and when to **synthesize the whole**.

2.2.3.3 Analytical Thinking vs. Systems Thinking

Strategists are constantly making decisions about which lens to use when evaluating a challenge. The table below breaks down the contrasting features of **analytical** and **systems thinking** across key dimensions. While analytical thinking excels in well-structured, controllable environments, systems thinking is indispensable for navigating dynamic, interdependent systems. The table below outlines the key differences between these two modes of thinking across several strategic dimensions:

Table 2.3. Analytical Thinking vs. Systems Thinking

Dimension	Analytical Thinking	Systems Thinking
Primary Focus	Isolating and improving individual parts of a system	Understanding relationships, feedback loops, and the whole system
Causality	Linear and dircct ("A causes B")	Nonlinear and circular ("A influences B, which loops back to influence A")
Time Perspective	Short-term, snapshot-oriented	Long-term, dynamic evolution and behavior over time
Change Orientation	Optimize what exists	Redesign the system structure to change behavior
Feedback Handling	Often ignored or simplified	Central, explicitly modeled to anticipate system behavior
Tools and Methods	KPIs, dashboards, cost-benefit analysis, regression models	Causal loop diagrams, stock-and-flow models, system dynamics simulations
Decision Logic	Control and correct variances	Influence system structure and leverage feedback effects
Assumptions	The system is stable, controllable, and decomposable	The system is dynamic, adaptive, and emergent
Failure Risk	Misses ripple effects or unintended consequences	Can become too abstract or broad without focus
Best Use Cases	Budgeting, pricing, performance reviews, efficiency metrics	Organizational learning, sustainability, innovation ecosystems, and crisis response
Mental Framing	"What is the problem and how can we fix it efficiently?"	"What is the system doing and why is it producing these outcomes?"

This comparison highlights why relying exclusively on analytical tools can lead to **blind spots** in strategic planning. For example, a company facing declining employee engagement might analyze turnover data, salary benchmarks, or exit surveys, yet fail to see how a toxic feedback loop between poor management, low morale, and rising workload is degrading the system itself.

Conversely, systems thinking can reveal **root causes and leverage points**, but without analytical support, it may lack the operational detail required to drive implementation.

The most effective strategists move fluidly between these modes, **using analysis to measure what matters** and **systems thinking to understand what drives it**. For students, this means learning to see data points not as isolated facts, but as part of larger systems that evolve, adapt, and respond to feedback.

Unilever's Waste Reduction Strategy: Where Analysis Meets Systems Thinking

Unilever, one of the world's largest consumer goods companies, set an ambitious goal to halve the environmental impact of its products. A major obstacle was plastic packaging waste, especially in emerging markets where recycling infrastructure is underdeveloped.

Analytical Approach:

Initially, the company used **analytical tools** to assess waste by product line, geography, and material type. This helped pinpoint high-impact categories (e.g., single-use sachets) and estimate cost-benefit trade-offs for biodegradable materials, lightweighting, and recycling investments.

* Dashboards and regression analysis were used to measure the relationship between packaging costs and volume reductions.
* Supply chain simulations helped optimize warehouse utilization as packaging density changed.

Systemic Insight:

But the problem turned out to be **structural**, not just material. Customers bought single-use sachets because of **income constraints**, **distribution limitations**, and a **lack of refill stations**. Even if biodegradable materials were introduced, they would still end up in landfills or waterways without behavior change and infrastructure.

Unilever shifted toward **systems thinking** by mapping:

* **Causal loops** between affordability, consumer behavior, waste volume, and brand equity.
* **Delays** in the impact of consumer education efforts.
* **Feedback loops** between retail formats, packaging size, and environmental outcomes.

Strategic Move:

Instead of just changing materials, Unilever partnered with local governments and NGOs to **co-develop reuse models**, such as mobile refill trucks and smart dispensers. The company also piloted digital incentive systems, giving loyalty points for reused containers, creating a **reinforcing loop of behavior change**.

STEM Integration:

* **Science**: Lifecycle analysis and material science-informed sustainable packaging design.
* **Technology**: Smart dispensers and data collection apps helped track behavior shifts.
* **Engineering**: Supply chain redesign allowed integration of bulk delivery and reuse models.
* **Mathematics**: Scenario modeling projected the ROI of infrastructure investments over 5–10 years.

By blending **analytical precision** with **systems-level insight**, Unilever didn't just reduce packaging waste; it redesigned the entire consumption system around sustainability. The result was both environmental and strategic: reduced costs, increased brand trust, and early-mover advantage in circular packaging.

This case exemplifies the power of integrating **analytics (to see the parts)** and **systems thinking (to reshape the whole)**. STEM-literate strategists must learn to ask not only *"How do we fix this?"* but *"What is the system that creates this need—and how can we redesign it?"*

2.2.3.4 Integrating Both: The Strategic Sweet Spot

The future of strategy doesn't lie in choosing between analytical thinking and systems thinking; it lies in **synthesizing both**. While each mindset offers a different lens, **the most impactful strategic decisions emerge at their intersection**. Analytical thinking gives us the tools to measure and optimize; systems thinking shows us what really matters and how everything is connected. Used together, they enable not just efficiency but **evolution**.

Strategists working in dynamic environments, such as tech ecosystems, healthcare systems, or sustainability transitions, must learn to **shift fluidly between micro-level analysis and macro-level synthesis**. This integration creates a strategic "sweet spot" where leaders can both diagnose problems with precision and design solutions that are resilient, adaptive, and aligned with long-term system health.

A Four-Step Integration Process for Strategists

1. **Frame the System First**
   Start with a systems view: What are the key actors, structures, and feedback loops involved? What patterns of behavior are recurring? This framing ensures that strategic actions don't solve the wrong problem.

Example: A transportation firm sees rising delivery delays. Rather than just rerouting trucks (analytical fix), leadership maps the full delivery ecosystem, revealing that warehouse bottlenecks and scheduling mismatches, not traffic, are the root causes.

2. **Use Analytics to Target Leverage Points**
   Once the system is understood, bring in analytical tools to **quantify impacts**, **prioritize interventions**, and **simulate trade-offs**. Analytics helps validate insights from systems mapping and supports decision-making with evidence.

   Example: After mapping emissions hotspots in a manufacturing system, analysts use regression and scenario models to test the ROI of solar energy retrofits versus supply chain redesign.

3. **Design Feedback-Aware Interventions**
   Avoid one-off fixes. Use systems thinking to design interventions that will **reinforce positive loops** or **dampen harmful ones** over time. Anticipate delays, unintended effects, and adaptive responses.

   Example: Rather than offering bonuses for short-term sales growth, a company designs a performance system that links bonuses to customer retention, creating a reinforcing loop between value creation and loyalty.

4. **Monitor, Learn, and Adapt Continuously**
   Blend analytical dashboards with systems diagnostics (e.g., feedback tracking, behavior modeling). Regularly revisit both the metrics and the system map to **adjust strategy in real time**.

   Example: In urban innovation, smart cities use both KPIs (e.g., traffic volume, power usage) and digital twins to simulate and adapt strategies before crises emerge.

Why Integration Matters

Organizations that rely solely on analytics risk becoming **efficient at the wrong things**. Those who rely solely on systems thinking may see the big picture but struggle to implement it. Only by integrating both can strategists:

* Achieve **clarity without oversimplification**
* Drive **performance without sacrificing resilience**
* Align **short-term wins with long-term transformation**

In a STEM-driven strategy world, this integration is not optional—it's essential.

Reconsider **Ping An's transformation**: Systems thinking revealed that financial, healthcare, and AI services were interdependent. Analytical tools were then optimized for each unit's performance. But the real genius was in designing **reinforcing feedback loops,** where data from one unit enhanced the others, creating compounding strategic value. Ping An didn't just operate businesses; it architected a living system.

Tesla's Integrated Thinking—Precision Meets System Strategy

Tesla is not just a car company; it's a vertically integrated energy and technology ecosystem. Its ability to dominate in electric vehicles (EVs), battery storage, and renewable energy stems from a unique ability to blend analytical precision with systems thinking**. From supply chains to software, Tesla leverages STEM tools not only to optimize operations but also to** redesign the entire value system **of mobility and energy.**

Analytical Precision in Action:

Tesla uses advanced analytics **to drive decisions across manufacturing, pricing, and customer experience:**

* **Predictive Maintenance**: Vehicle sensor data is analyzed in real time to predict part failures before they occur, reducing downtime and warranty costs.
* **Battery Performance Modeling**: Mathematical models simulate charging cycles, temperature effects, and material degradation—optimizing both cost and lifespan.
* **Gigafactory Efficiency**: Tesla's manufacturing dashboards track thousands of variables, from robot cycle times to supply throughput, allowing for lean operations and precision scaling.

Systems Thinking in Strategy:

Tesla's true innovation lies in seeing its business not as discrete silos, but as an interconnected energy system**:**

* **Reinforcing Loops**: As EV adoption increases, battery costs decrease due to scale. Lower battery costs fuel solar storage adoption, which reinforces the EV ecosystem.
* **Ecosystem Architecture**: Tesla's vehicle, solar, storage, and AI initiatives are designed to **feed data and value into each other,** a systems architecture rather than a product portfolio.
* **Time Delays and Policy**: Tesla anticipates regulatory shifts, infrastructure delays, and public adoption curves, positioning itself early in the system for long-term gain.

This strategic integration is powered by a deep application of **STEM disciplines**, each embedded into Tesla's core operations. The table below illustrates how science, technology, engineering, and mathematics are not just support functions, but **strategic enablers** of Tesla's systems-based approach:

Table 2.4. STEM as Strategy: Tesla's Integrated Innovation Model

STEM Element	Tesla Application
Science	Battery chemistry R&D, autonomous vehicle safety, energy storage materials
Technology	OTA (over-the-air) software updates, AI-based driving systems, and solar panel integration
Engineering	Vertical integration of gigafactories, design of modular power walls, and supercharger networks
Mathematics	Demand forecasting, supply chain modeling, risk optimization algorithms

By embedding STEM into its strategic DNA, Tesla transforms traditional business functions into a coordinated, adaptive system, one that senses, learns, and evolves in real time. Tesla's success is not simply about fast cars or green energy; it's about using analytical thinking to master the parts **and** systems thinking to reimagine the whole**. The company's ability to iterate quickly while shaping the future of mobility shows how** STEM-enabled integrated thinking **creates not just strategy, but strategic ecosystems.**

2.3 Modeling Strategic Systems

If systems thinking is the strategist's mindset, **modeling is the strategist's method**. It is how abstract insights become tangible, how complexity becomes visual, and how decisions become testable before they're made. Modeling allows leaders to make the **invisible structure of systems visible**, transforming complexity into clarity, not by simplifying reality, but by exposing how reality works beneath the surface.

Most traditional strategy tools (e.g., SWOT, Porter's Five Forces, PESTLE) help describe *what is*. Strategic modeling, in contrast, helps explore *what could be*. It enables the construction of **"living" models,** representations of real-world dynamics that evolve, adapt, and respond to change. These models are especially valuable in STEM-driven environments, where outcomes emerge from nonlinear interactions, where feedback loops distort cause and effect, and where delays and dependencies obscure strategic consequences.

Modeling is not about predicting the future with certainty; it is about creating a **decision space** where different futures can be explored. A well-built model helps answer questions like:

- *What will happen if we scale this initiative?*
- *Where are the hidden bottlenecks in our ecosystem?*
- *How will this policy ripple through our organization or market over time?*

In a STEM world, where decision velocity is increasing and the cost of error is rising, modeling provides the **simulation sandbox** where leaders can learn safely, act intelligently, and adapt continuously.

Most importantly, strategic modeling unlocks the ability to **test ideas before committing resources**, **visualize ripple effects**, and **understand the whole system, not just isolated parts**. It is what makes systems thinking operational, and what transforms strategic judgment from intuition to insight.

2.3.1 What Is a Strategic System Model?

A **strategic system model** is a visual and conceptual representation of how key elements (Table 2.3) in a strategic environment interact, evolve, and influence one another over time. Unlike static charts or linear frameworks, a system model captures the **dynamic structure** of a system: how its components are connected, how they behave under different conditions, and how outcomes emerge from the interplay of decisions, feedback loops, and time delays.

At its core, a strategic system model answers this question: **"How does this system work, and how will it change if we intervene?"**

It enables strategists to **see structure behind events**, identify **non-obvious leverage points**, and anticipate **unintended consequences** before they occur. Whether used to understand customer behavior, innovation diffusion, supply chain resilience, or workforce dynamics, these models shift strategy from reactive guesswork to **proactive design**. To build such models, strategists rely on a set of foundational elements that define how systems behave and evolve over time. These building blocks, such as stocks, flows, feedback loops, and delays, help map the dynamics of complex environments. The table below outlines these core components and illustrates how each contributes to strategic insight.

These components are not isolated; they interact constantly. A change in flow affects a stock, which in turn alters a loop, which feeds back to reshape the system. Strategic models **map these cascading interdependencies**, providing insights not visible through conventional analysis.

Let's consider a basic strategic system model for customer lifetime value (CLV) in a subscription-based business.

* **Stock**: Total active subscribers
* **Inflow (Flow)**: Monthly new signups
* **Outflow (Flow)**: Churn (monthly cancellations)
* **Reinforcing Feedback Loop**: Happy customers refer others → referrals increase signups → subscriber base grows → more word of mouth
* **Balancing Feedback Loop**: Poor service → increased churn → fewer subscribers → higher pressure on remaining service team → even poorer service

A simple intervention like increasing ad spend might grow the inflow, but without improving service quality, the **system structure will drive churn back up,** erasing gains. The model helps identify where to intervene **for lasting results**, not temporary spikes.

Table 2.5. Key Building Blocks of a Strategic System Model

Element	**Definition**	**Strategic Example**
Stocks	The accumulations or resources in the system, things that build up or diminish over time	Inventory levels, talent pool, and customer trust
Flows	The rates of change that increase or decrease stocks	Hiring rate, churn rate, sales rate
Feedback Loops	Circular cause-and-effect relationships that regulate system behavior	Faster delivery increases satisfaction, leading to more usage
Delays	Time lags between cause and effect	Training impacts performance after 6 months
Boundaries	The scope of the system being modeled, what's in, what's out	Market ecosystem vs. internal operations
Nonlinearities	Points where small changes produce disproportionate outcomes	Network effects, tipping points

Strategic system models are **ideal candidates for STEM integration**:

* **Science** informs how systems behave under stress (e.g., ecological resilience, epidemic spread, tipping points).
* **Technology** enables simulation platforms that visualize models dynamically (e.g., Vensim, Stella, AnyLogic).
* **Engineering** connects the model to real infrastructure—logistics networks, production systems, and cloud architecture.
* **Mathematics** powers the equations that govern flows, loops, and sensitivity analysis.

Together, these tools allow strategists not only to draw system maps, but to **run them**, **stress test them**, and **refine them** in real-time strategy development.

2.3.2 Types of Strategic Models

Strategic models are not one-size-fits-all. Different problems, systems, and strategic environments call for **different modeling approaches**, each suited to capturing particular structures, behaviors, and questions. Whether the goal is to map qualitative relationships, simulate quantitative dynamics, or anticipate emergent behaviors, choosing the right model type is critical.

Below are five essential types of strategic system models, with practical examples and strategic relevance.

2.3.2.1 Causal Loop Diagrams (CLDs)

Purpose:
To visually map feedback loops and understand how variables influence one another—positively or negatively—over time.

Structure:
Causal loop diagrams use arrows and polarity (+/– signs) to illustrate how changes in one variable affect others. They help reveal **reinforcing loops** (amplify change) and **balancing loops** (stabilize the system).

Use Case:
In innovation strategy, a CLD might show how R&D investment increases product differentiation, which increases customer adoption, which generates revenue for more R&D—a reinforcing loop. A balancing loop could show how market saturation limits future adoption.

Benefits:

* Clarifies dynamic relationships
* Exposes unintended consequences
* Ideal for early-stage strategic exploration

CLDs are useful when decision-makers have a mental model they want to **test and refine visually**, even without full data.

2.3.2.2 Stock and Flow Models

Purpose:
To track how **quantities (stocks)** accumulate or deplete over time through **rates of change (flows)**—providing a basis for dynamic simulation.

Structure:
These models include:

* Stocks (boxes): e.g., customers, inventory, talent
* Flows (pipes): e.g., acquisition rate, churn, hiring, exits
* Converters and connectors: to represent policies, rates, and nonlinear relationships

Use Case:
In workforce planning, stock and flow models can simulate how hiring, attrition, and training impact skill availability over time.

Benefits:

* Allows time-based simulation
* Captures delay effects
* Supports scenario analysis and forecasting

In supply chain strategy, firms use these models to forecast inventory turnover under different demand scenarios—revealing when and where stockouts might occur.

2.3.2.3 System Archetypes

Purpose:
To identify **recurring patterns of system behavior** that often lead to strategic failure—even when intentions are good.

Structure:
Archetypes are templates that reveal **deep structural flaws** in systems. Common ones include:

* "Fixes that Fail"
* "Shifting the Burden"
* "Limits to Growth"
* "Success to the Successful"

Use Case:
A company cutting R&D to boost short-term profits may face a "Fixes that Fail" archetype—initial gains lead to long-term erosion in innovation and market share.

Benefits:

* Diagnose root causes
* Helps break recurring failure patterns
* Encourages long-term thinking

Archetypes help strategists recognize when **short-term logic leads to long-term traps**.

2.3.2.4 Simulation Models

Purpose:
To **experiment with different scenarios** in a modeled environment and observe how system behavior changes over time.

Structure:
Built with software tools, simulation models can include dozens (or hundreds) of interacting variables, running over months or years of simulated time.

Use Case:
A multinational retailer might simulate the impact of introducing same-day delivery in new markets—testing different assumptions about demand, logistics costs, and workforce capacity.

Benefits:

* Supports "what-if" experimentation
* Identifies tipping points and thresholds
* Reduces risk before committing real-world resources

Many simulation models use discrete-event simulation, agent-based modeling, or system dynamics—all powered by STEM techniques like Monte Carlo simulations and optimization algorithms.

2.3.2.5 Digital Twins

Purpose:
To create a **real-time virtual replica** of a system, continuously synchronized with its physical counterpart.

Structure:
Digital twins combine IoT sensors, AI algorithms, and cloud computing to model behavior in real-time. They often overlay simulation capabilities onto live operational systems.

Use Case:
In smart manufacturing, a factory's digital twin monitors equipment health, predicts breakdowns, and optimizes production schedules before actual disruptions occur.

Benefits:

* Provides a live control panel for strategy execution
* Enables predictive decision-making
* Blends simulation and reality

Digital twins are foundational to **Industry 4.0**, allowing strategists to model, intervene, and adapt continuously—at scale and in real time.

To support model selection, **Table 2.6** summarizes the core strengths of each model type and the strategic questions they are best equipped to answer:

Table 2.6. Choosing the Right Model

Model Type	**Ideal For**	**Strategic Question Answered**
Causal Loop Diagram	Early-stage strategic mapping	*What influences what—and how?*
Stock & Flow Model	Time-based forecasting and simulation	*How do resources accumulate or deplete?*
System Archetype	Identifying failure patterns	*Are we trapped in a common strategic pitfall?*
Simulation Model	Scenario testing and stress-testing	*What happens if we scale or change strategy?*
Digital Twin	Real-time control and predictive insights	*How is our system behaving right now—and next?*

2.3.3 From Mental Models to Dynamic Simulations

Every strategic decision begins in the mind of a decision-maker. Whether a CEO is sketching out a new growth initiative or a team lead allocating resources, decisions are guided by **mental models,** internal representations of how the world works. These models draw on experience, intuition, and reasoning. But while often useful, **mental models are invisible, untested, and vulnerable to bias and oversimplification**.

Strategic system modeling transforms these internal beliefs into **external, testable simulations**. It enables leaders to surface hidden assumptions, identify gaps in logic, and run "what-if" experiments that sharpen judgment. This is where systems thinking becomes more than philosophy; it becomes **engineering for strategy**.

2.3.3.1 What Are Mental Models?

Mental models are the internal blueprints we use to understand how the world works. They help us simplify complexity, form predictions, and make decisions. In strategic contexts, mental models serve as **unspoken guides,** shaping how leaders perceive problems, interpret signals, and design solutions. Yet, because they are invisible and informal, mental models are often incomplete or flawed.

In essence, a **mental model is a personal system simulation**. It may not be drawn or calculated, but it still represents the strategist's belief about cause and effect: *"If we enter this market, we'll gain share", or "If we improve efficiency, our profits will rise."*

These intuitive judgments are essential, but if left unexamined, they can lead to **strategic blind spots**, particularly in dynamic or nonlinear systems.

Mental Models in Everyday Strategy

Mental models are simplified representations of how the world works. Strategists use them, often unconsciously, to make sense of complexity, anticipate outcomes, and guide decisions. These models shape how we interpret cause and effect, assess trade-offs, and design interventions.

While mental models can be abstract, they are deeply embedded in everyday strategic thinking across domains. The table below illustrates common mental models used in marketing, operations, sustainability, HR, and innovation, each capturing a chain of logic that guides action:

Table 2.7. Everyday Mental Models Across Strategic Domains

Domain	Example Mental Model
Marketing	*Increased ad spend leads to more visibility → more traffic → more conversions.*
Operations	*Faster production leads to lower unit cost.*
Sustainability	*Carbon offsetting can neutralize the impact of emissions.*
HR Strategy	*Higher pay leads to better retention and a stronger culture*
Innovation	*Fail fast → learn quickly → iterate to success.*

These models feel intuitive, and sometimes they are right. But without external validation, they **remain assumptions**, not evidence.

The Hidden Power (and Risk) of Mental Models

Mental models are powerful because they shape what we notice, prioritize, and act on. But their invisibility makes them risky:

* **They simplify complexity**—sometimes too much.
* **They are culturally shaped**—two leaders in the same firm may see the same system very differently.
* **They are sticky**—even in the face of contradictory evidence.

A senior executive might assume that "remote workers are less productive." This model, shaped by experience and bias, could lead to costly decisions about office policies—ignoring data that suggests the opposite.

Why Mental Models Need Modeling

In STEM-rich systems, like supply chains, data ecosystems, or energy grids, mental models can't keep up with real-world complexity. Strategists must translate their mental models into **formalized system models** that can:

* Be visualized and critiqued
* Be simulated and updated
* Reveal hidden feedback, time delays, and emergent behavior

From Model to Map: A mental model is like an idea of a city. A strategic system model is the city's actual map, with streets, congestion patterns, and traffic lights clearly laid out.

Mental Models as the Starting Point

While flawed, mental models are still valuable; they are where strategy begins. Modeling tools such as causal loop diagrams and system archetypes **start by capturing the strategist's internal logic** and making it explicit. This creates the basis for shared understanding, collaborative inquiry, and refinement through data or simulation.

Key Steps in Developing a Strategic Model from a Mental Model:

1. **Elicit**: Identify the mental assumptions shaping current decisions.
2. **Visualize**: Draw the feedback loops, dependencies, and causal pathways.
3. **Challenge**: Ask: What are we missing? Where's the evidence?
4. **Translate**: Build a formal system model to test the logic under various conditions.

By bridging the gap between mental models and dynamic simulations, strategists build **cognitive transparency,** a cornerstone of strategic intelligence in complex systems.

2.3.3.2 Why Mental Models Break Down

Mental models serve as shortcuts, efficient approximations of reality that help us make decisions quickly. But in complex, fast-moving, or highly interconnected environments, these shortcuts often **fail to capture the full picture**. They collapse under pressure when:

- Feedback is delayed or hidden
- Causes and effects are nonlinear
- Multiple variables interact across time and space
- External disruptions invalidate past assumptions

In such systems, even seasoned leaders fall into **mental traps,** believing they understand the system, when in reality, they are reacting to symptoms, not structure.

Below are five core reasons why mental models break down, each illustrated with a strategic lens:

1. Linear Thinking in a Nonlinear World

Mental models tend to assume straight-line causality: *"If we do X, Y will happen."* But most strategic systems are **nonlinear,** small inputs can create big effects (or no effects), and outcomes often depend on threshold points.

A price reduction may lead to higher sales—until it crosses a threshold where customers perceive the product as "cheap" or low-quality. Beyond that point, sales may decline.

Why it breaks down:
Linear thinking fails to anticipate tipping points, compounding feedback, or diminishing returns, leading to overconfidence in flawed solutions.

2. Static Assumptions in Dynamic Systems

Mental models often freeze systems in time: *"Our customers behave this way,"* or *"This process always works."* In reality, systems evolve, competitors adapt, customer preferences shift, and new technologies rewire incentives.

A hotel chain assumes its loyalty program will continue driving bookings. But as Gen Z travelers prioritize sustainability and local experiences, the model loses relevance.

Why it breaks down:
Static models can't respond to environmental changes, making yesterday's truths tomorrow's liabilities.

3. Cognitive Bias and Selective Attention

Human brains are wired to seek patterns, favor familiar data, and confirm prior beliefs. These **cognitive biases** shape how mental models are built—and how stubbornly they are held.

An executive may overestimate the impact of marketing spend because it aligns with their expertise, while ignoring product quality issues driving churn.

Why it breaks down:
Strategic blind spots form when leaders filter out contradictory evidence or overweight anecdotal experiences.

4. Fragmented Logic Across the Organization

Mental models are usually built in silos—marketing sees one part of the system, operations another, and finance yet another. Rarely do these models **synchronize across functions**.

Marketing pushes a campaign to increase demand, while operations haven't prepared for increased volume, and customer support is understaffed—resulting in backlash.

Why it breaks down:
Without shared models, teams optimize for local outcomes, not system-wide performance—creating uncoordinated or conflicting strategies.

5. Underestimation of Time Delays

Many strategic outcomes are delayed, such as talent investments, reputation building, cultural change, and R&D payoffs. Mental models often **ignore these lags**, expecting immediate results and misreading early feedback.

A company launches a sustainability initiative and expects brand loyalty to rise in six months. In reality, brand perception shifts over years, not quarters.

Why it breaks down:

Strategic failure often stems not from a lack of effort, but from flawed assumptions embedded in mental models. When effects don't appear quickly, leaders may abandon strategies prematurely, mistaking a delay for failure. But the problem often lies deeper: in how the system is understood, framed, and acted upon.

Table 2.8 outlines common flaws in mental models, the mechanisms through which they break down, and the strategic consequences that follow:

In STEM-rich environments, where cause and effect are complex, delayed, and interdependent, flawed mental models can have **exponential costs**. By identifying these breakdowns and modeling systems explicitly, strategists gain the **clarity and control** needed to design robust, future-ready strategies.

Table 2.8. Why Mental Models Fail

Flaw in Mental Model	Breakdown Mechanism	Strategic Consequence
Linear thinking	Misses tipping points and emergent behavior	Misguided resource allocation
Static assumptions	Ignores environmental evolution	Strategy becomes obsolete
Cognitive bias	Filters out key signals and feedback	Overconfidence in flawed conclusions
Functional silos	Misaligned departmental models	Poor cross-functional coordination
Delay blindness	Misreads early performance signals	Premature pivot or abandonment of strategy

2.3.3.3 From Maps to Models

The shift from intuitive thinking to dynamic modeling begins with a simple yet profound act: **externalizing what's inside your head**. Mental models, when kept internal, are limited by bias and invisibility. But once they are mapped visually and made testable, they become the foundation for rigorous, system-based strategic design.

The journey from maps to models transforms strategy from a guessing game into a **computational, data-driven discipline.** In this transition, diagrams evolve into simulations; sketches become dashboards; and mental heuristics are replaced by validated feedback loops and forecasts.

Step 1: Draw the Map

The first step in building a strategic system model is mapping the perceived logic **of the system. This often begins with:**

* **Causal Loop Diagrams (CLDs)**: These maps visually depict variables and the feedback relationships between them.
* **Influence Diagrams**: Broader maps that include stakeholders, boundaries, and external drivers.

A strategist might map how employee engagement drives productivity, which boosts customer satisfaction, reinforcing both financial performance and culture.

This step is not about getting the model "right" at first, but about **surfacing assumptions**, exposing gaps, and aligning team understanding.

Step 2: Define the Structure

Once the causal relationships are mapped, the next step is to specify the structure and quantify flows. This includes:

* Identifying **stocks** (e.g., number of customers, total talent pool)
* Defining **flows** (e.g., customer acquisition, employee attrition)
* Pinpointing **delays**, **nonlinearities**, and **thresholds**
* Establishing **units of measurement** and **timeframes**

In an HR strategy model, the stock might be "skilled employees," with inflows like "training completions per month" and outflows like "attrition rate." A delay might represent the lag between training investment and productivity improvement.

At this point, the model is transitioning from **qualitative insight to quantitative simulation**—a hallmark of STEM-based strategic modeling.

Step 3: Build the Simulation

Using platforms such as **Vensim**, **Stella**, **AnyLogic**, or **Python-based modeling tools**, strategists can now simulate system behavior over time. This enables:

- Scenario testing ("What happens if we increase hiring by 15%?")
- Sensitivity analysis ("Which variable most impacts performance?")
- Policy design ("Which combination of changes creates sustainable improvement?")

Amazon simulates its warehouse staffing and demand management systems using system dynamics and agent-based models—anticipating where breakdowns might occur if demand spikes or labor shortages arise.

Step 4: Run, Refine, and Learn

The simulation becomes a **strategic learning environment**:

* Test interventions before real-world deployment
* Explore unexpected behaviors and tipping points
* Refine assumptions and update logic continuously

This is where **mental models are no longer speculative—they are strategic assets** that evolve with new data and learning. Simulations make systems visible and tangible, enabling strategists to manage complexity instead of being overwhelmed by it.

To illustrate this shift, the table below contrasts traditional planning approaches with STEM-based strategic modeling. It highlights how data, feedback, and integration transform strategy from a static exercise into a dynamic, adaptive process:

Table 2.9. Evolving Strategy: Traditional vs. STEM-Based Modeling

Traditional Approach	STEM-Based Strategic Modeling
Intuition-based planning	Data-informed system simulation
Linear spreadsheets	Feedback-rich dynamic models
One-time analysis	Continuous scenario testing
Separate department logic	Integrated system-wide thinking
Guess-and-check	Simulation-before-execution

This transformation is not only about better strategy, it's about **better strategists**. It equips leaders to **see the future more clearly, intervene more intelligently, and learn faster than the system changes.**

2.3.3.4 The Power of Simulation

In today's volatile and interdependent world, strategy cannot rely on static plans or reactive thinking. Strategic environments, from global supply chains and innovation ecosystems to regulatory landscapes and workforce dynamics, are **complex systems**. They evolve over time, respond to internal and external feedback, and often behave in unexpected ways. To navigate this complexity, **simulation becomes a critical strategic tool.**

Simulation empowers strategists to go beyond intuition and spreadsheets. It allows them to **test, refine, and evolve strategies before the real world does it for them, often at great cost.** With simulation, the strategist becomes an experimental system designer, running virtual experiments in a controlled space where learning is fast, failure is cheap, and insight is high.

What Simulation Enables

Strategic simulation tools allow organizations to model, test, and iterate in a way that accelerates learning. Specifically, simulation enables several capabilities highlighted in Table 2.10:

Table 2.10. Capabilities of Simulation Enable

Capability	Strategic Benefit
Scenario testing	Simulate different futures (e.g., policy shifts, technological changes, disruptions)
Sensitivity analysis	Identify which variables or inputs have the greatest influence on outcomes.
Stress-testing interventions	Anticipate unintended consequences or system shocks.
Learning through modeling	Build shared mental models across teams and stakeholders.
Anticipating delays and tipping points	Better timing of initiatives and avoiding overreactions.

A manufacturing company used simulation to test its global sourcing strategy. When they modeled a disruption in one region, they found a cascade of bottlenecks across inventory, logistics, and customer satisfaction, well before it happened in the real world.

In STEM-intensive industries, simulation has become not just a tool but a **strategic infrastructure**:

* **Science** contributes frameworks for modeling behavior (e.g., thermodynamics, epidemiology, systems biology)
* **Technology** enables real-time data integration, visual dashboards, and cloud-based modeling platforms
* **Engineering** connects simulation to process design, supply chain flows, and optimization logic
* **Mathematics** drives the underlying logic, differential equations, probability distributions, and feedback dynamics

This STEM integration turns strategic simulations into **living models** that evolve with data, respond to inputs, and support continuous recalibration. Table 2.11 shows strategic applications of simulations in terms of different industries.

Table 2.11. Strategic Applications of Simulation

Industry	Use of Simulation
Healthcare	Modeling patient flow, disease spread, and capacity management.
Retail	Demand forecasting, inventory optimization, and pricing strategies.
Logistics	Route optimization, last-mile delivery models, contingency planning.
Urban Planning	Simulating transportation networks, zoning changes, and public safety interventions.
Public Policy	National energy transitions, pandemic responses, and tax policy forecasting.
Tech Firms	Modeling cloud usage demand, algorithm adoption, and feature rollouts.

For example, the city-state of Singapore used dynamic simulation models during COVID-19 to test different lockdown and vaccination strategies. Their scenario-based modeling allowed policymakers to balance health outcomes with economic activity, adjusting interventions in near real-time with extraordinary precision.

Simulation doesn't guarantee better decisions, but it guarantees **better decision-making processes**. It helps strategists learn faster, fail earlier, and align actions with system-level consequences. In an era where complexity defines the strategic game, **simulation is not optional; it's foundational.**

2.3.4 STEM Tools for System Modeling

Strategic modeling gains its true power when supported by robust tools drawn from science, technology, engineering, and mathematics. These tools transform ideas into structured simulations, allowing strategists not only to understand complex systems but to intervene in them intelligently. In this section, we explore key STEM tools used in strategic system modeling, not as abstract concepts, but as **practical instruments** for diagnosing system behavior, testing interventions, and shaping desired futures.

STEM tools are not confined to laboratories; they're increasingly embedded in **corporate strategy offices, think tanks, tech firms, and public policy labs**. These tools support a new generation of leaders who blend strategic insight with computational power.

Let's walk through six foundational tools that have redefined what it means to model complex strategic systems.

1. System Dynamics

System Dynamics (SD) is a method developed by Jay Forrester at MIT to model and simulate feedback-driven systems over time. It uses **stocks, flows, delays, and feedback loops** to understand how changes in one part of a system influence others.

A strategist at a telecom company might use SD to model the feedback loop between customer churn, service quality, and employee morale, revealing that rapid customer acquisition without parallel investment in support infrastructure leads to long-term performance degradation.

Why it matters: SD captures **nonlinear behavior, tipping points, and unintended consequences**. It's especially valuable for long-term planning in domains like sustainability, public health, innovation, or workforce development.

Tools: Vensim, Stella Architect, InsightMaker

2. Agent-Based Modeling (ABM)

Agent-Based Modeling simulates systems by modeling the behaviors and interactions of individual "agents", which can represent people, firms, machines, or even ideas. These agents follow rules, learn, adapt, and interact, often creating **emergent behaviors** not visible in traditional models.

A tech startup uses ABM to explore how different user personas respond to pricing changes in a platform ecosystem. The model shows how early adopters influence mainstream users, creating feedback loops in adoption patterns.

Why it matters: ABM excels in modeling **heterogeneity, micro-level behavior, and emergent complexity**—perfect for strategic settings like platform competition, innovation diffusion, consumer behavior, and policy implementation.

Tools: NetLogo, AnyLogic, GAMA, Mesa (Python)

3. Bayesian Modeling

Bayesian modeling is a statistical approach that allows strategists to **update predictions as new data becomes available.** Rather than treating probabilities as fixed, Bayesian models revise beliefs dynamically, making them highly adaptable to volatile environments.

A logistics firm forecasts demand spikes during global disruptions using a Bayesian model that incorporates real-time signals like fuel prices, port delays, and weather changes. The model improves over time as new data is fed in.

Why it matters: In fast-changing systems where uncertainty is high and data streams are continuous, Bayesian tools offer **adaptive intelligence**, helping leaders refine forecasts with each new input.

Tools: Stan, PyMC, BayesFusion, WinBUGS

4. Monte Carlo Simulation

Monte Carlo simulation uses randomness and repeated sampling to **simulate thousands of possible outcomes**, providing a probability distribution of results. It helps assess **risk and uncertainty**, especially when exact predictions are impossible.

A multinational uses Monte Carlo methods to evaluate the financial risks of entering a new market. By simulating currency volatility, regulatory changes, and demand fluctuations, the firm gains insight into the range and likelihood of potential profit outcomes.

Why it matters: Monte Carlo enables **decision-making under uncertainty**, showing not just what could happen, but how likely each outcome is. This is especially useful in financial planning, supply chain resilience, and innovation strategy.

Tools: @Risk, Crystal Ball, Python (NumPy), R (mc2d, simEd)

5. Optimization Algorithms

Optimization tools seek to **identify the best possible solution** from a set of alternatives, subject to constraints. In strategy, this might mean optimizing resource allocation, scheduling, logistics, or pricing decisions.

An airline uses linear and nonlinear optimization to determine optimal fleet assignments across routes, maximizing profit while minimizing fuel use and delay penalties.

Why it matters: Optimization shifts strategy from *what-if* to *what's best*—making it indispensable in operations, marketing, capital budgeting, and digital infrastructure.

Tools: Gurobi, CPLEX, SciPy.optimize (Python), Excel Solver

6. Digital Twins

A **Digital Twin** is a real-time, digital replica of a physical system, such as a factory, city, supply chain, or even a business unit. It integrates sensors, analytics, and simulation models to monitor and predict system behavior.

A retail company builds a digital twin of its supply network, enabling managers to simulate disruptions (e.g., a regional shutdown) and instantly test rerouting options in response.

Why it matters: Digital twins support **continuous strategic experimentation**, allowing for real-time stress-testing, scenario planning, and AI-assisted decision-making. They are the **frontier of dynamic strategy**.

Tools: Siemens MindSphere, Azure Digital Twins, IBM Maximo, TwinCAT

Integrated STEM Strategy in Action

Consider this layered example: a global renewable energy firm combines several of these tools in a unified strategic modeling environment.

* **System Dynamics** helps model long-term energy adoption trends and climate impact.
* **Agent-Based Modeling** explores how households adopt smart-grid technologies.
* **Monte Carlo** simulates policy and climate risk scenarios.
* **Optimization algorithms** fine-tune capital investments across projects.
* **Digital twins** allow real-time monitoring of wind farm performance.

This is **STEM-powered strategic foresight,** a modeling toolkit that not only predicts the future but shapes it.

STEM tools transform strategy into a dynamic, testable, and adaptive discipline. They **move strategy from the boardroom to the simulation lab,** where ideas are tested before they are executed, and foresight becomes infrastructure.

As we move to the next section, we'll see how these tools aren't just technical; they're cognitive extensions of how great strategists think and act in systems.

STEM Spotlight: Singapore — A Nation That Models Everything

Singapore is widely recognized not just for its economic success but for how it **uses STEM tools to model and manage itself as a strategic system.** The city-state applies a uniquely holistic approach to national governance by integrating system dynamics, agent-based modeling, optimization, and digital twins across sectors.

Urban Planning with Digital Twins

Singapore's **Virtual Singapore** initiative is a fully functioning **3D digital twin of the entire country**. Built in collaboration with Dassault Systèmes, this twin integrates real-time data from traffic sensors, population movement, and energy usage. Urban planners use it to:

* Simulate new construction and infrastructure development
* Anticipate how building design affects heat distribution and traffic flow
* Optimize land use in one of the world's most space-constrained cities

This allows Singapore to **stress-test city-wide decisions** before implementation, reducing waste, improving quality of life, and enhancing climate resilience.

Pandemic Response with Agent-Based and System Dynamics Models

During the COVID-19 pandemic, Singapore employed **agent-based models** to simulate individual behavior in schools, transport, and workplaces. These models were layered with **system dynamics frameworks** to assess long-term public health interventions and economic impacts.

* Models informed decisions on lockdown timing, mask mandates, and vaccination rollouts.
* Strategic simulations helped balance economic reopening with health protection.

Singapore's **multi-tool modeling approach** enabled it to maintain one of the lowest death rates and fastest economic recoveries globally.

Education and Workforce Foresight

Using **Bayesian models** and **optimization algorithms**, Singapore's Ministry of Education continuously aligns school curricula and university funding with future labor market needs. Workforce planning models forecast talent gaps 10–15 years out, dynamically adjusting national education strategy.

Energy and Climate Systems

The Singapore Energy Market Authority uses **Monte Carlo simulations** and **system dynamics** to test energy transition scenarios, exploring risks in solar expansion, carbon pricing, and regional grid integration. Digital twins of the energy grid allow predictive maintenance and adaptive capacity planning.

What Strategists Can Learn

Singapore treats strategy not as planning, but as **adaptive modeling of a living system.** Its success lies not in perfect foresight, but in building models that learn, update, and simulate possibilities before action is taken.

* It combines **quantitative rigor** with **qualitative judgment**
* It runs **real-time simulations** across policy domains
* It turns **modeling into governance infrastructure**

Whether you're designing corporate strategy, sustainability plans, or digital infrastructure, **Singapore provides a blueprint** for modeling complex systems with precision, speed, and agility.

2.4 Strategic Modeling in Practice: Ethics, Governance, and Cross-Disciplinary Insights

As strategic modeling becomes more powerful and accessible through STEM tools, a critical layer of responsibility emerges: **how we model, who models, and to what end**. Strategic systems modeling is not just a technical task; it's a **social and ethical act**, one that shapes decisions, influences behavior, and redistributes power across stakeholders. Whether applied in governments, corporations, or AI-driven platforms, modeling choices carry implicit assumptions that affect real lives and outcomes.

This section explores the **ethical boundaries**, **governance structures**, and **cross-disciplinary lessons** that must accompany modeling practices in STEM strategy. It challenges us to think not only about *what models can do*, but also *what they should do,* and how to ensure they reflect values like transparency, accountability, equity, and inclusivity.

By drawing from disciplines such as public policy, computer science, urban planning, and biomedical engineering, we begin to see that models are more than analytical tools; they are **strategic artifacts** that must be designed with care and governed with rigor.

2.4.1 Ethical Considerations in Strategic Modeling

As the power of modeling increases, so too does its potential to amplify **unintended consequences, institutional biases, and ethical blind spots.** Strategic models are not ethically neutral—they reflect choices about what gets included, what is left out, and whose perspectives shape the logic. These choices have implications for **fairness, accountability, and justice**, particularly in high-stakes domains such as healthcare, urban planning, AI-driven hiring, or national policy. When strategists build models without critical reflection, they risk embedding assumptions that may reinforce inequality, exclude marginalized voices, or lead to system-wide failures masked as optimization.

At its core, **ethical strategic modeling** means asking not only *"Is this model technically correct?"* but also *"Is it just, inclusive, and aligned with our values?"*

1. The Ethics of Assumptions

Every model begins with assumptions: what variables matter, how they interact, what counts as success. These assumptions reflect the worldview of the modeler—and if left unchecked, they can encode narrow logics that overlook the complex realities of affected communities.

A predictive policing model might assume that past arrest data reflects true crime rates. But if historical arrests are racially biased, the model will reinforce over-policing in marginalized neighborhoods, an ethically hazardous feedback loop.

Strategic models must be **interrogated for the ethical implications of their structure.** Who benefits from the model? Who bears the risk? What voices are missing from the design?

2. Transparency and Explainability

Ethical modeling requires **transparency,** making model logic visible, explainable, and reviewable by both technical and non-technical stakeholders. In opaque models (e.g., proprietary black-box AI tools), strategic decisions may be made without a full understanding of how outcomes are generated.

Transparency matters not only for trust, but for **accountability**. If a model drives layoffs, denies insurance, or excludes communities from infrastructure investment, affected individuals deserve to understand why.

Strategy Implication: Ethical strategists should document model assumptions, data sources, and logic pathways. Explainability isn't a technical afterthought; it's a **core leadership responsibility**.

3. Inclusion and Participatory Design

Modeling processes that exclude frontline workers, communities, or underrepresented groups often fail, not because the math is wrong, but because the **human logic is incomplete**. Inclusive modeling involves:

* Co-designing models with those affected by the outcomes
* Validating assumptions through community or stakeholder feedback
* Testing scenarios that reflect diverse lived experiences

In Nairobi, planners used participatory system mapping to design a flood resilience strategy. By including informal settlement residents in the modeling process, the city uncovered critical drainage patterns and resource gaps invisible to traditional top-down models.

Inclusivity makes models **more accurate, legitimate, and robust.** It also reinforces a culture of ethical foresight, where modeling becomes a shared language for dialogue, not a technocratic tool for control.

4. Data Ethics and Privacy

Strategic models often rely on large-scale data, employee records, consumer behavior, health profiles, and mobility data. But data is not just a resource; it's **an expression of human identity**. Unethical data use, such as collecting without consent, anonymizing poorly, or repurposing data against user intent, can erode trust and violate rights.

A smart city simulation that tracks individual movement patterns without disclosure or opt-out options may optimize traffic flow, but at the cost of civil liberties.

Ethical modeling requires **data governance principles**:

* Informed consent
* Secure data storage
* Clarity on how and why data will be used
* Mechanisms for redress if harm occurs

In strategic terms, data ethics is a **reputational asset**. Ethical handling of data reinforces stakeholder trust and reduces regulatory and legal risk.

5. Model Lifecycle Responsibility

Models are not static; they evolve as systems change. But many organizations treat models as one-time tools, failing to update them as conditions shift. Ethical modeling includes a **duty of care** over the model's lifecycle:

* Regular audits and recalibration
* Monitoring for drift or new biases
* Sunset clauses for outdated models
* Clear accountability for misuse or misinterpretation

Strategic Risk: A workforce optimization model that no longer reflects current labor laws or social expectations can lead to discriminatory outcomes, despite working as originally intended. Table 2.7 presents ethical risks and strategic responses. Sustainable modeling requires an **ethics of maintenance,** keeping models honest, current, and reflective of new knowledge and norms.

Table 2.12. Ethical Risks and Strategic Responses

Ethical Risk	Strategic Response
Biased assumptions	Conduct multi-stakeholder model reviews
Black-box logic	Prioritize transparency and explainability
Exclusion of affected groups	Use participatory modeling approaches
Misuse of personal data	Enforce strong data governance and privacy policies
Model obsolescence	Institute model update cycles and performance monitoring

The Ethical Strategist's Mindset

In a STEM-driven strategy environment, power lies not only in data and tools, but in the **values encoded within them.** Ethical modeling is not about perfection; it's about reflection, accountability, and humility in the face of complexity. Ethical strategists recognize that **models shape systems, and systems shape lives.** Their task is not only to optimize but to **steward** the future with integrity.

2.4.2 Governance and Transparency in Model Use

Ethics in modeling cannot be sustained without effective **governance structures,** the institutional systems that define how models are developed, deployed, monitored, and held accountable. While ethics emphasizes individual and collective responsibility, **governance brings structure, oversight, and enforceability.** Without governance, even well-intentioned models can drift into opacity, irrelevance, or misuse.

As strategic modeling becomes embedded in decision-making, from public infrastructure to algorithmic hiring to global climate policy, **transparency and governance become the dual pillars** that determine whether models build legitimacy or sow distrust.

Why Governance Matters in Modeling

Strategic models often **guide real-world outcomes**: where a city builds housing, who receives a loan, how a company manages layoffs, or how nations allocate vaccines. These are not internal simulations; they are **decision engines**. And as with any powerful system, decision engines must be governed.

Strategic Consequence: When a model influences people's access to resources or rights, lack of governance can lead to public backlash, litigation, or loss of license to operate.

Governance ensures:

* Models are reviewed and approved by independent experts
* Decision-makers understand the model logic, not just outputs
* Organizations can explain and defend model-driven actions
* Feedback loops exist for refining models based on outcomes

2. Governance Mechanisms: From Protocols to Policies

A mature modeling governance system includes:

* **Documentation Standards**: Clear explanation of variables, assumptions, limitations, and intended use cases

* **Version Control**: Models evolve; each version must be tracked, labeled, and timestamped
* **Audit Trails**: Records of who changed what, when, and why, ensuring accountability
* **Oversight Committees**: Interdisciplinary panels to review high-impact models
* **Redress Mechanisms**: Systems to report harm or error, and to update or suspend problematic models

At a major financial institution, credit scoring models must be reviewed quarterly by a Risk and Ethics Board. The board has the authority to veto models that show evidence of discriminatory impact, even if they're profitable.

3. The Strategic Case for Transparency

Transparency is more than a compliance requirement; it is a **strategic asset**. Organizations that explain how and why their models work can:

* Earn trust from customers, employees, regulators, and the public
* Reduce resistance to change, especially in automation or restructuring
* Build credibility in fast-moving fields like AI, biotechnology, and logistics
* Attract top talent who value open, accountable workplaces

Transparency is particularly crucial when models are deployed in **public-facing or ethically sensitive domains**—e.g., hiring, education, healthcare, policing, or social benefits.

Behavioral Insight: Studies show that people are more likely to accept algorithmic decisions, even unfavorable ones, when they understand the rationale behind them.

4. Global Trends in Model Governance

Governments and international bodies are beginning to treat algorithmic models as **governable artifacts**, subject to legal, ethical, and technical scrutiny.

* **EU AI Act**: Requires transparency and risk categorization for all AI systems. High-risk models (e.g., in education, law enforcement, and finance) must include human oversight, explainability, and regular evaluation.
* **OECD AI Principles**: Recommend transparency, robustness, and accountability in AI and model-driven systems.
* **U.S. Executive Order on AI (2023)**: Mandates safety testing and independent audits for federally funded AI models.

These shifts signal a new era in strategy, where **model governance is as important as corporate governance**.

5. The Role of STEM Leaders in Model Governance

STEM-trained strategists are uniquely positioned to **bridge technical complexity and organizational accountability.** They can:

* Translate model logic for executives and boards
* Anticipate misuse or bias before deployment
* Design governance frameworks that evolve with technology
* Promote a modeling culture grounded in *precision with humility*

Governance is not about slowing innovation; it's about **making innovation trustworthy, testable, and sustainable.** In fast-moving, STEM-driven environments, strategic governance ensures that models, algorithms, and decisions remain transparent, accountable, and aligned with evolving standards. The table below summarizes key components of governance and their strategic purposes, offering a practical framework for embedding governance into innovation workflows:

Table 2.13. Governance in Action: A Summary Framework

Governance Component	Strategic Purpose
Documentation & Disclosure	Ensures transparency and stakeholder understanding
Version & Change Tracking	Maintains model integrity over time
Audit & Oversight Panels	Provides independent review and risk mitigation
Feedback Loops & Redress	Enables continuous improvement and harm mitigation
Alignment with Regulation	Anticipates legal compliance and ethical scrutiny

2.4.3 Cross-Disciplinary Lessons from System Design

Strategic modeling doesn't belong to any one discipline; it's an evolving practice enriched by insights from **engineering, biology, public policy, computer science, architecture, and beyond.** While business strategy traditionally borrows from economics or operations research, today's complex environments demand **cross-disciplinary thinking** that goes beyond linear models and static planning.

This section explores what strategists can learn from how other disciplines model complex systems. These fields offer not only novel techniques but also **mental models, values, and failure stories** that can help strategic thinkers build more resilient, adaptive, and inclusive systems.

1. From Engineering: Modular Design and Resilience

Engineering disciplines, from aerospace to software architecture, teach the value of **modularity and fault tolerance**. Instead of trying to model everything in one monolithic system, engineers often break down complexity into **interoperable modules** that can be adjusted or replaced without collapsing the whole.

Strategic Lesson: Design strategic models that can evolve over time. Don't build "one-shot" models; create adaptable systems that can scale, swap components, and integrate new information.

Engineering also emphasizes **stress testing**, boundary conditions, and system redundancy, tools often underused in business modeling but essential in volatile environments.

2. From Biology: Feedback, Adaptation, and Emergence

Biological systems, from neural networks to ecosystems, model **adaptive complexity**. These systems don't operate with rigid hierarchies or single-point control. Instead, they thrive through **feedback loops, decentralized control, and continuous learning**.

Strategic Analogy: Think of an entrepreneurial ecosystem like a rainforest, not a machine. It's shaped by co-evolution, mutual dependencies, and niche specialization. Models should reflect this emergent logic, not force a top-down blueprint.

Biology reminds us that complexity isn't noise, it's **information**. Strategic models can benefit from **evolutionary logic**, where strategies are tested, adapted, and selected for fitness over time.

3. From Public Policy: Stakeholder Complexity and Political Feasibility

Public policy modeling grapples with **messy trade-offs**, competing interests, and limited resources. Tools like **cost-benefit analysis, multicriteria modeling, and participatory scenario building** are used not to "solve" problems but to **illuminate tensions and negotiate better outcomes**.

Strategic Insight: Strategic models must reflect not only technical efficiency but also **political and cultural viability.** Just because a strategy works on paper doesn't mean it will survive stakeholder pushback, cultural misalignment, or regulatory scrutiny.

Policy modeling also excels in **scenario planning,** anticipating divergent futures shaped by weak signals, structural shifts, and black swan events. Business strategists can learn to move beyond linear forecasts toward **narrative-rich, multi-path modeling.**

4. From Architecture and Urban Design: Spatial Thinking and Human-Centered Design

Architects and planners use modeling not just for technical optimization but for **experiential understanding of** how people interact with systems over time and space. These fields prioritize **design thinking, inclusivity, and the lived experience** of stakeholders.

Strategy Application: Models should not only be analytically robust but also **visually intuitive and emotionally resonant.** Strategy maps, dashboards, and simulation environments should communicate meaning, not just data.

Spatial models like **systems maps, causal loop diagrams, and storyboards** help align diverse stakeholders around complex initiatives, from digital transformation to global expansion.

5. From Computer Science: Algorithmic Transparency and Agile Modeling

Computer science brings **algorithmic precision, iterative refinement, and computational ethics**. Agile software methods, building, testing, and refining in short sprints, mirror the principles of **agile strategy modeling**.

Strategic Takeaway: Rather than build the "perfect" model in isolation, deploy **minimum viable models (MVMs)** early and often. Let data and feedback refine the model iteratively.

Computer science also emphasizes **explainability in AI,** a lesson critical to strategic modeling. Models that stakeholders can't understand are **models that can't be trusted.** But explainability doesn't emerge from one discipline alone—it's the result of integrating insights from multiple fields. The table below highlights how different disciplines contribute unique strengths to strategic modeling, helping leaders build models that are not only technically sound but also transparent, adaptive, and stakeholder-aware:

Table 2.14. What Each Discipline Offers Strategic Modeling

Discipline	Key Contribution to Strategic Modeling
Engineering	Modular design, stress testing, resilience
Biology	Adaptive learning, feedback loops, emergence
Public Policy	Stakeholder complexity, scenario analysis, and political feasibility
Architecture	Spatial logic, stakeholder experience, and visual modeling
Computer Science	Iterative development, algorithmic transparency, minimum viable models

The Power of Cross-Disciplinary Modeling

Complex strategy problems, climate change, digital disruption, global health, and ethical AI require more than linear business logic. They require **models built with the wisdom of many fields**. Cross-disciplinary modeling does not mean abandoning core strategy; it means **enhancing it with richer lenses, smarter assumptions, and deeper human insight.**

Strategists of the future are not siloed specialists. They are **systemic thinkers fluent in the languages of biology, data science, engineering, and ethics.** Their models are not only smart, but wise.

STEM Spotlight: Singapore – A Nation Modeled as a Strategic System

Singapore is more than a city-state; it is a **living model of system strategy in action**. From urban design to economic development, healthcare, climate resilience, and digital infrastructure, Singapore applies **STEM-powered systems thinking** at a national scale. Its strategic success is not accidental; it is modeled, tested, governed, and continuously refined.

Systems Thinking at the Core

Since the 1960s, Singapore has modeled its transformation from a vulnerable post-colonial port to a global hub using principles from engineering, urban planning, and systems design. Instead of fragmented policymaking, it integrates strategy through centralized foresight units like the **Centre for Strategic Futures (CSF)** and **Smart Nation and Digital Government Group (SNDGG)**.

The Urban Redevelopment Authority (URA) uses system dynamics models and spatial analytics to simulate land use, transportation flow, and population density over 50-year horizons, long before decisions are implemented.

STEM Tools in Practice

Singapore blends traditional planning with modern modeling techniques:

* **Digital Twins** of city infrastructure simulate traffic, energy use, flood risk, and emergency response.
* **AI-augmented policy simulations** forecast housing needs, social services, and economic scenarios.
* **Participatory systems maps** are co-developed with citizens to guide urban renewal and public health.

Ethics and Governance

Singapore doesn't simply deploy models—it governs them with a focus on **public trust and social legitimacy**. Data protection laws, transparency portals, and citizen consultation platforms accompany most model-driven policies.

* *TraceTogether*, a COVID-19 contact tracing app, was paired with public disclosures and data access constraints to balance effectiveness with privacy.

Insight: Modeling is not just technical, it's cultural. Singapore succeeds because it integrates **STEM tools with ethical governance, inclusive design, and long-range systems vision.**

Strategic Takeaway

For students of strategic management, Singapore proves what's possible when a nation models itself as a system:

* It **tests policies before they fail.**
* It **balances efficiency with equity.**
* It **scales STEM tools with ethical oversight.**
* And it **teaches future leaders** to model not only for profit, but for people and the planet.

2.5. Chapter 2 Wrap-Up: Modeling Strategy as a System

As we close this chapter, it becomes clear that in a STEM-driven world, strategy is no longer a static roadmap; it is a dynamic system. Today's complex, interdependent environments require leaders to think in loops, not lines; to anticipate ripple effects, not just first-order outcomes. Modeling strategy as a system enables decision-makers to uncover hidden leverage points, visualize consequences, and design feedback-rich responses.

Throughout this chapter, we explored how **systems thinking** reshapes the strategic mindset. We moved beyond traditional analysis to examine how causal maps, feedback loops, and simulations allow organizations to respond adaptively rather than reactively. We looked at the power of dynamic modeling in anticipating change and in governing complexity with clarity. We also examined the ethical responsibilities and governance challenges that arise when system models influence high-stakes decisions.

From understanding the difference between analytical and systems thinking, to applying tools like system dynamics, digital twins, and Bayesian simulations, one message stood out: **the strategist of the future must be a systems modeler,** someone who thinks in terms of flows, feedback, uncertainty, and iteration. These are not abstract ideas; they are the foundations of real-world decision-making in industries as diverse as healthcare, technology, logistics, and finance.

Importantly, we also addressed **the human side of modeling,** from cross-disciplinary collaboration to ethical foresight and inclusive governance. Strategic models are not just technical tools; they are reflections of our values, assumptions, and intentions. A well-designed model can empower smart decision-making; a poorly designed one can create blind spots, biases, or worse, systemic harm.

As you prepare to model strategy in your own organization or field, remember: it's not just about simulating outcomes. It's about designing systems that adapt, learn, and lead ethically.

2.5.1. Discussion Questions

1. What is the difference between a system map and a strategy model?
2. How can feedback loops be used to improve long-term strategic planning?
3. Why is causal thinking essential for identifying leverage points?
4. What are the risks of over-reliance on static strategy frameworks?
5. How does simulation enhance strategic foresight?
6. In what ways do ethical considerations affect how models are designed and used?
7. How does cross-disciplinary modeling strengthen decision-making?
8. What distinguishes systems thinking from analytical thinking in practice?
9. How do governance and transparency influence stakeholder trust in models?
10. How could you begin to model your organization's current strategy as a system?

2.5.2. Micro-Case: Zara's Strategy Loop

Zara, long admired for its fast-fashion supply chain, is now facing a new frontier: how to maintain speed and global responsiveness while navigating rising sustainability pressures, digital transformation, and regional regulatory shifts.

Recognizing these systemic challenges, Zara's leadership team initiates a **strategy modeling sprint** using systems thinking principles. The goal? To uncover systemic tensions and design a strategy that learns and adapts in real time.

They begin by building a **causal loop diagram** of Zara's global design-to-retail process. Key variables include fabric sourcing timelines, production cycle speed, CO_2 emissions, digital design input, consumer behavior data, and regional labor conditions. The model reveals a reinforcing loop: faster turnaround increases product freshness and sales, but also increases energy consumption and waste.

The team introduces a balancing loop by incorporating **algorithmic fabric optimization**, using machine learning to forecast material needs and reduce overproduction. They simulate how increased use of digital twins and AI-assisted design can shorten sampling cycles by 30%, reducing emissions and cost.

To embed learning into the system, Zara creates **feedback dashboards** that show real-time inventory fluctuation, customer sentiment by region, and energy consumption per unit sold. This looped information system enables faster, more ethical decisions, balancing speed with sustainability.

But challenges remain. The system model surfaces ethical concerns, such as data privacy in AI personalization and supplier exploitation risks. To address these, Zara sets governance protocols that require human oversight in all automation decisions and institutes transparency scoring for third-party suppliers.

Zara's approach exemplifies how strategy, when modeled as a dynamic system, becomes a **living process,** one that senses, adapts, and evolves continuously.

Discussion Questions

1. What loops (reinforcing or balancing) were key to Zara's system design?
2. How did Zara identify and test leverage points?
3. What were the ethical risks of automation, and how were they addressed?
4. If you were advising Zara, what feedback loop would you prioritize next?
5. How would you visualize Zara's strategic system using one of the tools from this chapter?

2.5.3. Interactive Exercise: Modeling Organizational Strategy

Scenario:
You are part of a strategic innovation team at a multinational retail company aiming to model its entire supply chain to enhance resilience and sustainability. The CEO has asked your team to use a systems approach to simulate the impacts of delays, demand fluctuations, and policy changes.

Your Task:
Using any two tools or frameworks from this chapter (e.g., causal mapping, system dynamics, simulation modeling, or digital twins), design a one-page system model prototype that:

* Maps key variables and feedback loops across the supply chain
* Identifies at least two leverage points for reducing emissions and improving speed
* Proposes ethical safeguards for transparency and algorithmic decision-making

Bonus Challenge:
Use a digital tool (e.g., Vensim, Loopy, or Miro) to visualize your causal map. Reflect on what surprises you discover when mapping feedback across functions.

2.5.4. Recommended HBR Case Studies

❖ **Systems Thinking**

- **Overview**: The Harvard Business Review case study *"Systems Thinking"* by Brian Trelstad and Preeti Varma explores how systems thinking can be applied to understand and address complex, interdependent challenges in organizations and society. It introduces the core concepts of systems thinking, such as identifying root causes, recognizing feedback loops, and mapping interconnections within a system. The case emphasizes that many persistent problems cannot be solved through linear thinking or isolated interventions. Instead, it encourages a holistic approach that considers the broader system dynamics and long-term consequences of actions. This case is particularly useful for leaders, social innovators, and students seeking to develop strategies that lead to sustainable, systemic change.
- **Citation**: Trelstad, B., & Varma, P. (2021). *Systems Thinking* (Case No. 322030-PDF-ENG). Harvard Business School Publishing.

❖ **Maple Leaf Foods: Changing the System**

- **Overview**: This case explores how Maple Leaf Foods (MLF), a major Canadian food processor, redefined its corporate purpose to become the world's most sustainable protein company. Rather than focusing solely on financial performance, MLF began integrating environmental and social considerations into its core strategy. The case examines the company's shift toward sustainability, the internal and external challenges it faced, and how it engaged with various stakeholders to drive systemic change. The main case is accompanied by five supplemental cases (A–E), each offering a different stakeholder perspective, highlighting how systemic change within an organization affects and is influenced by the broader ecosystem in which it operates.
- **Citation**: Bansal, P., Montgomery, W., & MacMillan, K. (2020). *Maple Leaf Foods: Changing the System* (Case No. W20831-PDF-ENG). Ivey Publishing.

❖ **Operation Walk Canada: Improving Capacity in Guatemala**

- **Overview**: This case, authored by Nicole R.D. Haggerty, Hannah Litchfield, Margarita Amoranto, Gabriel Yallen, Christo Bosco, Sajee Maheswaran, and Horodjei Yaya, was published on December 7, 2022. It focuses on Operation Walk Canada, a not-for-profit orthopedic surgical mission based in London, Ontario. The organization provides hip and knee replacement surgeries to patients in low- and middle-income countries like Guatemala, where access to such care is extremely limited. After a two-year pause due to the COVID-19 pandemic, the organization faced a growing backlog of patients and increasing pressure on its limited resources. The case challenges readers to consider how Operation Walk can scale its impact sustainably, address systemic healthcare gaps, and prepare for its return to Guatemala in 2022.

- **Citation**: Haggerty, N. R. D., Litchfield, H., Amoranto, M., Yallen, G., Bosco, C., Maheswaran, S., & Yaya, H. (2022). *Operation Walk Canada: Improving Capacity in Guatemala* (Case No. W31633-PDF-ENG). Ivey Publishing.

2.5.5. Strategy-in-Film: Learn by Watching

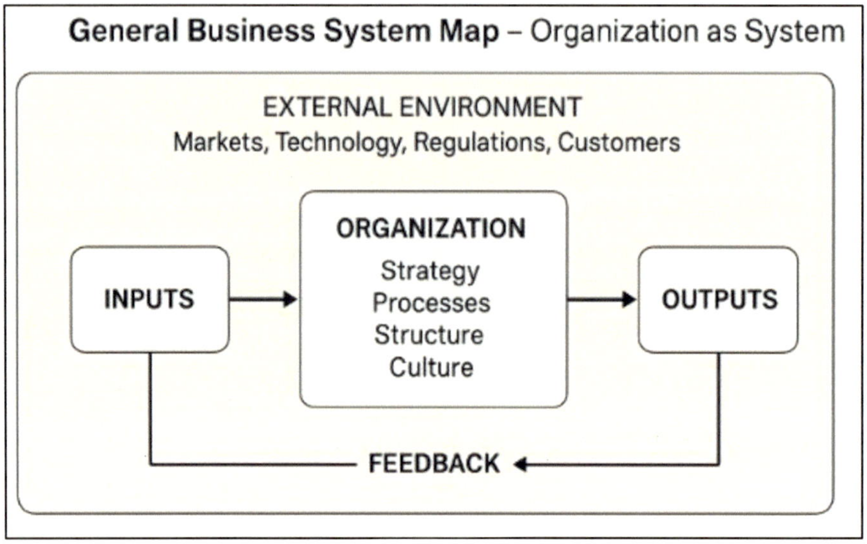

Strategy unfolds within interconnected systems, where decisions trigger ripple effects far beyond immediate goals. The films below capture how leaders confront complexity, navigate feedback loops, and manage unintended consequences. As you deepen your understanding of systems thinking in this chapter, use these narratives to observe how strategic decisions shape, and are shaped by, dynamic environments.

Movie / Documentary	Strategic Theme Highlighted
Apollo 13	Crisis systems modeling, interdependencies
The Martian	Feedback loops, adaptive problem-solving
Moneyball	Systemic decision-making in sports analytics
Arrival	Systems communication and time-loop reasoning
Contagion	Modeling risk, feedback, and epidemic systems
Tenet	Cause-effect inversion, nonlinear timelines
Inside Job	Ethical failures in financial system modeling
Westworld (Series)	Dynamic system design, AI modeling, and simulation
The Great Hack	Systems manipulation through digital feedback loops
Iron Man 2	STEM systems integration in strategic innovation

References

Arnold, R. D., & Wade, J. P. (2015). A definition of systems thinking: A systems approach. *Procedia Computer Science*, *44*, 669-678.

Baghaei Lakeh, A., & Ghaffarzadegan, N. (2015). Does analytical thinking improve understanding of accumulation?. *System Dynamics Review*, *31*(1-2), 46-65.

Bethune, K. G. (2022). *Reimagining design: Unlocking strategic innovation*. MIT Press.

Bolívar, M. P. R., Galera, A. N., & Muñoz, L. A. (2015). Governance, transparency and accountability: An international comparison. *Journal of Policy Modeling*, *37*(1), 136-174.

Cabrera, D., & Cabrera, L. (2023). What is systems thinking?. In *Learning, design, and technology: An international compendium of theory, research, practice, and policy* (pp. 1495-1522). Cham: Springer International Publishing.

Cosenz, F., & Noto, G. (2016). Applying system dynamics modelling to strategic management: A literature review. *Systems Research and Behavioral Science*, *33*(6), 703-741.

Dennis, J., & Tesser, M. (2025). *Lean Blockchain Systems Thinking: Reinventing Value Streams*. CRC Press.

Eisenreich, A., Füller, J., Stuchtey, M., & Gimenez-Jimenez, D. (2022). Toward a circular value chain: Impact of the circular economy on a company's value chain processes. *Journal of Cleaner Production*, *378*, 134375.

Forliano, C., De Bernardi, P., Rozsa, Z., & Bertello, A. (2024). Systems dynamics research in management and organization studies: Overview and research agenda. *Journal of Innovation & Knowledge*, *9*(3), 100512.

Forrester, J. W. (1995). The beginning of system dynamics. *McKinsey Quarterly*, 4-17.

Fowler, A. (2003). Systems modelling, simulation, and the dynamics of strategy. *Journal of Business Research*, *56*(2), 135-144.

Frank, M. (2000). Engineering systems thinking and systems thinking. *Systems Engineering*, *3*(3), 163-168.

Grewatsch, S., Kennedy, S., & Bansal, P. (2023). Tackling wicked problems in strategic management with systems thinking. *Strategic Organization*, *21*(3), 721-732.

Groesser, S. N., & Schaffernicht, M. (2012). Mental models of dynamic systems: taking stock and looking ahead. *System dynamics review*, *28*(1), 46-68.

Haniff, A. P., & Galloway, L. (2022). Modeling strategic alignment in project networks. *International Journal of Project Management*, *40*(5), 517-530.

Hayes, A. S. (2024). Time, ties, transactions: temporality and relational work in economic exchange. *Theory and Society*, *53*(3), 625-651.

Hermans, L. M., & Cunningham, S. W. (2018). *Actor and strategy models: Practical applications and step-wise approaches*. John Wiley & Sons.

Iacovidou, E., Hahladakis, J. N., & Purnell, P. (2021). A systems thinking approach to understanding the challenges of achieving the circular economy. *Environmental Science and Pollution Research*, *28*, 24785-24806.

Jackson, M. C. (2019). *Critical systems thinking and the management of complexity*. John Wiley & Sons.

Jackson, M. C. (2001). Critical systems thinking and practice. *European Journal of Operational Research*, *128*(2), 233-244.

Kaplan, R. S., & Norton, D. P. (1996). Strategic learning & the balanced scorecard. *Strategy & Leadership*, *24*(5), 18-24.

Kaplan, R. S., & Norton, D. P. (2003). *Strategy maps: Converting intangible assets into tangible outcomes*. Harvard Business Press.

Kleijnen, J. P. (2001). Ethical issues in modeling: Some reflections. *European Journal of Operational Research*, *130*(1), 223-230.

Kleijnen, J. (2011). Ethical issues in engineering models: an operations researcher's reflections. *Science and Engineering Ethics*, *17*, 539-552.

Kondrashkina, I. I., Manin, P. V., & Kelejnikova, S. V. (2024). Systems Thinking in Strategic Business Management. In *Sustainable Cooperation for the Creation of Green Supply Chains Based on Environmental Technologies and Responsible Innovations* (pp. 429-436). Cham: Springer Nature Switzerland.

Kunc, M. (2024). The Systems Thinking Approach to Strategic Management. *Systems*, *12*(6), 213.

Kunc, M. (2008). Using systems thinking to enhance strategy maps. *Management Decision*, *46*(5), 761-778.

Liu, S., Wu, Y., Jiang, G., & Dong, J. (2024). System dynamics modeling of collaborative governance in smart cities: a case study of Dongguan, China. *Scientific Reports*, *14*(1), 31758.

Lyneis, J. M., Cooper, K. G., & Els, S. A. (2001). Strategic management of complex projects: a case study using system dynamics. *System Dynamics Review: The Journal of the System Dynamics Society*, *17*(3), 237-260.

Lyons, M. H., Adjali, I., Collings, D., & Jensen, K. O. (2003). Complex systems models for strategic decision making. *BT Technology Journal*, *21*(2), 11-27.

Lyytinen, K., Weber, B., Becker, M. C., & Pentland, B. T. (2024). Digital twins of organization: implications for organization design. *Journal of Organization Design*, *13*(3), 77-93.

Mao, H., Liu, S., & Gong, Y. (2024). Balancing structural IT capabilities for organizational agility in digital transformation: a resource orchestration view. *International Journal of Operations & Production Management*, *44*(1), 315-344.

Martinsuo, M., & Anttila, R. (2022). Practices of strategic alignment in and between innovation project portfolios. *Project Leadership and Society*, *3*, 100066.

Meijer, A. (2013). Understanding the complex dynamics of transparency. *Public Administration Review*, *73*(3), 429-439.

Morecroft, J. D. (2015). *Strategic modelling and business dynamics: A feedback systems approach*. John Wiley & Sons.

Nowak, A., Rychwalska, A., & Borkowski, W. (2013). Why simulate? To develop a mental model. *Journal of Artificial Societies and Social Simulation*, *16*(3), 12.

Palakshappa, N., Venkateswar, S., & Ganesh, S. (2023). Broadening the circle: creativity, regeneration and redistribution in value loops. *Social Responsibility Journal*, *19*(10), 1870-1884.

Paucar-Caceres, A., & Pagano, R. (2009). Systems thinking and the use of systemic methodologies in knowledge management. *Systems Research and Behavioral Science: The Official Journal of the International Federation for Systems Research*, *26*(3), 343-355.

Pavlov, O. V., Cavana, R. Y., Wheat, I. D., Saeed, K., Radzicki, M. J., & Dangerfield, B. C. (2025). Introducing Feedback Thinking and System Dynamics Modeling in Economics Education. *System Dynamics Review*, *41*(2), e70001.

Pil, F. K., & Holweg, M. (2006). Evolving from value chain to value grid. *MIT Sloan Management Review*, *47*(4), 72.

Price, R., Román, F. J., & Rountree, B. (2011). The impact of governance reform on performance and transparency. *Journal of Financial Economics*, *99*(1), 76-96.

Rojas-Córdova, C., Williamson, A. J., Pertuze, J. A., & Calvo, G. (2023). Why one strategy does not fit all: a systematic review on exploration–exploitation in different organizational archetypes. *Review of Managerial Science*, *17*(7), 2251-2295.

Schaffernicht, M., & Groesser, S. N. (2011). A comprehensive method for comparing mental models of dynamic systems. *European Journal of Operational Research*, *210*(1), 57-67.

Schiuma, G., Carlucci, D., & Sole, F. (2012). Applying a systems thinking framework to assess knowledge assets dynamics for business performance improvement. *Expert Systems with Applications*, *39*(9), 8044-8050.

Shafique, A., Saeed, M. S., Javeed, D., & Qureshi, K. N. (2024). AI Models and Ethical Considerations in Research. In *Next Generation AI Language Models in Research* (pp. 170-190). CRC Press.

Simatupang, T. M., Piboonrungroj, P., & Williams, S. J. (2017). The emergence of value chain thinking. *International Journal of Value Chain Management*, *8*(1), 40-57.

Singh, A., Dwivedi, A., Agrawal, D., Bag, S., & Chauhan, A. (2024). Can sustainable and digital objectives synchronize? A study of ESG activities for digital supply chains using multi-methods. *Business Strategy and the Environment*, *33*(8), 8413-8435.

Vinayavekhin, S., & Phaal, R. (2023). Roadmapping for strategic alignment, integration, and synchronization. In *Next Generation Roadmapping: Establishing Technology and Innovation Pathways Towards Sustainable Value* (pp. 1-24). Cham: Springer International Publishing.

Voulvoulis, N., Giakoumis, T., Hunt, C., Kioupi, V., Petrou, N., Souliotis, I., & Vaghela, C. J. G. E. C. (2022). Systems thinking as a paradigm shift for sustainability transformation. *Global Environmental Change*, *75*, 102544.

Walters, D., & Lancaster, G. (2000). Implementing value strategy through the value chain. *Management Decision*, *38*(3), 160-178.

Weiser, A. K., & Laamanen, T. (2022). Extending the dynamic equilibrium model of paradox: Unveiling the dissipative dynamics in organizations. *Organization Theory*, *3*(3), 26317877221090317.

Zajac, E. J., Kraatz, M. S., & Bresser, R. K. (2000). Modeling the dynamics of strategic fit: A normative approach to strategic change. *Strategic Management Journal*, *21*(4), 429-453.

CHAPTER 3

Strategic Analytics and Decision Intelligence

"Data is not strategy but without it strategy is just guessing with confidence"

by MAK

Learning Objectives:

By the end of this chapter, you will be able to:

Wisdom is not in seeing the obvious, but in perceiving the patterns hidden in the dark. Analytics gives strategy its night vision.

- ❖ Explain the evolution of strategic decision-making from intuition-based to data-driven models within the context of STEM-enabled environments.
- ❖ Identify and interpret key strategic analytics tools, such as XGBoost, SHAP values, PDP, Monte Carlo Simulation, Bayesian Optimization, and Digital Twins, and describe their roles in enhancing strategic foresight and decision quality.
- ❖ Apply predictive techniques to augment traditional strategic tools, including SWOT analysis and VRIO frameworks, transforming static assessments into dynamic, data-informed insights.
- ❖ Integrate environmental scanning and internal/external analysis with real-time analytics to support opportunity recognition and competitive positioning.
- ❖ Assess the ethical implications of automated or semi-automated decision-making tools in strategic management, including issues of bias, transparency, and accountability.
- ❖ Construct strategy scenarios and simulations using analytics to support resource allocation, market entry timing, or risk management under uncertainty.
- ❖ Bridge classical strategy models with modern STEM capabilities, showing how machine learning and decision intelligence enhance core strategic thinking.

In today's fast-moving, data-rich environment, strategy is no longer just about big-picture vision or gut instinct; it's about making smart decisions in real time, often in the face of uncertainty and complexity. While foundational strategy tools like SWOT, VRIO, and environmental scanning remain essential, they now operate alongside powerful new technologies that allow strategists to forecast, simulate, and optimize choices in ways never before possible.

This chapter explores the emerging field of **strategic analytics and decision intelligence**, a fusion of data science, systems thinking, and strategic reasoning. These tools go beyond dashboards or reports; they empower organizations to ask smarter questions, uncover hidden patterns, and respond dynamically to shifting market conditions.

We will examine how modern analytics tools like **XGBoost**, **SHAP values**, **Partial Dependence Plots**, **Bayesian Optimization**, **Monte Carlo Simulation**, and **Digital Twins** are reshaping how strategy is developed, tested, and executed. These tools not only provide answers, they help leaders understand *why* certain decisions matter and *how* resources create value.

Just as Chapter 2 introduced systems thinking for understanding strategic complexity, this chapter brings decision-making into sharper focus. The goal is not to turn managers into data scientists, but to help future strategists become literate in the tools that drive modern intelligence, and to use them to make better strategic calls.

Opening Case: Ping An: Reengineering Strategy with Intelligence

In the competitive world of financial services, Ping An once resembled a traditional insurance company, offering policies, processing claims, and managing investments. But over the past decade, it has transformed into one of the world's most analytically advanced corporations, blurring the lines between finance, healthcare, and technology.

The pivot began with a question asked at the board level: *What if strategic decision-making could be digitized, accelerated, and continually improved using advanced analytics?* For Ping An, the answer was not to merely digitize processes but to reengineer how strategy itself was formulated and executed.

The company began embedding machine learning models into everyday strategic decisions. Its underwriting division deployed gradient boosting algorithms to evaluate risk more accurately than human analysts. Over time, these models were combined with SHAP values, tools that not only delivered predictions but also explained the "why" behind each decision. This transparency allowed Ping An's managers to trace risk drivers and adjust their strategic offerings in real time.

Meanwhile, the company's healthcare subsidiary, Ping An Good Doctor, used AI to simulate patient diagnoses, forecast medical resource needs, and optimize city-level health systems using digital twin technologies. By integrating patient behavior data, hospital infrastructure, and epidemiological patterns, the firm ran thousands of simulations to explore strategic scenarios—much like a digital version of urban strategic planning.

Ping An didn't discard traditional strategic tools like SWOT or VRIO. Instead, it enhanced them. Predictive models update risk and opportunity profiles daily. VRIO analyses became dynamic, using data-informed SHAP rankings to determine whether a capability was truly rare or imitable. Partial dependence plots uncovered nonlinear relationships between internal resources and future market success, information that static dashboards often miss.

This is not just a story of technology adoption. It is a story of how an organization can reimagine **strategic intelligence,** the process of turning data into insight, and insight into lasting advantage.

Play the Strategist

Ping An built its competitive advantage not by reacting to change, but by rethinking how strategic decisions are made in the first place. Imagine you're advising a leadership team that wants to invest in analytics but isn't sure how to align it with its strategic planning process.

* What traditional tools (like SWOT or VRIO) would you recommend enhancing with analytics, and how?
* Which strategic questions should be prioritized for predictive modeling or simulation?
* What risks might arise if the organization depends too heavily on automated insights?

3.1 From Insight to Intelligence

Strategic thinking has always required insight, the ability to see patterns, recognize opportunities, and make sound judgments in the face of complexity. For much of the 20th century, these insights came from experience, industry intuition, and periodic planning retreats. But in the 21st-century strategy landscape, insight alone is no longer enough. Organizations must move from insight to **intelligence,** a more dynamic, data-enhanced form of decision-making powered by models, simulations, and real-time feedback loops.

This transformation mirrors a broader shift: strategy is no longer episodic; it's continuous. It is no longer solely based on human intuition; it's amplified by algorithms. Decision intelligence represents this new frontier. It combines traditional strategic frameworks with advanced analytics, allowing managers to identify emerging trends, test options before committing resources, and simulate future states with remarkable precision.

In this new reality, the strategist is less of a static planner and more of a **real-time sensemaker,** someone who navigates uncertainty by interpreting signals, not just scanning headlines. Tools like **XGBoost** allow firms to build models that detect hidden market dynamics. **SHAP values** explain the "why" behind strategic outcomes, giving executives a clearer view of what drives success. **Partial dependence plots (PDPs)** uncover how different variables interact, making it easier to spot nonlinear opportunities or threats. **Monte Carlo simulations** and **digital twins** help simulate a wide range of strategic futures before any irreversible decision is made.

But these tools do not replace strategic thinking; they enhance it. Data is not a substitute for vision; it is a partner in clarity. STEM-enabled strategy doesn't remove the human element. Instead, it raises the bar for how humans use their judgment by equipping them with better foresight, sharper interpretation, and deeper understanding of strategic trade-offs.

This section lays the conceptual foundation for strategic analytics by answering two core questions:

1. **Why does strategy require intelligence in a data-driven world?**
2. **How does data move from insight to intelligence, and then into action?**

As we continue, you will see how data becomes not just a mirror of past performance, but a lens through which future advantage is discovered and designed.

3.1.1 Why Strategic Decisions Fail Without Intelligence

Most strategic failures are not the result of poor intentions or even poor ideas; they are often the result of **incomplete thinking**. A strategy may look promising on paper, pass through multiple executive layers, and even align with a firm's mission. Yet, when implemented, it falters. Why? The decision-making process failed to anticipate hidden risks, emerging variables, or unintended consequences.

In traditional settings, strategic planning has relied heavily on periodic reviews, expert judgment, and historical data. While useful, this approach assumes a world that is relatively stable and knowable. But in today's dynamic environment, marked by volatility, digital disruption, and complex interdependencies, **strategic blind spots multiply**. What used to be a two-dimensional chess game has become a fast-paced simulation involving thousands of moving parts.

Without decision intelligence, firms fall prey to several common strategic pitfalls:

* **Overconfidence in historical success**: A classic trap. Companies assume what worked before will work again. Blockbuster failed to see Netflix coming, not because it lacked vision, but because it lacked the data capabilities and insight architecture to recognize behavioral shifts early enough.
* **Failure to anticipate interactions**: Strategic decisions are rarely linear. A pricing change may affect customer loyalty, supply chain efficiency, and brand perception simultaneously. Without modeling these interdependencies, such as with **partial dependence plots (PDPs)** or **digital twin simulations,** strategists are left guessing.

* **Neglecting low-probability, high-impact risks**: The 2008 financial crisis, COVID-19, and supply chain breakdowns remind us that improbable events can carry catastrophic consequences. **Monte Carlo simulations** help quantify these risks, but without them, firms often rely on best-case assumptions.
* **Static analysis in a dynamic world**: SWOT or VRIO frameworks remain useful, but if they are only revisited once a year, they lose relevance. Predictive analytics and **real-time SHAP-based resource assessments** transform these tools from snapshots into live strategy dashboards.
* **Cognitive and confirmation biases**: Executives, like all humans, filter information through personal experiences. Decision intelligence systems, when designed ethically, can surface contrarian patterns and challenge intuition with evidence. This does not eliminate human judgment, but it expands its boundaries.

These failure modes aren't merely operational; they're strategic. A strategy that ignores fast-changing variables or underestimates the interconnected nature of decisions is not simply "risky"; it is structurally unsound.

What modern organizations need is a **feedback-rich, forward-looking, analytically empowered strategy system**. One that doesn't just validate what executives already believe, but helps them explore what they haven't yet considered.

As we move forward, we will explore how decision intelligence tools help shift strategy from static assumptions to adaptive learning, where decisions are tested, tuned, and evolved before the cost of being wrong becomes too high.

3.1.2 Intelligence as a Strategic Capability

In a world where the speed and complexity of change outpace traditional decision-making, **intelligence is no longer a function; it is a strategic capability**. Like brand equity or operational efficiency, an organization's ability to convert data into meaningful, actionable insight is now a core source of competitive advantage.

But what exactly do we mean by "intelligence" in a strategic context?

At its core, **strategic intelligence** refers to the organization's ability to gather, process, interpret, and act upon relevant information in a way that enhances strategic outcomes. It is not merely data accumulation, nor is it limited to analytics dashboards. Strategic intelligence is the **institutional capacity to learn, adapt, and decide better than competitors,** faster, with fewer blind spots, and with a deeper understanding of cause and effect.

Building this capability involves more than hiring data scientists or investing in software. It requires embedding intelligence into the very architecture of strategic thinking:

* **Sensing**: The ability to detect weak signals and emerging trends, often buried in unstructured data. Tools like natural language processing and machine learning classifiers can scan news, patents, customer reviews, and social media for early signs of change.
* **Sensemaking**: Interpreting patterns from noisy or incomplete information. This is where models like **XGBoost** and interpretability tools like **SHAP values** come in, translating complex statistical output into understandable insights for strategic decision-makers.
* **Simulation**: Forecasting outcomes before acting. **Monte Carlo simulations** and **digital twins** allow firms to test different strategic moves under various scenarios, transforming planning from an art into a science of calculated trade-offs.
* **Learning**: Intelligence compounds when organizations treat each decision as a data point in a longer-term feedback loop. By integrating model performance reviews, decision post-mortems, and strategy audits, firms build a **learning infrastructure** that strengthens their intelligence muscle over time.

When these components are woven into the culture and processes of a firm, intelligence becomes a **repeatable advantage**. Consider Amazon's ability to launch, test, and refine new services at scale, or Unilever's use of AI-driven insights to localize product strategies across regions. These organizations are not simply more "data-savvy"; they are **strategically intelligent by design**.

Importantly, intelligence is not just about reacting. It's about **anticipating,** moving from lagging indicators to leading ones. Rather than asking, "What happened?" intelligent firms ask, "What's about to happen, and how can we shape it?"

This evolution elevates the role of the strategist. In traditional models, strategic planning was episodic and hierarchical. In intelligent enterprises, strategy is **fluid, cross-functional, and model-informed**. Leaders no longer wait for quarterly reports; they interact with live dashboards, explore scenario outputs, and co-create options with analysts and algorithms.

Strategic intelligence is not about removing the human element; it's about **augmenting human judgment** with tools that uncover what intuition alone might miss. In the chapters ahead, we will explore how organizations can operationalize this capability and what tools, systems, and mindsets are required to do so effectively.

3.1.3 From Pattern Recognition to Pattern Simulation

Strategic leaders have long been valued for their ability to recognize patterns, identify trends, spot opportunities, and sense threats before they become obvious. This skill remains vital. But in today's data-rich, technology-intensive environment, pattern recognition is no longer the endpoint of strategic thinking. It is the **starting point**.

The frontier of strategy now lies in **pattern simulation,** the ability not just to notice trends, but to test how they will evolve, interact, and respond to different strategic choices. This shift transforms strategy from an observational art to a dynamic, experimental science.

Let's illustrate the difference.

A traditional strategist might recognize that consumer behavior is shifting toward sustainability. This is pattern recognition. A strategist equipped with simulation tools goes further: they model how different sustainability investments will influence market share, customer loyalty, and regulatory exposure under various conditions. They use tools like **Monte Carlo simulations** to explore dozens, or thousands, of possible futures, and **digital twins** to recreate how the business system behaves under change.

This move from recognition to simulation redefines what it means to think strategically:

* **From static to dynamic**: Instead of viewing the business landscape as a fixed environment with stable variables, simulation embraces complexity. It accounts for feedback loops, time delays, and nonlinear interactions, features often missed by spreadsheets or static strategy frameworks.
* **From single-scenario planning to multi-path forecasting**: Rather than building one strategic plan based on assumed certainty, simulations allow strategists to test multiple scenarios—best case, worst case, and everything in between.
* **From intuition-driven response to evidence-backed exploration**: While strategic insight still starts with a hypothesis, simulation tools provide a way to test assumptions before irreversible commitments are made.

Consider the airline industry. A strategist may notice an emerging preference for low-carbon travel. A pattern simulator can build models that test how pricing changes, fuel innovations, and carbon taxes might affect route profitability over the next decade. The insights from that simulation are not merely descriptive; they are **decision-enabling**.

Tools like **XGBoost** help build these models quickly and accurately. **SHAP values** reveal which variables most influence future outcomes. **PDPs** visualize how changes in key inputs alter predicted results. Together, these tools shift the strategist's role from interpreting the past to designing the future.

This transition does not diminish the value of strategic experience; it enhances it. The strategist becomes a conductor of insight, using data not to replace judgment but to stretch it, to test more ideas, under more conditions, with better evidence.

In the chapters to follow, this mindset will guide how we approach tools and techniques. Not just as passive analytical instruments, but as **engines of exploration,** bridging imagination with computation to shape a strategy that is not only smart, but forward-ready.

3.1.4 Strategic Intelligence in Action: The New Decision Blueprint

The strategist of the 21st century no longer operates with a linear checklist of analysis, planning, and execution. Instead, the new decision blueprint is **cyclical, iterative, and intelligence-driven**. Insight, simulation, and feedback are not sequential; they are continuous, forming a learning loop that refines strategy in real time.

This transformation rests on a fundamental shift: **from deterministic planning to probabilistic decision-making**. Rather than attempting to predict one future, today's strategist prepares for many possible futures, using data not as a fixed report but as a source of adaptive learning.

Strategic intelligence, in this sense, becomes a **capability system,** a blend of tools, data, human judgment, and organizational design. Let's break down this blueprint into its four essential layers:

1. Strategic Sensing

Modern organizations must detect weak signals early, be it a regulatory shift, a viral customer complaint, or an emerging technology. Tools like natural language processing, web scraping, and real-time data feeds enable organizations to listen at scale. This replaces anecdotal "gut feeling" with evidence-based vigilance.

2. Intelligence Generation

This is where the raw data is transformed into insight. Using tools like **XGBoost**, organizations model how key variables predict strategic outcomes, such as market demand, employee turnover, or brand perception. **SHAP values** and **partial dependence plots (PDPs)** add transparency, helping executives see not just the "what," but the "why" behind predictions.

3. Scenario Simulation

Rather than guessing how a decision might play out, simulations allow firms to explore alternative futures. **Monte Carlo simulations** test strategies under uncertainty. **Digital twins** simulate real-world systems, factories, cities, or even customer journeys, to stress-test assumptions before acting. This is strategy rehearsal at machine speed.

4. Strategic Adaptation

Finally, intelligence systems must learn. As new data flows in, predictive models are retrained, assumptions are recalibrated, and decisions are adjusted. Organizations embed this feedback loop into their planning processes, moving from annual strategy reviews to continuous adjustment cycles.

This intelligence-powered approach **does not discard classical strategy models;** it refines them. SWOT, VRIO, Porter's Five Forces, and scenario planning are still valuable. But in intelligent organizations, these tools are connected to live data, refreshed in real time, and embedded within an adaptive architecture.

The strategist, in turn, becomes a **designer of learning systems,** someone who orchestrates tools, interprets signals, and builds a resilient strategy pipeline that adjusts as the world changes.

As we transition into Section 3.2, we will explore the **analytics continuum,** from descriptive statistics to prescriptive optimization, and how each stage supports smarter strategic decisions. But the mindset should already be clear: strategy is no longer something you decide once. It is something you learn, simulate, and evolve intelligently.

3.2 From Data to Strategy – The Analytics Continuum

Strategic intelligence, as discussed in the previous section, hinges on an organization's ability to move beyond raw data toward informed, effective decisions. To navigate this journey, firms rely on what is often called the **analytics continuum,** a progression of analytical capabilities that evolve from simply describing what happened to prescribing what should be done.

Understanding this continuum is crucial for today's strategist. It outlines the **levels of analytical maturity** and shows how each step contributes uniquely to strategic decision-making. This section explores the three major stages of the continuum, **descriptive, predictive, and prescriptive analytics,** and how each one supports more intelligent and actionable strategy design.

3.2.1 Descriptive Analytics: Understanding What Has Happened

At the base of the continuum lies **descriptive analytics,** the process of summarizing historical data to understand past events and trends. This is the most familiar level of analytics, often manifested in dashboards, reports, charts, and key performance indicators (KPIs).

In a strategic context, descriptive analytics provides clarity. It tells managers what happened, how key metrics have shifted, and where problems or opportunities have historically emerged. Retail firms use it to track seasonal sales patterns. Airlines analyze on-time performance. Banks review churn rates.

While valuable, descriptive analytics is **retrospective**. It reveals symptoms, not causes; outcomes, not actions. That's why intelligent strategy treats this stage not as a decision tool, but as a **diagnostic foundation**. It frames the conversation but cannot close it.

In intelligent organizations, descriptive analytics feeds directly into model inputs. It informs **baseline assumptions**, shapes trend detection, and serves as the first layer of real-time sensing. For example, Intel uses descriptive analytics to understand manufacturing yields across global plants. Patterns in defect rates, maintenance schedules, and environmental conditions become input variables for deeper analysis in predictive and prescriptive systems.

3.2.2 Predictive Analytics: Anticipating What Could Happen

Moving up the continuum, **predictive analytics** leverages historical data and statistical models to forecast future outcomes. Rather than just knowing what happened, organizations begin to ask: *What is likely to happen next?*

This stage represents a major leap. By using machine learning models like **XGBoost**, firms can uncover patterns that are too complex or subtle for human analysis. These models do not just identify correlations; they make **probabilistic inferences** about what is likely to occur under similar conditions.

Strategically, this allows firms to:

* Forecast demand across regions
* Predict employee turnover or customer churn
* Anticipate the financial outcomes of different investment scenarios
* Identify early warning signs of disruption

But predictive analytics alone is not enough. The challenge lies in **interpretability**. Without knowing why a prediction occurs, decision-makers struggle to trust it. That's where tools like **SHAP values** and **PDPs** come in. These techniques explain the drivers behind predictions, revealing how specific inputs, like marketing spend, product features, or geographic factors, influence outcomes.

For example, Spotify uses predictive analytics to anticipate listener behavior and playlist engagement. Models evaluate how song characteristics, user habits, and social trends will shape listening patterns, allowing the company to make strategic content investments ahead of time.

Predictive analytics shifts the strategist's role from reaction to anticipation. It does not eliminate uncertainty, but it **clarifies likelihoods**, enabling leaders to focus attention and resources where they matter most.

3.2.3 Prescriptive Analytics: Recommending What Should Be Done

At the highest level of the continuum is **prescriptive analytics,** the use of optimization models, simulations, and algorithms to recommend actions that are most likely to deliver desired outcomes. Here, the question becomes: *Given what we know, what should we do?*

Prescriptive analytics builds on prediction but adds **decision logic**. It explores not only what might happen, but also how different decisions influence those outcomes. It accounts for trade-offs, constraints, and goals, turning complex problems into computable strategies.

Key methods include:

* **Bayesian optimization**: Used to find optimal solutions when testing is expensive or time-consuming. It's commonly applied in marketing budgets, pricing models, and product feature design.
* **Monte Carlo simulations**: Used to explore strategic choices under uncertainty. This helps firms understand not just the average outcome, but the full range of potential risks and rewards.
* **Digital twin simulations**: These are virtual representations of real-world systems, such as factories, logistics networks, or even customer service ecosystems. They allow firms to experiment with strategic changes before making them real.

UPS uses prescriptive analytics in route optimization. Rather than simply forecasting delivery times, its ORION system recommends the exact path each driver should take, saving millions of miles annually.

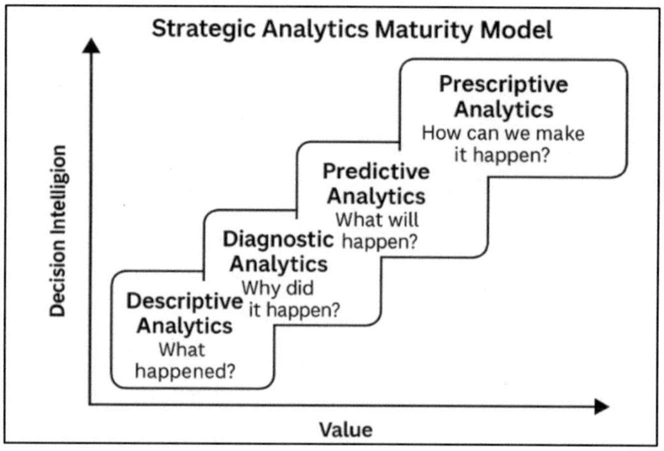

At this level, analytics is no longer just an advisor; it becomes a **partner in decision-making**. But to use it responsibly, organizations must address ethical implications: Who is accountable for machine-driven recommendations? How are biases in data managed? How transparent are the assumptions behind optimization?

In a STEM-informed strategy, prescriptive analytics is powerful, but not autonomous. It **enhances judgment** by providing simulated foresight, not replacing human wisdom.

3.2.4 Integrating the Continuum into Strategic Decision-Making

While each level of the analytics continuum offers value on its own, they become transformative when integrated into a continuous loop of decision-making. Descriptive insights guide predictive modeling. Predictive outcomes feed into prescriptive recommendations. Each level informs the next, and collectively they shape a **real-time, adaptive strategy system**.

This integration enables firms to shift from rigid strategic planning cycles to a living system of analytics-powered decisions. Smart strategists don't stop at asking what happened; they anticipate what's next and explore what actions will generate the best possible outcomes under uncertainty.

The table below summarizes the continuum and illustrates how each stage contributes uniquely to strategic intelligence:

Table 3.1. The Analytics Continuum in Strategic Decision-Making

Analytics Level	Key Question	Strategic Purpose	Core Methods/Tools	STEM-Enabling Examples
Descriptive Analytics	What has happened?	Summarize past performance	Dashboards, KPIs, summary statistics	Intel uses factory-level dashboards to monitor production efficiency and detect anomalies.
Predictive Analytics	What could happen?	Forecast trends and emerging risks	XGBoost, SHAP, PDPs, time-series models	Spotify predicts user listening patterns to guide content and investment decisions.
Prescriptive Analytics	What should we do about it?	Recommend optimal strategic actions.	Bayesian Optimization, Monte Carlo Simulation, Digital Twins	UPS's ORION system prescribes optimal delivery routes to minimize time and fuel costs.

The analytics continuum illustrates how strategic decision-making evolves from understanding the past to simulating the future. Each stage supports progressively deeper insights, enabling smarter, more agile strategy execution in STEM-driven organizations.

Together, these analytics capabilities form the backbone of decision intelligence. In the chapters ahead, we shift from the abstract to the actionable, introducing the specific tools that enable each level of analysis and showing how they power competitive strategy across industries.

Next, we enter the strategist's toolbox, beginning with **XGBoost**, a predictive engine that combines speed, accuracy, and strategic flexibility in ways traditional models never could.

3.3 Strategic Analytics Toolkit – Tools That Power Intelligent Strategy

The analytics continuum defines the **strategic logic** behind intelligent decision-making. But to put that logic into action, organizations need more than vision—they need tools. This section introduces the **strategic analytics toolkit**: a curated set of advanced techniques and technologies that allow strategists to extract insights, model futures, and optimize choices with scientific precision.

These tools are not generic statistical instruments; they are **strategy-enabling engines**. Each one plays a specific role in converting data into foresight, uncertainty into simulations, and scenarios into informed action. Just as finance uses discounted cash flow or marketing relies on segmentation analysis, strategic management in a STEM world requires its own **analytical canon,** one that matches the complexity and pace of modern competition.

In this section, we focus on six tools that together power the strategist's evolving role:

* **XGBoost** – for high-performance predictive modeling
* **Bayesian Optimization** – for intelligent experimentation and decision refinement
* **SHAP** – for interpretability and strategic transparency
* **Partial Dependence Plots (PDPs)** – for isolating and visualizing key relationships
* **Monte Carlo Simulation** – for modeling risk under uncertainty
* **Digital Twins** – for simulating business systems and stress-testing strategies

Each subsection that follows will not only explain how the tool works but also demonstrate how it creates value in strategic contexts, from product innovation to resource allocation, market expansion, crisis response, and long-term planning.

More importantly, these tools do not replace strategic thinking; they **amplify it**. They allow strategists to challenge assumptions, simulate alternatives, and design resilient pathways in increasingly complex environments.

We begin with a foundational workhorse of modern predictive strategy: **XGBoost**.

3.3.1 XGBoost – The Predictive Engine for Strategy

In today's data-intensive strategic environments, leaders need fast, reliable, and scalable ways to anticipate what lies ahead. This is where **XGBoost** (Extreme Gradient Boosting) emerges—not just as a technical model, but as a **strategic engine** for making better predictions with real-world complexity.

Originally designed for machine learning competitions, XGBoost has quickly become a favorite among data scientists and business strategists alike. Its popularity is not accidental. XGBoost is known for its **exceptional accuracy**, **speed**, and **robustness** when working with messy, nonlinear, high-dimensional data—all characteristics that define strategic decision contexts.

At its core, XGBoost is a **gradient boosting algorithm**. Rather than building one single model, it builds many small decision trees sequentially, with each tree correcting the errors of the ones before it. This process leads to a highly refined ensemble model that is exceptionally good at predicting outcomes based on numerous interacting variables.

Why Strategists Should Care

While XGBoost operates under the hood with sophisticated math, its value to strategists lies in its **output**—predictions about customer behavior, operational bottlenecks, market dynamics, and risk exposure, among others. More importantly, XGBoost can handle:

* **Heterogeneous data** (structured, categorical, missing values)
* **Nonlinear relationships** that traditional regression would miss
* **Large-scale datasets** without sacrificing computational speed

In strategic settings, XGBoost becomes a **predictive microscope**—revealing the underlying drivers of outcomes such as:

* Which new markets are likely to generate the highest return?
* Which customer segments are most at risk of churn?
* What factors predict a product's success or failure in the next quarter?
* How likely is a disruption in the supply chain under various scenarios?

A global logistics firm uses XGBoost to forecast delivery delays based on weather, geopolitical risk, and fleet conditions. These predictions inform real-time route adjustments, inventory buffers, and dynamic pricing, all core elements of its competitive strategy. What makes XGBoost particularly valuable in strategic contexts is not just its predictive power, but the way its technical features translate into tangible strategic advantages. The table below summarizes how XGBoost's capabilities support faster, smarter, and more resilient decision-making:

But the power of XGBoost does not stop at prediction. When combined with **interpretability tools** such as **SHAP values** (Section 3.3.3), it becomes not only a black-box forecaster but a **transparent strategic partner**, helping executives understand which inputs matter most and how changes in those variables might shift the outcome.

Table 3.2. Strategic Advantages of XGBoost

Feature	Strategic Benefit
High prediction accuracy	Supports more confident decisions in uncertain environments
Fast training time	Enables rapid scenario testing and iteration
Handles complex interactions	Captures real-world complexity better than linear models
Built-in regularization	Reduces overfitting, increasing the generalizability of strategic models

Strategic Use Case

Imagine a multinational hotel chain aiming to predict which cities will offer the best return on investment for expansion. Traditional analysis might rely on average income or tourist traffic. An XGBoost model can incorporate **dozens of variables,** from regional labor costs to online review sentiment, regulatory risk, and competitive saturation, and identify nonlinear patterns that no spreadsheet could capture.

Armed with these insights, executives can create data-backed expansion strategies, allocate resources more effectively, and anticipate obstacles before they arise.

As we will see in the next sections, XGBoost is only one piece of the strategic analytics puzzle. Once predictions are made, the next step is to optimize decisions based on those predictions. That's where Bayesian Optimization comes in, offering a powerful framework for choosing the best course of action under uncertainty.

3.3.2 Bayesian Optimization – Intelligent Exploration for Strategic Decisions

Making the right strategic choice often means navigating trade-offs in the face of uncertainty. Whether it's deciding how to allocate a marketing budget, choosing the right pricing structure, or balancing speed versus quality in a supply chain, leaders must **optimize outcomes without perfect information**. This is where **Bayesian Optimization (BO)** becomes an invaluable ally.

Unlike traditional optimization techniques that require exhaustive search or trial-and-error, Bayesian Optimization offers a **smarter, faster, and more data-efficient** way to find the best solution—particularly when the objective function is expensive, noisy, or unknown. In strategic settings, this means discovering the best action without needing to test every possibility.

How It Works

Bayesian Optimization uses two components:

1. **A surrogate model** (often a Gaussian Process) to approximate the unknown function being optimized, such as revenue growth, cost reduction, or NPS improvement.
2. **An acquisition function** to determine the next best point to sample, balancing two goals:
    * **Exploration** – testing new areas where data is scarce
    * **Exploitation** – refining what's already known to improve outcomes

> Think of Bayesian Optimization as a chess player who learns from every move, refining their strategy with each decision. Rather than playing thousands of games to find the best opening, they use smart logic to identify which variations are most promising.

In essence, the algorithm builds a **belief** about the relationship between inputs (e.g., investment levels, pricing tiers) and the outcome (e.g., profit), and it **updates this belief** after every test or data point. Over time, it zeroes in on the optimal configuration.

Why It Matters for Strategists

Bayesian Optimization is especially powerful when:

* **Data is limited or expensive to acquire** (e.g., A/B tests, simulations, pilots)
* **The solution space is large and nonlinear** (e.g., dynamic pricing models)
* **Multiple trade-offs must be balanced** (e.g., cost vs. innovation speed)

For strategic decision-makers, this translates into a tool that doesn't just support decisions, it actively **guides experimentation**. It enables:

* Intelligent exploration of new markets or business models
* Dynamic adjustment of investment portfolios
* Simulation-based learning in uncertain environments

A SaaS firm wants to optimize its pricing across different global regions. Instead of running hundreds of price experiments (which risk customer loss), it uses Bayesian Optimization to find the best price points with minimal trial-and-error. This leads to higher profitability without overexposing the brand to pricing risk.

What makes Bayesian Optimization strategically powerful is its ability to balance exploration and efficiency, even in data-scarce or high-stakes environments. The table below outlines the key features of Bayesian Optimization and the strategic value each brings to decision-making:

Table 3.3. Strategic Benefits of Bayesian Optimization

Feature	Strategic Value
Data-efficient learning	Supports decision-making even when data is limited or expensive
Dynamic exploration	Encourages discovery of novel options rather than repeating old patterns
Trade-off optimization	Balances multiple strategic objectives without oversimplifying
Embedded uncertainty modeling	Helps leaders understand confidence levels in each decision

Application in Strategy Labs and Innovation Units

Many forward-thinking organizations are building **"strategy labs",** spaces where simulation, modeling, and experimentation drive planning. In such labs, Bayesian Optimization becomes a **central decision engine**, helping teams test innovation strategies before market rollout, optimize new service delivery configurations, or even tune supply chain resilience under disruption.

It also fits seamlessly into **closed-loop decision systems**, where predictions (from models like XGBoost) are followed by optimization (via BO), and outcomes feed back into learning systems.

As we move forward, we address a common challenge in machine-driven strategy: **interpretability**. Predictions and optimizations are valuable, but only if leaders can understand, trust, and explain them. In the next section, we explore how **SHAP values** provide that strategic transparency.

3.3.3 SHAP – Making Strategy Models Transparent and Trustworthy

As machine learning tools like XGBoost and Bayesian Optimization become more powerful, they also become more complex. For many strategists, the question is no longer *"Can we predict it?"* but rather *"Can we explain it?"* In high-stakes environments, such as financial forecasting, regulatory compliance, healthcare strategy, or ESG planning, **black-box models** pose a risk. If decision-makers cannot interpret why a model makes a recommendation, they may be reluctant to trust or act on it.

Enter **SHAP (SHapley Additive exPlanations),** a tool designed to bring **clarity and accountability** to even the most complex predictive models.

What Is SHAP?

SHAP is an explainability framework based on a concept from game theory called the **Shapley value**. In simple terms, it quantifies how much each input variable contributes to a specific prediction, **for each individual case**. Rather than giving generic importance scores, SHAP shows exactly how each feature increases or decreases the predicted value in a given context.

For example, if a model predicts a customer will churn, SHAP can explain that the low net promoter score (NPS) and long customer service wait times were the biggest contributing factors, while recent upgrades helped reduce the risk.

This level of **granular insight** is essential for strategic decision-making because it shifts machine learning from abstract math to **actionable intelligence**.

A global insurance firm uses SHAP to explain why certain policyholders are predicted to file claims. By understanding the top contributors, like geographic risk, claim history, and policy structure, they can refine pricing models and tailor risk-mitigation strategies across different markets.

Strategic Applications of SHAP

* **Model auditability**: SHAP provides evidence for how a model reached a decision, which supports regulatory and board-level transparency.
* **Root cause analysis**: SHAP pinpoints which variables are truly driving outcomes, guiding corrective actions or strategy pivots.
* **Personalized strategy**: SHAP explanations can be tailored to individual customers, products, or markets, enabling micro-strategy at scale.
* **Executive trust**: Transparent models increase adoption of analytics-driven decisions among senior leaders, avoiding resistance rooted in opacity.

SHAP in Strategic Conversations

Imagine presenting a predictive model to a board of directors without being able to explain how it works. SHAP makes it possible to say:

"This model predicts a 35% decline in market share under Scenario A. The primary drivers are projected R&D underspending and weakened brand equity, while favorable cost ratios mitigate some of the decline."

Such interpretability turns analytics into **narratives**, enabling leaders to make decisions not just *with* data, but *through* it. The table below summarizes how SHAP enhances strategic modeling by bridging the gap between technical accuracy and executive clarity:

Table 3.4. SHAP Summary Table

SHAP Feature	Strategic Value
Local interpretability	Explains individual predictions for specific customers, markets, or products
Global interpretability	Summarizes how features impact predictions across the entire model
Visual explanation tools	Enhances communication of insights in executive presentations
Auditable AI	Supports transparency and ethical governance in data-driven strategy

SHAP brings **strategic accountability** to machine learning. It closes the gap between model performance and organizational adoption. With it, data science becomes not only powerful but also persuasive and actionable.

In the next section, we build on this by exploring **Partial Dependence Plots (PDPs),** another interpretability tool that enables strategists to visualize how changing one input alters strategic outcomes.

3.3.4 PDP – Visualizing Strategic Sensitivity

Understanding which factors influence outcomes is essential, but strategy often demands more than just identifying key drivers. Leaders want to know: *How much does each variable matter? What happens if we increase investment in R&D? What's the tipping point for customer dissatisfaction? How sensitive is our outcome to changes in pricing, talent, or innovation?*

Partial Dependence Plots (PDPs) answer these questions by offering a **clear visual window** into how changes in one variable, while holding others constant, affect the predicted outcome. They transform abstract model mechanics into **strategic sensitivity curves** that reveal how a particular input influences performance.

What Is a Partial Dependence Plot?

A PDP shows the **marginal effect** of a selected feature (such as cost, time, or policy strength) on the model's prediction. It allows strategists to isolate the impact of that variable across its range, assuming all other inputs remain unchanged.

Think of it as a **simulation dial**: as you slide the variable from low to high, the PDP shows how your predicted outcome (e.g., sales growth, mortality reduction, employee attrition) rises, falls, or plateaus.

A university admissions office uses a PDP to understand how predicted enrollment changes with increased scholarship amounts. The plot reveals that gains flatten after $5,000, providing an evidence-based ceiling for financial aid optimization.

Strategic Uses of PDPs

PDPs are especially valuable in strategy contexts where decision-makers need to:

* **Test sensitivity** to key levers (e.g., pricing, investment, training hours)
* **Quantify trade-offs** in complex decisions
* **Simulate scenarios** without running full-scale experiments
* **Identify nonlinear tipping points** and diminishing returns

These visualizations empower teams to **move beyond gut instinct** and engage in informed debates backed by empirical insights.

Real-World Strategy Application

A digital health startup wants to understand how patient engagement changes with different app design features, like notification frequency. A PDP shows that increasing notifications from 1 to 3 per day improves engagement, but beyond 4, engagement drops sharply due to notification fatigue. This insight directly shapes product and marketing strategy.

In operations, a manufacturing firm may use PDPs to analyze how downtime hours impact delivery performance. The result? A clear threshold: delays rise exponentially beyond 4 hours, guiding preventive maintenance schedules and resource buffers.

What makes PDPs especially valuable is their ability to translate complex model behavior into **clear, visual insights** that support strategic planning. The table below summarizes the key capabilities of PDPs and how they enhance strategic decision-making:.

Table 3.5. PDP Summary Table

PDP Capability	Strategic Impact
Visualizes variable impact	Helps leaders see how individual factors shape strategic outcomes
Reveals nonlinearity	Identifies where additional investment yields more or less return
Supports scenario analysis	Empowers simulation-based planning before making costly decisions
Strengthens strategy communication	Makes model results accessible to non-technical decision-makers

While SHAP helps explain why a prediction was made, PDPs help answer: *What if we change something?* Together, they enable **interpretability and interactivity**, providing strategists with tools not just to trust the model, but to challenge, explore, and act on it.

Next, we'll move from predictive visualization to **risk simulation,** where uncertainty is not avoided but modeled. In the following section, we turn to **Monte Carlo Simulation**, a method that equips leaders to evaluate the full range of possible futures.

3.3.5 Monte Carlo Simulation – Modeling Uncertainty for Strategic Resilience

In strategic decision-making, few things are certain. Market demand fluctuates. Supply chains break. Costs spike. Consumer behavior shifts. When the future is uncertain, deterministic forecasts are no longer sufficient. Leaders need a **range of possible outcomes,** not just a single estimate. This is where **Monte Carlo Simulation** becomes a powerful tool for building **strategic resilience**.

Monte Carlo Simulation (MCS) is not just about modeling randomness; it's about **embracing uncertainty to make better decisions**. By generating thousands (or even millions) of scenarios based on variable probability distributions, it provides a full picture of potential outcomes and how likely each one is to occur.

Rather than asking, "What will happen?" Monte Carlo asks, "What could happen, and with what probability?"

How Monte Carlo Simulation Works

The core idea behind MCS is simple:

1. **Identify the uncertain inputs** (e.g., customer demand, raw material prices, lead times).
2. **Define probability distributions** for those variables (e.g., normal, uniform, triangular).
3. **Run thousands of simulations** where each input is randomly sampled from its distribution.
4. **Aggregate the results** to visualize the full outcome space, mean, median, volatility, percentiles, and risk zones.

A solar energy firm uses Monte Carlo Simulation to model how weather variability and government incentives affect project ROI. Instead of relying on a single forecast, it models thousands of possible futures, guiding more robust investment decisions.

Strategic Power of Monte Carlo Simulation

For strategists, Monte Carlo is not just a tool for finance or engineering; it's a **strategic foresight engine**. It helps answer:

* What's the probability we'll fall below a critical performance threshold?
* How much risk does a new product launch introduce to our portfolio?
* Under what conditions does our growth strategy break down?

It turns risk into a **quantifiable variable,** allowing leaders to stress-test assumptions, build contingency plans, and design strategies that are not only optimized but **resilient**.

Table 3.6. Use Cases Across Strategic Domains

Domain	Monte Carlo Application
Financial Strategy	Forecasting earnings volatility under economic shocks
Operations Management	Modeling production delays and inventory stockouts
Strategic Planning	Stress-testing expansion plans across demand and regulatory scenarios
Product Innovation	Estimating time-to-market under R&D uncertainty
Crisis Response	Evaluating resilience under geopolitical, climate, or health disruptions

A multinational consumer goods company uses Monte Carlo Simulation to model how variations in consumer sentiment, supplier reliability, and freight costs affect its quarterly profitability. The simulation shows that under a worst-case scenario (5th percentile), profits would drop by 18%, triggering a supply chain diversification plan.

This approach highlights how Monte Carlo methods enhance strategic foresight by embedding uncertainty directly into the modeling process. Rather than relying on single-point forecasts, leaders can explore a full range of possible outcomes and prepare accordingly.

The table below outlines the core features of Monte Carlo simulation and the strategic value each brings to high-stakes decision-making:

Table 3.7. Strategic Benefits of Monte Carlo

Feature	Strategic Value
Captures uncertainty	Provides realistic, risk-aware decision support
Quantifies downside scenarios	Enables early planning for worst-case outcomes
Enhances stakeholder confidence	Supports transparency and defensibility in forecasts and board reports
Complements predictive analytics	Converts point forecasts into probability distributions.

Monte Carlo Simulation pushes strategy beyond best guesses. It introduces a **probabilistic mindset,** helping leaders make decisions not despite uncertainty, but because they've modeled it. In volatile times, it turns analytics into **adaptive preparedness**.

In the next and final tool of our analytics toolkit, we move from simulations to **mirrors of reality**. We explore **Digital Twins**, a technology that replicates real-world systems, allowing strategists to rehearse decisions in virtual space before committing in the real world.

3.3.6 Digital Twins – Simulating Strategy in a Living System

Imagine being able to rehearse your company's next big strategic move before making it, testing its impact on operations, finance, customer satisfaction, and even the competitive landscape. **Digital Twins** make that possible. Once confined to engineering and manufacturing, Digital Twin technology has evolved into a transformative strategic tool, allowing firms to create dynamic, data-driven replicas of real-world systems to simulate decisions in real time.

A **Digital Twin** is a virtual model of a physical system, be it a factory, supply chain, business unit, or entire enterprise. It continuously receives real-time data from the actual system, learns from it, and evolves alongside it. More than just a simulation, a Digital Twin becomes a **living laboratory** for strategic experimentation.

What Makes a Digital Twin Strategic?

While traditional simulations are static and scenario-limited, Digital Twins are:

* **Real-time** – updated continuously with operational, market, or customer data
* **System-integrated** – reflect cross-functional interdependencies (e.g., how a pricing change affects supply chain, customer satisfaction, and revenue)
* **Feedback-enabled** – evolve based on actual outcomes and external signals

A global port authority builds a Digital Twin of its logistics ecosystem—including shipping lanes, truck routes, customs operations, and labor availability. It uses this virtual environment to test expansion strategies, optimize ship docking schedules, and simulate responses to weather or geopolitical disruptions.

Strategic Applications of Digital Twins

Digital Twins allow decision-makers to:

* **Model complex systems** without shutting down operations
* **Run what-if scenarios** without real-world risk
* **Understand systemic ripple effects** before they occur
* **Align long-term strategy** with dynamic operational realities

They are especially powerful for high-stakes, high-cost decisions, such as entering a new market, launching a product, reconfiguring a supply network, or reorganizing service delivery.

From Forecasting to Foresight

Digital Twins shift strategic thinking from forecast-based planning to **foresight-based strategy**. By interacting with a live simulation, leaders don't just ask *"What might happen?"*. They can **see it, test it, and adapt in real time**.

Strategic Use Case: A global hotel chain creates a Digital Twin of its core markets, incorporating data on tourism flows, competitor pricing, consumer sentiment, and energy usage. It simulates how green investments, pricing adjustments, and loyalty program changes affect not just financials, but brand equity and ESG metrics across geographies.

Table 3.8. Digital Twin Strategic Summary

Capability	Strategic Benefit
Real-time system modeling	Enables responsive strategy development
Cross-functional impact analysis	Reveals ripple effects of decisions across the business
Risk-free scenario testing	Reduces uncertainty and decision cost
Continuous strategic alignment	Keeps long-term planning in sync with fast-changing data

Digital Twins are not a future luxury; they are an emerging **strategic necessity**. In volatile and complex environments, they allow leaders to develop, stress-test, and refine strategies before they are implemented, bridging the gap between planning and execution, vision and velocity.

With this final tool, we complete the **STEM Strategist's Toolkit**. But these tools are only powerful when used within a broader decision architecture. In the next section, we bring them together into an integrated framework, a system for applying these tools across the strategic cycle.

3.3.7 Selecting the Right Tool – A Strategic Comparison

In the rapidly evolving world of strategy analytics, having a powerful toolkit is only half the battle. Knowing **which tool to use, when, and why** is what differentiates a strategist from a technician. Each tool in the STEM Strategist's Toolkit brings unique capabilities, but they are not interchangeable. Some excel in prediction, others in interpretation or simulation. The right choice depends on the **strategic question, the data environment, the urgency of the decision, and the level of uncertainty involved**.

This section provides a comparative lens to help decision-makers **map tools to problems**. Are you seeking deep accuracy or faster iteration? Is your challenge about uncertainty, experimentation, explanation, or system dynamics? The table below serves as a strategic compass, guiding the selection of analytical methods based on purpose, strengths, trade-offs, and application domains.

Table 3.9. Strategic Analytics Toolkit Comparison Table

Tool	Best Use Case	Strategic Strength	Key Limitation	Ideal Application Context
XGBoost	Predicting outcomes with messy, high-dimensional data	High accuracy; handles nonlinearity and missing data	Limited transparency without explainability tools	Forecasting customer behavior, market trends, and financial risk
Bayesian Optimization	Optimizing decisions with limited data or high testing costs	Efficient learning with fewer trials; exploration + exploitation	Requires a well-defined objective function	Pricing strategies, marketing budget allocation, and innovation search
SHAP	Explaining individual or global model decisions	Model transparency; localized insights for trust	Doesn't optimize or predict by itself	Board reporting, regulatory strategy, model interpretation
PDP	Visualizing sensitivity to individual variables	Shows marginal effects clearly; easy to communicate	Assumes other variables held constant	Scenario testing, resource allocation thresholds, tipping points
Monte Carlo Simulation	Modeling risk and uncertainty across scenarios	Probabilistic foresight; stress-testing	Requires assumptions about distributions	Financial strategy, crisis planning, scenario risk forecasting
Digital Twin	Simulating real-time, system-level decisions	Dynamic interaction, system feedback, holistic view	High setup complexity; data-intensive	Supply chain strategy, market entry planning, ESG simulation

Strategic Decision Guide: How to Choose the Right Tool

* **If your goal is to predict future outcomes with high fidelity across complex data…** start with **XGBoost** and enhance with **SHAP** for interpretability.
* **If you're navigating a trade-off-heavy landscape with costly experimentation…** use **Bayesian Optimization** to converge on optimal strategies efficiently.

* **If decision-makers need clarity and trust in what drives your models…** pair **SHAP** with **PDPs** to explain and visualize strategic levers.
* **If you need to plan for the unknown and prepare for volatility…** turn to **Monte Carlo Simulation** to understand risk distributions.
* **If you're testing big, system-level moves (e.g., expansion, reorganization)…** leverage **Digital Twins** to simulate and rehearse strategies in a virtual, low-risk environment.

Together, these tools provide a **layered and adaptive architecture for strategic thinking**. Rather than forcing every problem into the same statistical mold, modern strategy demands a **flexible toolkit**, each component brought to bear when the problem calls for it.

In the next section, we explore how these tools can **reinvent classic strategy frameworks,** not replacing their logic, but upgrading their execution through analytics. We begin with VRIO, now supercharged by SHAP.

3.4. Reinventing Classic Strategy with Analytics

Strategy, at its core, has always been about frameworks, timeless mental models like SWOT, VRIO, environmental scanning, and strategic fit that guide leaders in navigating uncertainty, evaluating advantage, and allocating resources. These frameworks have endured not because they are perfect, but because they offer **intellectual structure in an unpredictable world**. However, in the age of data and machine intelligence, these models need more than wisdom; they need **reinforcement through analytics**.

This section explores how classic strategic tools can be **augmented, reimagined, and reinvented** through advanced analytics. Instead of discarding what has worked, we inject these familiar frameworks with **data-driven precision and algorithmic insight,** turning them from static checklists into dynamic systems of diagnosis, prediction, and action.

We show how:

* **VRIO**, when combined with SHAP values, moves beyond "do we have a valuable resource?" to "what is driving value creation in real-time?"
* **SWOT** becomes more than brainstorming—it becomes **predictive scanning**, using trend data and machine learning to quantify strengths, weaknesses, opportunities, and threats.
* **Environmental scanning** shifts from static PEST analysis to real-time **AI-powered sensing**, where algorithms flag emerging signals and surface hidden disruptions.
* And **Strategic Fit** transforms from a conceptual ideal into a **modeling problem**, where alignment between internal capabilities and external conditions is continuously optimized through simulation and feedback.

These hybrid models represent not a departure from traditional strategy, but an evolution. They blend the **intuition of human insight** with the **rigor of computational logic**, enabling strategic decisions that are not only theoretically sound but empirically validated, real-time adaptive, and future-ready.

We begin with VRIO, infused with SHAP, to demonstrate how analytics can dynamically identify, explain, and amplify sources of competitive advantage.

3.4.1 VRIO + SHAP – Dynamic Value Identification

The VRIO framework, standing for **Value, Rarity, Imitability, and Organization,** has long served as a foundational tool for internal strategic analysis. It helps leaders assess whether a firm's resources and capabilities can deliver sustained competitive advantage. However, in today's dynamic and data-intensive business environment, VRIO's traditional checklist approach is no longer sufficient on its own. What leaders now need is a **way to measure and monitor strategic value in real time**, at scale, and with diagnostic clarity.

This is where the integration of **SHAP (SHapley Additive exPlanations)** into the VRIO framework fundamentally transforms the nature of strategic analysis. SHAP enables us to **quantify the contribution of each resource or capability to a firm's competitive performance** using interpretable machine learning. Rather than speculating on value, strategists can **measure it directly** and adjust strategy based on empirical insights.

Rethinking "Value" with SHAP: From Assumption to Evidence

Traditionally, the "Value" component of VRIO asks: Does a resource or capability help the firm exploit opportunities or neutralize threats? With SHAP, we go further. We ask: **To what extent does this variable explain performance, right now, and in which direction?**

For example, suppose a firm uses XGBoost to predict its Net Promoter Score (NPS), a proxy for customer loyalty, and includes variables such as R&D investment, digital capabilities, brand strength, and employee retention. SHAP can then isolate the contribution of each variable to NPS, both across the model and for specific customer segments.

- If SHAP reveals that **digital capability consistently contributes most to high NPS scores**, it signals that this resource is truly valuable.
- If another variable, say traditional advertising spend, consistently shows a **negative SHAP impact**, then its strategic value should be reevaluated, even if it was previously thought to be a core asset.

This is **VRIO in motion,** evaluating "V" not once a year in a planning retreat, but continuously and with measurable precision.

Rarity and Imitability: From Narratives to Benchmarking

SHAP also supports a **more grounded analysis of Rarity and Imitability**. Once the high-impact features (i.e., resources) are identified, strategists can benchmark their availability across competitors or industries. For example:

- A hotel chain finds through SHAP analysis that **proprietary guest personalization algorithms** are driving customer satisfaction and revenue growth.
- By scanning competitors' technological infrastructure (via public filings, vendor partnerships, or AI-based web scraping tools), the company can assess whether this resource is **rare** or becoming **commoditized**.
- Similarly, if SHAP indicates that a high-impact capability stems from **deep cross-training and a unique company culture**, the firm can classify it as **costly to imitate**—satisfying the "I" condition.

This creates a bridge between **quantitative model outputs and strategic storytelling,** bringing empirical teeth to what was once abstract analysis.

Organizational Support: Closing the Loop

The final VRIO component, **Organization,** asks whether the firm is structured to fully exploit a given resource. SHAP offers a way to test this by examining the **interaction effects** within models. If SHAP values show that a resource's impact is conditional on another variable (e.g., leadership alignment, investment levels, or regional market strength), this suggests **organization-dependent performance**.

For instance, in a global retail firm:

- SHAP might reveal that **advanced analytics platforms** only create significant value when paired with **data-literate middle management**. In branches without this support, their value is neutral or even negative due to misuse.
- This indicates the resource is only strategic **if the organization is aligned** to deploy it effectively, fulfilling the "O" requirement.

In this way, SHAP doesn't just help identify high-impact resources; it shows **when, where, and why they work**, linking value creation directly to internal alignment and capability deployment.

Strategic Use Case: Tech-Enabled Retailer

Consider a fast-growing omnichannel retailer analyzing the drivers of its regional profitability. A model is built using XGBoost, incorporating data on logistics efficiency, mobile app usage, store design, labor costs, customer density, and AI-based pricing tools. SHAP analysis reveals:

* The **AI pricing engine** is the single most consistent driver of high profitability, with SHAP values significantly outperforming all other features.
* The model also shows that its impact is highly dependent on **mobile app adoption rates,** a proxy for customer engagement with price alerts and personalization.
* Based on this, leadership classifies the pricing engine as **Valuable, Rare, and Costly to Imitate,** but **only valuable when organizational enablers are in place** (mobile integration, training, cross-channel marketing).

The result? A data-driven VRIO scorecard, one that evolves with the business, identifies competitive leverage and guides investment in scaling the capability where organizational alignment is strongest. Table 3.3 summarizes the logic.

Table 3.10. SHAP-Enhanced VRIO Logic

VRIO Dimension	SHAP-Enriched Insight	Strategic Benefit
Value	Quantifies the real-time impact of each resource	Validates strategic assumptions empirically
Rarity	Compares high-impact features across market benchmarks	Identifies true differentiators
Imitability	Flags complex, embedded resources with nonlinear interactions	Guides the protection and scaling of unique capabilities
Organization	Reveals dependencies and conditional impacts	Ensures internal alignment before resource scaling

By combining VRIO with SHAP, strategists shift from **static inventory-taking** to **dynamic competitive diagnosis**. Strategy becomes a **living system**, constantly updated by empirical evidence and explained through transparent narratives. The outcome? Resources that are not only valuable in theory, but validated, explainable, and actionable in practice.

In the next section, we apply this same logic to another timeless tool, **SWOT analysis,** and explore how predictive analytics can convert brainstorming into **quantified foresight**.

Implementing VRIO + SHAP: A Strategic Execution Roadmap

Integrating SHAP analysis into VRIO assessments requires both **data infrastructure** and **strategic intent**. Below is a practical, staged approach to implementation, suitable for strategy teams, internal consultants, or innovation units within organizations.

Step 1: Define Strategic Performance Outcomes

Start by selecting a **strategic performance variable** that reflects competitive advantage. This becomes the dependent variable in your predictive model. Examples include:

* Market share growth
* Customer retention rate
* Profit margin by region
* Net Promoter Score (NPS)
* ESG performance
* Innovation output or time-to-market

The variable should be **measurable**, **relevant to your value proposition**, and **sensitive to internal capabilities**.

Step 2: Inventory Strategic Resources and Capabilities

Next, compile a list of **firm-specific resources and capabilities**—these will become the features (independent variables) in your model. Consider both tangible and intangible assets, such as:

* Proprietary technology
* Brand equity
* R&D intensity
* Organizational culture metrics (e.g., engagement scores)
* Leadership tenure
* Operational efficiency ratios
* AI maturity or analytics readiness
* Patents, partnerships, or licenses

This is your **modern VRIO inventory**, structured for modeling rather than only qualitative judgment.

Step 3: Build the Predictive Model (e.g., with XGBoost)

Using historical data, build a predictive model (like XGBoost) to estimate the chosen outcome variable based on the features collected. Ensure the data is cleaned, preprocessed, and balanced.

* Split data into training and validation sets

* Tune the model using cross-validation
* Confirm model accuracy with metrics like R^2, RMSE, AUC, etc.

This model will become the foundation for SHAP analysis.

Step 4: Generate SHAP Values for Interpretability

Apply SHAP analysis to your model to **interpret the importance of each resource or capability,** both globally (across all observations) and locally (for specific units, regions, or customer segments).

Key outputs include:

* SHAP summary plots: Which features drive outcomes most?
* SHAP force plots: Why did a prediction look the way it did?
* SHAP dependence plots: How do feature effects vary over their range?

These outputs translate model predictions into **strategic narratives**.

Step 5: Map SHAP Insights to VRIO Dimensions

Using **SHAP values**, strategists can evaluate resources not just qualitatively, but quantitatively, grounding the VRIO framework in data-driven insights. This approach allows leaders to assess how each resource contributes to strategic advantage, based on actual model behavior.

The table below outlines how SHAP can be applied to each of the four VRIO pillars, turning abstract criteria into measurable, interpretable components:

Table 3.11. SHAP-Enhanced VRIO Framework

VRIO Element	SHAP-Driven Criteria
Value	High positive SHAP contribution to the desired outcome
Rarity	Resource not widely available among peer firms
Imitability	SHAP indicates embeddedness or complexity (e.g., interaction effects)
Organization	The conditional SHAP effects value is only unlocked when other structures are in place.

This converts SHAP outputs into a **VRIO scorecard with empirical grounding**.

Step 6: Visualize and Communicate Results

Transform the analysis into actionable insights for stakeholders:

- Create a **VRIO heatmap** with SHAP scores annotated
- Build **data-driven resource dashboards**
- Prepare **scenario briefs**: "If we increase capability X, how does that affect outcome Y?"
- Present strategic options backed by SHAP-based diagnostics

The goal is to **embed VRIO+SHAP into regular strategic planning cycles**—not as a one-off analysis, but as a continuous feedback tool.

Step 7: Act on Findings and Reassess Regularly

Use the resulting insights to:

* Allocate resources more precisely
* Divest from non-contributing capabilities
* Double down on high-SHAP, high-VRIO assets
* Design training or restructuring to improve "O" (organizational support)

As the environment and data evolve, **rerun models quarterly or annually** to detect shifts in resource value. This turns the static VRIO analysis into a **dynamic system of strategic sensing and response**.

Optional: Automation and Scaling

For advanced firms, consider embedding this approach into a **strategic analytics platform** or **corporate data warehouse**, allowing for:

* Live SHAP value dashboards for key performance drivers
* Strategy labs with interactive resource simulation
* Integration with Digital Twins for full-cycle planning

3.4.2 SWOT + Predictive Trend Data: Augmenting Strategic Awareness with Foresight

For decades, the SWOT framework, **Strengths, Weaknesses, Opportunities, and Threats,** has helped organizations summarize internal and external factors that shape strategic direction. Yet in today's dynamic, data-driven environment, traditional SWOT analyses often fall short: they are static, opinion-based, and locked in the present. What if we could **infuse SWOT with predictive insight**, making it forward-looking, evidence-based, and responsive?

By integrating **Predictive Trend Data** with SWOT, strategists move from speculative diagnostics to **proactive intelligence**. This fusion empowers organizations to not only identify internal capabilities and external forces but also **forecast how they will evolve**, interact, and impact future performance.

3.4.2.1 From Descriptive to Predictive SWOT

The SWOT framework has long served as a foundational tool in strategic planning. Its appeal lies in its simplicity, providing a structured way to think about internal capabilities (Strengths and Weaknesses) and external conditions (Opportunities and Threats). However, in an age defined by volatility, complexity, and rapid change, **traditional SWOT analysis has critical limitations**. It is often static, subjective, backward-looking, and detached from empirical data. To remain relevant in the STEM-powered strategy domain, SWOT must evolve from a descriptive summary to a predictive intelligence system.

Descriptive SWOT: Static and Subjective

In its traditional form, SWOT captures what stakeholders *perceive* to be the organization's current internal and external conditions. For example:

* Strength: "We have strong brand recognition."
* Weakness: "Our IT systems are outdated."
* Opportunity: "There's growth in emerging markets."
* Threat: "Competition is intensifying."

While useful for brainstorming, this approach is inherently subjective and anchored in the present moment. It lacks:

* **Evidence-based validation** (Are the threats real? How likely are the opportunities?)
* **Temporal dimension** (Will these factors change in 6 months? 2 years?)
* **Sensitivity modeling** (Which items have the most strategic leverage?)
* **Scenario consideration** (What happens if X changes? What's the tipping point?)

The result is a SWOT chart that can easily become a checklist, a political negotiation, or a consensus artifact, rather than a strategic catalyst.

Predictive SWOT: Dynamic and Data-Enriched

The shift to **Predictive SWOT** transforms this classic framework into a forward-looking decision-support engine. Rather than relying on anecdotal input or managerial intuition, each quadrant is now informed by **empirical trend data, predictive models, and simulations**. The focus moves from "what we think" to "what is likely to happen," enabling leaders to anticipate how internal capabilities and external forces may evolve and interact over time.

The table below illustrates how this transformation reframes each SWOT element through a predictive lens:

Table 3.12. From Descriptive to Predictive SWOT

SWOT Element	Descriptive Approach	Predictive Enhancement
Strength	"Strong R&D team"	"Patent output forecasted to grow 18% YoY; talent retention probability = 93% over next 2 years"
Weakness	"Slow delivery system"	"Delivery delays predicted to increase 24% during peak seasons under current capacity models"
Opportunity	"AI growth in logistics"	"AI-driven logistics projected to account for 38% of market by 2028, with early adopters gaining 3.4x ROI"
Threat	"New market entrants"	"Startup density in the target region growing 12% annually; price war risk estimated at 0.7 likelihood in 12 months"

With this predictive logic, SWOT becomes a **living system,** capable of evolving as new data arrives and assumptions shift.

The Strategic Value of Predictive SWOT

A Predictive SWOT enables a range of high-value strategic capabilities:

1. **Foresight over hindsight**: Instead of relying on current-state analysis, firms can model future states and position themselves in advance.
2. **Probability-informed threat and opportunity evaluation**: Forecasts and simulations assign likelihoods and magnitudes, sharpening focus on what matters most.
3. **Resource allocation alignment**: Predictive insights clarify where strengths must be reinforced, which weaknesses are most urgent, and which opportunities/threats warrant investment.
4. **Agile scenario development**: As environmental signals shift, SWOT elements are updated dynamically—making the framework responsive rather than static.

Example: Retail Expansion Strategy

Imagine a mid-sized retail chain evaluating expansion into three regions. A traditional SWOT might note "strong brand equity" and "opportunity in suburban areas" based on internal surveys and anecdotal trends. A predictive SWOT, however, shows:

* **Brand equity** among Gen Z in Region A is projected to decline by 14% due to low digital presence (based on social sentiment analysis).
* **Suburban growth** in Region B is forecasted to peak in 18 months and decline afterward (based on housing and mobility trends).
* **Logistical weaknesses** in Region C could result in a 22% increase in delivery costs due to supplier volatility (from Monte Carlo simulations).

These findings completely change the decision. Instead of acting on generic assumptions, the firm now has **directional clarity and probabilistic intelligence**—enabling confident action with risk-adjusted expectations.

Implementation Tip

To implement Predictive SWOT:

1. Start by identifying traditional SWOT elements through workshops or strategic discussions.
2. For each item, ask: *Can we validate this with data? Can we project its trajectory?*
3. Integrate predictive tools (e.g., SHAP, PDPs, market trend forecasts, simulations) to model each item's evolution.
4. Reframe your SWOT quadrants into **evidence-backed insights** rather than labels.

3.4.2.2 Data Sources for Predictive SWOT

Moving SWOT analysis from subjective brainstorming to predictive insight requires a transformation in how we **source, validate, and structure information**. A Predictive SWOT is only as powerful as the data that feeds it. Thus, one of the most critical enablers of this shift is the integration of high-quality, relevant, and timely data from both internal and external environments.

From Perception to Prediction: The Data Imperative

In traditional SWOT exercises, data is often anecdotal: managers' opinions, team reflections, or static reports. These inputs lack depth, consistency, and a forward-looking perspective. By contrast, a Predictive SWOT requires data that is:

* **Structured and unstructured**
* **Real-time or recent**
* **Quantitative and qualitative**
* **Internal and external**
* **Historical and forward-looking (forecasts or simulations)**

The key is not simply more data, but *strategically aligned data,* selected for its ability to inform the evolution of each SWOT quadrant.

Mapping the Right Data to Each Quadrant

To make **Predictive SWOT** actionable, it's essential to align each quadrant with the right data sources. This ensures that strategic insights are grounded in evidence, not assumptions. By mapping internal and external data streams to each element of the SWOT framework, organizations can continuously refresh their analysis with real-time intelligence.

The table below provides a structured guide to sourcing relevant data for each quadrant:

Table 3.13. Data Mapping for Predictive SWOT

SWOT Quadrant	Internal/External	Primary Data Sources
Strengths	Internal	KPIs, product performance dashboards, internal talent data, IP filings, process benchmarking, employee retention, internal survey data
Weaknesses	Internal	Error logs, customer complaints, downtime reports, churn analysis, cost-overrun data, HR turnover, capacity utilization
Opportunities	External	Market trend databases, innovation diffusion reports, demographic projections, emerging tech indexes, ESG scoring trends, venture funding data
Threats	External	Competitive intelligence feeds, geopolitical risk indices, regulatory tracking platforms, social sentiment analysis, inflation, and supply chain volatility data

Let's explore these in more detail.

Internal Data for Strengths and Weaknesses

Strengths and weaknesses are grounded in an organization's own resources, capabilities, and performance. Predictive data here enhances self-awareness and future-readiness.

- **Performance Metrics**: Time-series data on sales growth, delivery speed, defect rates, or Net Promoter Score (NPS) helps uncover trends behind current capabilities.
- **Human Capital Analytics**: Predictive models using retention scores, upskilling success rates, and team innovation rates illuminate workforce strengths or gaps.

- **Operational Dashboards**: Real-time visualizations of bottlenecks, downtime, or energy usage inform whether process strengths are durable or declining.

A manufacturing firm models machine downtime using internal IoT sensor data. When overlayed with maintenance logs and delivery performance, it predicts production reliability under various growth scenarios, a key input into both strengths and weaknesses.

External Data for Opportunities and Threats

Opportunities and threats live in the external environment—making accurate, forward-looking scanning essential.

* **Market Signals**: Data from services like PitchBook, Statista, or CB Insights can identify growing sectors, customer demands, or startup disruptors.
* **Policy and Regulatory Trends**: Tools like FiscalNote, LexisNexis, or WTO databases help forecast changes in compliance burdens or trade dynamics.
* **Environmental and ESG Trends**: Climate risk dashboards and ESG databases (e.g., Refinitiv, MSCI) provide early signals about emerging green markets or social backlash.
* **Competitor Modeling**: Web scraping, public filings, and NLP tools can uncover shifts in competitive strategy, pricing patterns, or innovation pipelines.

A hospitality chain uses climate vulnerability forecasts and GDP growth projections across regions to determine where to allocate sustainability investments—treating climate resilience not as a reactive threat, but as a growth-oriented opportunity.

Using Predictive Models and Simulations

Data alone is not enough; Predictive SWOT requires **analytical engines** to transform raw information into strategic foresight.

* **Time-Series Forecasting** (e.g., ARIMA, Prophet): To project trendlines for internal KPIs or external demand.
* **Predictive ML Models** (e.g., XGBoost): To score the probability of customer churn, product failure, or market entry success.
* **SHAP & PDPs**: To unpack why key variables are driving predicted shifts in strengths or threats.
* **Monte Carlo Simulation**: To model uncertainty ranges around emerging opportunities or strategic risks.
* **Digital Twin Data Feeds**: For continuously updating the SWOT based on real-time business system changes.

Qualitative Data and Human Intelligence Still Matter

Even in STEM strategy, **strategic intuition, narrative insight, and expert judgment** are indispensable, especially for identifying "weak signals" and interpreting cultural or behavioral patterns. Integrating structured qualitative inputs strengthens SWOT's depth.

- **Executive Interviews**: To assess internal perception gaps.
- **Customer Voice Mining**: Using NLP on reviews, social media, or call center transcripts.
- **Scenario Workshops**: With experts modeling implications of ambiguous signals.

Strategic Integration Tip

To operationalize these data sources within the SWOT process:

1. **Build a Strategic Data Library** aligned with SWOT quadrants, updating continuously.
2. **Assign SWOT Ownership** to cross-functional teams (e.g., HR owns talent data; marketing owns brand metrics).
3. **Run Quarterly SWOT Refreshes** using updated data pipelines.
4. **Visualize SWOT Trends** over time to show movement—not just snapshots.

3.4.2.3 Predictive Threat and Opportunity Modeling

In a volatile, uncertain, complex, and ambiguous (VUCA) world, the most powerful SWOT quadrant transformation happens in **Threats and Opportunities**. These are the external variables that either constrain or expand strategic options—and they're no longer static elements described in hindsight. Predictive analytics empowers organizations to forecast, score, and monitor threats and opportunities as **dynamic signals**, not just passive environmental descriptors.

This section explores how predictive modeling tools can be used to **quantify, track, and simulate** these two quadrants, helping leaders become anticipatory, not reactive.

Why Prediction Matters in Threats and Opportunities

Traditional strategy workshops often treat threats and opportunities as abstract bullet points:

* Opportunity: "Emerging demand in Southeast Asia"
* Threat: "New regulatory burden in Europe"

But these statements raise questions:

* How **likely** is the demand surge? What is the **expected ROI** if we act?

* How **severe** is the regulatory threat? What's the **window** for response?

Predictive modeling replaces ambiguity with precision. It allows strategists to say:

"The Southeast Asia opportunity has a projected 22% YoY growth with a 90% confidence interval between 18% and 25%, based on customer sentiment trends, digital adoption rates, and import policy signals."

"The proposed EU law has a 60% chance of passage by Q4, with expected compliance costs rising by $1.2M annually."

By attaching probabilities, ranges, and risk/return calculations, predictive SWOT reframes threats and opportunities as quantifiable decision variables.

Modeling Opportunities

Opportunities are typically high-potential external changes or trends that could be harnessed for competitive advantage. Predictive analytics identifies them early and prioritizes them by impact.

1. Trend Acceleration Models

* **Purpose**: Forecast the pace and trajectory of emerging phenomena.
* **Tools**: Time series forecasting (Prophet, ARIMA), adoption S-curves, topic modeling on patent/citation data.
* **Example**: A mobility company models electric vehicle adoption across Latin America using infrastructure rollout rates, fuel cost volatility, and Twitter sentiment to identify regional investment priorities.

2. Market Entry Predictors

* **Purpose**: Evaluate the success likelihood of entering new markets or segments.
* **Tools**: Ensemble classifiers (e.g., XGBoost), clustering algorithms for market segmentation.
* **Example**: A fintech startup predicts cross-border payment growth in underserved African markets, integrating GDP, mobile penetration, regulation favorability, and competitor activity into a machine learning model.

3. ESG Opportunity Indexing

* **Purpose**: Identify profit-aligned sustainability initiatives.
* **Tools**: SHAP-enabled predictive scoring of sustainability investments (e.g., energy efficiency, diversity programs).

* **Example**: A hotel chain predicts revenue boosts from solar investment using regional energy price forecasts, ESG fund flow trends, and customer green preference scores.

Modeling Threats

Strategic threats are not only about disruption, they're about **risk velocity** and **response preparedness**. Predictive modeling uncovers when threats become probable, and what triggers them to intensify.

1. Disruption Scoring Models

* **Purpose**: Rank threats by proximity and potential damage.
* **Tools**: Random Forest or Gradient Boosting models trained on past disruptions.
* **Example**: A global supply chain platform scores ports and routes by disruption risk using weather patterns, geopolitical indicators, labor strike data, and local policy volatility.

2. Regulatory Risk Forecasting

* **Purpose**: Anticipate the likelihood and impact of legal changes.
* **Tools**: Natural Language Processing (NLP) on government databases, sentiment trend analysis from lawmaker speeches.
* **Example**: A pharma firm uses text mining on FDA press releases and congressional hearings to forecast regulation intensity on opioids—modeling impact on portfolio viability.

3. Competitive Threat Modeling

* **Purpose**: Predict competitor moves and their potential impact.
* **Tools**: Web scraping, public financial analysis, merger pattern detection, NLP on earnings calls.
* **Example**: A cloud services company detects early signs of a competitor entering its territory through patent filings, job postings, and increased marketing spend, quantified into a threat heatmap.

Implementation Framework: From Signal to Action

Predictive threat and opportunity modeling follows a cycle:

1. **Signal Detection**: Use real-time feeds (e.g., news, pricing, regulations, funding rounds) to detect weak signals.
2. **Feature Engineering**: Identify relevant variables and metrics (e.g., risk indicators, growth accelerators).

3. **Predictive Scoring**: Apply ML models to estimate likelihood, timing, and impact.
4. **Explainability and Visualization**: Use SHAP, PDP, or simulation dashboards to communicate insights to decision-makers.
5. **Strategic Response Mapping**: Assign owners to each high-risk/high-opportunity area and develop preemptive action plans.

The table below illustrates how this cycle can be applied in practice, using a heatmap-style template to align predictive insights with strategic responses:

Table 3.14. Predictive Threat and Opportunity Heatmap

External Force	Prediction Tool	Likelihood	Strategic Impact	SHAP-Backed Drivers	Action Plan
Carbon tax in the EU	NLP + Monte Carlo	75%	High (↑cost by $1.8M)	Regional emissions, lobbying activity	Accelerate carbon offset partnerships
Gen Z travel surge	Time series forecast	90%	High (↑sales in urban stays)	TikTok sentiment, booking pattern shifts	Launch city-hosted digital detox retreats
Tech competitor expansion	XGBoost on funding + patents	60%	Medium (5% share loss)	Hiring spikes, product mentions in reviews	Preemptive pricing and retention campaigns

Opportunities and threats must be reframed as **data-driven hypotheses,** with predictive modeling providing the evidence base for strategic prioritization. When integrated into a Predictive SWOT process, these models allow organizations to shift from reacting to trends to shaping them, from defending against threats to outmaneuvering them.

Predictive modeling doesn't replace strategy; it weaponizes it with foresight, simulation, and agility.

3.4.2.4 Visualizing the Predictive SWOT Dashboard

As strategic management enters the STEM era, the traditional SWOT matrix must evolve from a static quadrant to a **live dashboard,** a visual, data-rich, and interactive interface that transforms analysis into real-time strategic command. Predictive SWOT dashboards are not just summary visuals; they are dynamic **decision platforms** that allow leaders to scan, simulate, and act.

This section explains how to design, interpret, and leverage a Predictive SWOT Dashboard using real-world data streams, machine learning outputs, and interactive visualizations.

From 2×2 to Strategic Intelligence System

In its original form, the SWOT matrix was a 2×2 box filled with managerial intuition. In its predictive version, the box becomes a canvas populated by **model outputs**, **impact scores**, **confidence levels**, and **visual trend indicators**.

Each quadrant becomes a **strategic lens**:

* **Strengths**: Measured by internal performance drivers (e.g., brand equity score, innovation velocity index)
* **Weaknesses**: Quantified through KPIs underperformance and internal inefficiencies (e.g., churn rate, margin pressure)
* **Opportunities**: Identified by external trend signals and scored by forecasted ROI or market potential
* **Threats**: Simulated using predictive risk models, disruption scoring, or compliance probability ranges

The dashboard overlays these with filters (region, business unit, time horizon), enabling granular drill-downs or enterprise-wide views. The following table outlines the key features that make a predictive SWOT dashboard both dynamic and strategically valuable:

Table 3.14. Key Features of a Predictive SWOT Dashboard

Feature	Function	Toolset	Strategic Benefit
Live Data Feeds	Pulls from internal systems (ERP, CRM) and external APIs (news, sentiment, regulation)	Power BI, Tableau, Python Dash	Keeps SWOT current and reactive
Quadrant Scoring Engine	Assigns numeric strength/opportunity/threat scores via models	XGBoost, SHAP, PDP, BO	Moves SWOT from narrative to math
Confidence Bands	Displays forecast reliability (e.g., 80% CI)	Monte Carlo Simulation, Bayesian Estimators	Quantifies uncertainty for better risk planning
Scenario Switcher	Test different environments (e.g., high inflation, competitor entry)	Simulation + Dashboard interface	Guides proactive strategy shaping
Drill-Down Explorer	Allows strategic slicing (by geography, segment, product line)	Filterable visual layers	Supports localized or unit-specific actions
Strategic Alerts	Flags changes in threat/opportunity scores	Alert rules via threshold models	Enables real-time executive response

Sample Visualization: Predictive SWOT Dashboard

Here's a conceptual representation of how the dashboard might appear on a strategist's screen:

```
|-----------------------|-------------------------|
|    STRENGTHS          |      OPPORTUNITIES      |
|-----------------------|-------------------------|
| ✓ Brand Index: 87      | ↑ Gen Z Travel Demand    |
| ✓ Tech Leadership: 9.2| ↑ Sustainable Tourism +24% |
| ✓ Loyalty NPS: 75      | ↑ AI Concierge Trend     |
| [Confidence: 95%]     | [ROI Forecast: $8.6M]   |
|-----------------------|-------------------------|
|   WEAKNESSES          |       THREATS           |
|-----------------------|-------------------------|
| ✗ Rising Churn: 12.5% | ↓ Regulatory Uncertainty |
| ✗ Cybersecurity Lag   | ↓ Greenwashing Backlash |
| ✗ Cost Pressure: $3M  | ↓ Inflation-Driven Deferral |
| [Trend: +12% QoQ]     | [Risk Score: 8.7/10]    |
 -----------------------|-------------------------
```

* Clicking on **"Rising Churn"** opens a SHAP-backed model showing the top five factors: customer service response time, pricing tier, competitor offer exposure, app UX score, and recent feature bugs.
* Selecting **"AI Concierge Trend"** shows its PDP curve: revenue lifts plateaus after $2M in investment.
* Hovering on **"Greenwashing Backlash"** reveals a social media sentiment drop and pending ESG legislation flagged by NLP.

Implementing the Dashboard

Technical Stack:

* **Backend**: Python/R for modeling, SQL for data storage, APIs for live feeds
* **Analytics Layer**: SHAP for interpretability, XGBoost for forecasting, BO for optimization
* **Frontend**: Tableau, Power BI, Streamlit, or Dash for interactive UI

User Roles:

* **Strategic Planners**: Use quadrant scores to prioritize initiatives
* **C-Suite**: Review confidence intervals, worst-case scenarios, and risk-adjusted forecasts
* **Business Units**: Track performance shifts and external exposure at operational levels

Strategic Impact

A Predictive SWOT Dashboard does not replace strategy; it **enables strategy at the speed of complexity**. Instead of waiting for quarterly workshops, leaders have **live radar** on evolving strengths, rising risks, and unfolding opportunities. It institutionalizes vigilance, agility, and evidence-based action.

More importantly, it turns SWOT into what it was always meant to be: **a strategic cockpit**, not a static form.

3.4.3 Environmental Scanning Powered by ML and NLP

Environmental scanning has long been a cornerstone of strategic planning, yet in its traditional form, it is increasingly outdated. Once reliant on periodic reports, expert opinions, and selective news monitoring, environmental scanning was more of an art than a science. It helped firms understand their external context, but it was slow, reactive, and subjective. In today's complex, fast-moving world, where change is continuous and uncertainty is the norm, such an approach is insufficient.

Strategic management in a STEM-driven world demands more than just awareness of external trends; it requires **real-time, systemic sensing** across multiple information streams, regions, languages, and formats. The modern strategist must not only **see change coming** but also **model its trajectory, understand its sentiment, and estimate its strategic impact** before it reaches the core business.

This transformation is made possible by the integration of **Machine Learning (ML)** and **Natural Language Processing (NLP)** into the scanning process. Together, these tools move environmental scanning from a human-limited discipline to a computationally augmented strategic capability. ML enables pattern recognition across massive datasets, spotting subtle anomalies and weak signals. NLP enables machines to "read" and interpret unstructured content, such as news articles, financial filings, executive speeches, social media discussions, and regulatory briefings, at a scale and speed impossible for humans.

What emerges is a **new form of strategic intelligence**: automated, scalable, and adaptive environmental scanning that can surface risks, detect opportunities, and monitor stakeholder sentiment across the full landscape of external forces.

Rather than reacting to disruption, firms can now anticipate it. Rather than waiting for consultants to summarize trends, they can develop in-house scanning engines. And rather than relying solely on human judgment, they can combine intuition with machine-verified evidence to craft faster, sharper, and more adaptive strategies.

This section explores how ML and NLP revolutionize environmental scanning by turning fragmented data into foresight. It offers not just tools and techniques, but also strategic frameworks for embedding intelligent scanning into the firm's decision-making architecture, so that leaders are not just informed, but advantageously positioned in an uncertain world.

3.4.3.1 From Passive Monitoring to Intelligent Sensemaking

In the past, environmental scanning functioned largely as **passive monitoring,** a structured, but often shallow, collection of information about the external environment. Teams compiled data on competitors, regulations, consumer preferences, or geopolitical trends and presented them as static reports. These outputs were informative, but rarely transformative. The role of scanning was to observe, not to explain, predict, or integrate.

However, as strategic uncertainty has intensified and the volume of external data has exploded, **passive monitoring is no longer sufficient**. Today's organizations face what might be called a "strategic signal problem": there is too much noise and too little actionable insight. In this environment, **intelligent sensemaking,** the ability to extract, synthesize, and interpret dynamic signals from complex data sources, is becoming a defining capability of strategic success.

Intelligent sensemaking is about **moving from data to decision advantage**. It requires not just collecting information, but structuring it into patterns, testing hypotheses, detecting emerging themes, and evaluating the relevance and credibility of each signal in real time. And crucially, it demands **automation** and **augmentation,** the use of machine learning (ML) and natural language processing (NLP) to extend human cognition and enable continuous, adaptive scanning.

Why Passive Monitoring Fails in a STEM Strategy Environment

1. **Latency**: Traditional methods are slow, offering lagging insights when real-time responsiveness is needed.
2. **Cognitive Bias**: Human analysts often seek confirmation of assumptions rather than discovery of disruptions.
3. **Siloed Sources**: Conventional scanning pulls from a narrow range of sources, missing peripheral or early-stage indicators.
4. **Static Output**: Reports are snapshots, not living systems; they fail to update or evolve with new data.

In contrast, intelligent sensemaking treats environmental data as **a dynamic ecosystem,** one that must be modeled, simulated, and interpreted continuously.

How Machine Learning and NLP Enable Intelligent Sensemaking

* **Automated Data Ingestion**: NLP engines can continuously scrape and parse millions of documents, social media feeds, and public reports across multiple languages and domains.

* **Entity Recognition and Topic Modeling**: ML can identify companies, trends, events, or risks mentioned in news, earnings calls, or regulatory documents, then cluster them into themes.
* **Sentiment Analysis**: NLP evaluates the tone, polarity, and urgency of text to quantify how stakeholders (consumers, investors, regulators) feel about a topic or firm.
* **Trend Detection and Anomaly Scoring**: Algorithms track how the frequency or framing of issues changes over time—flagging unusual spikes in attention or sentiment.

These capabilities shift scanning from a reactive task to a **predictive and strategic function**—alerting decision-makers not just to what is happening, but **what is changing**, **why**, and **what that means** for the firm's trajectory.

Strategic Example: Pharmaceutical Supply Chain

A global pharmaceutical firm deploys intelligent environmental scanning to monitor over 10,000 global sources daily, ranging from scientific journals to government health alerts and supplier news. An NLP system flags a rising volume of concern around a rare plant-based compound used in one of the firm's core treatments. The alerts show a drop in availability due to deforestation policies in its sourcing region. The company responds by proactively diversifying its supply chain six months ahead of competitors, avoiding cost shocks and production delays.

Organizational Design for Sensemaking

To implement intelligent sensemaking, firms must rethink how they organize their strategic intelligence functions. This includes:

* Creating **Strategic Foresight Cells** or **AI-Augmented Strategy Labs**
* Embedding **ML/NLP dashboards** into executive decision routines
* Establishing **feedback loops** from strategic actions to data signals for continuous learning
* Training analysts not only to interpret insights, but also to validate, challenge, and communicate them to diverse stakeholders.

From passive scanning to real-time intelligence, this shift redefines how environmental data informs strategy. The strategist's role is no longer to compile what happened last quarter; it is to **orchestrate machine-human ecosystems** that monitor, interpret, and act on what's emerging *right now*.

3.4.3.2 From Information Overload to Strategic Signal Detection

In an age where organizations are flooded with external data, from social media chatter and customer reviews to policy documents, patent filings, and satellite imagery, **the challenge is no longer data scarcity but signal scarcity**. Amid the noise of millions of datapoints, only a small fraction offers truly strategic value. Yet most firms still treat all inputs as if they are equally relevant. This leads to analysis paralysis, poor prioritization, and reactive rather than anticipatory decision-making.

Strategic signal detection is the art and science of identifying the weak signals that precede major shifts. These signals are often **subtle, cross-domain, and weakly structured,** but when detected early, they offer disproportionate advantage. They allow firms to pivot ahead of competitors, prepare for disruption, or accelerate innovation in a favorable direction. Unlike reporting, signal detection is a forward-facing capability; it is about what *might become important*, not just what *already is*.

Why Traditional Information Processing Falls Short

1. **Volume Obscures Meaning**: Human teams cannot keep up with the exponential growth in external content. Important indicators get buried.
2. **Relevance is Contextual**: A signal that is irrelevant to one firm may be mission-critical to another—depending on strategy, market position, and risk profile.
3. **Temporal Blindness**: Traditional reports miss the "emerging trajectory" of an issue—how it's gaining momentum or spreading across geographies and industries.
4. **Lack of Feedback**: Static analysis does not learn from outcomes. If a signal was missed or a false alarm raised, there's no system to recalibrate filters.

These limitations highlight the need for **machine-assisted signal detection systems,** ones that are capable of dynamic filtering, cross-domain pattern recognition, and contextual ranking of relevance. Key Components of Strategic Signal Detection Systems are presented in Table 3.15.

Table 3.15. Key Components of Strategic Signal Detection Systems

Component	Strategic Function
Signal Library	A curated repository of predefined signal types relevant to the firm's domain (e.g., sustainability disruptions, regulatory shifts, AI breakthroughs)
AI-Driven Filtering	Uses NLP and ML classifiers to continuously evaluate whether incoming data points match known or new signal profiles
Strategic Relevance Scoring	Assigns a contextual importance score based on organizational goals, historical impact, and current exposure
Temporal Trend Modeling	Detects acceleration or deceleration of attention over time—early indicators of potential tipping points
Human Verification Loop	Analysts validate high-confidence signals and provide interpretation, narrative framing, or action recommendations.

Together, these elements create a feedback-driven ecosystem where strategic foresight is constantly refreshed.

Strategic Example: Automotive Electrification

A leading automotive firm tracks emerging EV battery technologies using an AI-based signal detection system. In early 2023, the system flagged a pattern of increasing venture capital flows and scientific citations in a niche subfield, solid-state sodium-ion batteries. Though not yet commercially viable, these signals indicated a brewing shift in the innovation frontier. Within weeks, the company's R&D and partnership arms began exploratory initiatives, giving it a first-mover advantage in a now rapidly expanding field.

Signal Detection vs. Noise Amplification

Not every trending topic is a signal. Some are mere noise—short-lived hype cycles, media exaggerations, or misleading correlations. Effective signal detection systems must be trained to:

- **Ignore noise**: Filter out media buzz that lacks structural evidence or relevance.
- **Detect cross-domain convergence**: For example, an uptick in AI regulation discussion + insurance risk modeling publications = a potential regulatory inflection point.
- **Link to scenarios**: Attach signals to possible strategic futures, so they can be embedded into scenario planning and resilience modeling.

Strategic Action from Signals

Signals are only valuable if they inform action. Leading firms build **Signal-to-Strategy Pipelines**, where insights flow into:

* Portfolio diversification strategy
* Market entry timing
* Resource reallocation or investment freeze
* Early engagement with regulators or partners
* Trigger-based decision matrices (e.g., "If Signal X passes intensity threshold Y, activate Strategic Option Z")

From drowning in data to detecting inflection points, this section marks a critical evolution in the strategist's toolkit. Signal detection transforms the external environment from an unknowable storm into a manageable landscape of early warnings, hidden opportunities, and competitive insights.

3.4.3.3 Predictive Topic Modeling and Issue Clustering

In the era of information explosion, organizations face an overwhelming flow of qualitative data: media coverage, analyst reports, stakeholder feedback, regulatory changes, and scientific publications. Hidden within this unstructured ocean are early signals of disruption, latent customer needs, emergent policy shifts, or industry realignments. Traditional environmental scanning methods, even if digitized, struggle to cope with this scale, velocity, and semantic nuance. That's where **predictive topic modeling** and **issue clustering** enter the strategy toolkit.

From Text to Topics: What Topic Modeling Does

Topic modeling refers to a class of **unsupervised machine learning** techniques designed to automatically discover the abstract "topics" that occur in a collection of documents. Rather than requiring human coders or pre-defined categories, these models infer patterns in word usage and group them into coherent themes. The most commonly used models in strategic contexts include:

* **Latent Dirichlet Allocation (LDA):** Detects probabilistic themes across large corpora—ideal for identifying what issues are being talked about.
* **BERTopic:** Uses transformer-based embeddings (like BERT) to identify semantically rich clusters in short-form text like news headlines, tweets, or executive updates.
* **Non-Negative Matrix Factorization (NMF):** Effective for spotting additive themes in specialized datasets such as R&D portfolios, sustainability reports, or customer complaints.

Instead of asking, "What are people saying about AI regulation?" a topic model reveals that:

* Topic 1: Ethics, bias, and fairness in algorithms
* Topic 2: Legal liability and data privacy
* Topic 3: Government procurement and AI in defense
  Each topic is composed of key terms and ranked documents, making it a live intelligence stream rather than a manual summary.

Strategic Use Case: Early Foresight in Healthcare Innovation

Consider a biotech firm tracking breakthroughs in gene therapy. Instead of reading thousands of abstracts, it uses LDA to surface evolving research clusters:

* Cluster A: Delivery methods for CRISPR
* Cluster B: Rare disease targeting
* Cluster C: Regulatory and ethical discussions in the EU
  Armed with this structure, the firm can assign strategic monitors, form cross-functional task forces, or initiate partnerships before rivals detect the signal.

From Topic Detection to Strategic Clustering

Predictive topic modeling is often just the first layer. The second is **issue clustering,** where topics are grouped into broader thematic areas aligned with strategic relevance. For instance:

* Topics on "water scarcity," "carbon credit volatility," and "regulatory taxonomies" may be clustered under **Climate Risk Exposure**.
* Topics on "customer churn," "low NPS," and "competitor app launches" may fall under **Digital Experience Threats**.

Clustering makes unstructured data legible to strategy. It enables pattern recognition over time, comparison across competitors, and integration into scenario planning. When paired with dynamic dashboards, these clusters evolve live, much like weather maps, alerting leaders to new storms before they arrive. The following toolchain equips strategy teams to extract, visualize, and act on these insights in real time:

Table 3.16. Toolchain for Strategy Teams

Tool/Technique	Strategic Role
LDA (Latent Dirichlet Allocation)	Extracts structured topics from long-form reports
BERTopic	Maps emerging issues from short-form sources like social media
NMF (Non-Negative Matrix Factorization)	Analyzes additive trends in niche datasets
UMAP / t-SNE (for visualization)	Visualizes issue maps for workshops and dashboards
DBSCAN / K-Means	Group topics into clusters for foresight mapping

Strategic intelligence platforms are only as powerful as the capabilities they enable. The following table outlines core capabilities that transform raw data into actionable foresight, each delivering distinct strategic value across the enterprise:

Table 3.17. Strategic Intelligence Capabilities and Their Enterprise Value

Capability	Strategic Value
Dynamic foresight	Early detection of industry-relevant disruptions
Semantic scaling	Analyze thousands of documents without human fatigue
Strategic sensemaking	Turns noisy signals into board-ready narratives
Cross-functional alignment	Links R&D, marketing, policy, and sustainability under shared issues

3.4.3.4 Competitive Intelligence from Open Sources

In a hyperconnected economy, strategic surprises often arise not from secrecy, but from overlooked signals hidden in plain sight. Today, some of the most powerful insights into competitor behavior, market shifts, and innovation trajectories reside in open-access data, company websites, patent filings, investor briefings, social media, app updates, regulatory disclosures, job postings, and more. The discipline of **Competitive Intelligence (CI)** is being reborn, not as static benchmarking, but as a dynamic, tech-enabled, open-source-driven engine of foresight.

With the right tools and techniques, environmental scanning can harness these fragmented public signals and synthesize them into actionable insight. Rather than waiting for quarterly reports or market rumors, modern strategists can observe competitive moves as they happen, anticipate responses, and map strategic intent.

From Passive Monitoring to Active Signal Detection

Traditional CI relied heavily on human monitoring and periodic summaries, often lagging behind real-time developments. In contrast, **open-source intelligence (OSINT)** tools use web scraping, natural language processing (NLP), and network analysis to continuously track, analyze, and visualize competitor activities.

Examples of Open Source Competitive Signals:

* **Job postings:** Reveal new R&D priorities, location expansions, or internal restructuring.
* **Patent databases:** Uncover emerging technological bets and IP strategies.
* **App update logs and release notes:** Signal feature rollouts, integration with new platforms, or UX redesign.

* **Social media posts and employee LinkedIn activity:** Reflect shifting employer branding, internal culture, or hiring sentiment.
* **Government procurement portals:** Reveal contract wins or bids that indicate revenue strategy shifts.

For instance, if a logistics company notices its rival is hiring 25 machine learning engineers in Singapore and recently filed three patents related to AI-powered route optimization, this suggests a strategic pivot toward automated supply chain intelligence in that region. That's not just information, it's foresight.

Table 3.18. Toolchain for Open Source Competitive Intelligence

Source Category	Tools & Methods	Strategic Insight Example
Job Postings	Web scraping, NLP job clustering	Detect new organizational capabilities
Patent Filings	Google Patents, Lens.org, keyword clustering	Spot innovation themes by region or assignee
Investor Briefings	NLP sentiment analysis, topic extraction	Assess leadership tone and strategic confidence
App Stores	Review mining, version comparison tools	Identify UX pain points or upcoming releases
Social Media & LinkedIn	Text mining, network analysis	Uncover talent flow and team restructures
Government Portals	Contract mining, timeline visualization	Anticipate geographic or public-sector expansion

By combining these sources, strategists can build **live CI dashboards** that track competitor movement across dimensions: capability building, geographic presence, innovation areas, brand positioning, and regulatory stance.

Real-World Example: Telecom Intelligence

A telecom provider uses NLP to analyze job listings across five rivals, mapping them by skills, functions, and cities. A surge in AI-focused roles in Latin America signals a latent price war powered by network automation. The provider responds with targeted investments in AI-driven customer service and builds a partnership pipeline in the same region.

This example illustrates how open-source intelligence (OSINT) can drive competitive intelligence (CI) that is both timely and actionable. The table below outlines the strategic benefits of OSINT-driven CI and how they directly impact decision-making:

Table 3.19. Strategic Benefits of OSINT-Driven Competitive Intelligence

Benefit	Impact on Strategic Decision-Making
Real-time competitor awareness	Enables preemptive strategy shifts
Non-intrusive and ethical sourcing	Reduces compliance risks while gathering insight
Pattern recognition across competitors	Supports trendspotting and opportunity mapping
Cross-functional relevance	Aligns product, marketing, HR, and operations

The New CI Mindset: From Surveillance to Simulation

Modern CI is not just about observing what competitors do; it's about **simulating how they think**. OSINT-driven frameworks allow strategy teams to reverse-engineer a rival's decision-making logic, stress-test their likely responses under different scenarios, and map opportunity white spaces that have not yet been contested.

In this view, **competitive intelligence becomes a strategic counterpoint,** a discipline that challenges assumptions, sharpens positioning, and prepares the firm not just to react, but to outmaneuver.

3.4.3.5 Strategic Foresight as Machine-Augmented Insight

In a world where yesterday's best practices rapidly become today's outdated routines, **strategic foresight** is no longer optional; it is mission-critical. Yet the nature of foresight itself has changed. Once the domain of expert panels, long-range planning documents, and intuition-driven scenario workshops, foresight today must be **augmented by machines**, powered by data, and embedded into decision systems that evolve as fast as the world they seek to predict.

This is the rise of **machine-augmented foresight,** a paradigm where artificial intelligence (AI), big data, and simulation engines extend human vision and enable organizations not only to anticipate change but to shape it.

From Forecasting to Foresight

Traditional forecasting answers the question: *What is most likely to happen based on past data?* Strategic foresight reframes the question: *What could happen, why, and what should we do about it?*

Forecasting focuses on extrapolating trends. Foresight focuses on identifying weak signals, disruptive inflections, and paths not yet taken. By combining human creativity with machine precision, organizations move beyond trend-following to **opportunity-creation**.

Tools That Extend Strategic Vision

The machine-augmented foresight toolkit draws from several disciplines, enabling strategy teams to move beyond static analysis and into dynamic, future-oriented thinking. For example, **NLP-based trend mining** helps detect emerging themes across media, patents, and blogs—surfacing weak signals before they become mainstream. To explore future possibilities, teams can use **scenario generation via GANs**, which creates plausible future worlds for stress-testing strategies.

When simulating how different stakeholders might behave in complex systems, **agent-based modeling** becomes invaluable. Meanwhile, **temporal topic modeling** allows strategists to track how key themes evolve over time, offering a longitudinal view of strategic narratives. For decision-making under uncertainty, **predictive analytics combined with what-if modeling** helps forecast outcomes under various strategic moves. Finally, **generative AI for visioning** supports ideation around future business models, products, or risks—turning foresight into tangible innovation.

These tools and their strategic applications are summarized below:

3.20. Machine-Augmented Foresight Toolkit

Tool or Technique	Strategic Use Case
NLP-based Trend Mining	Detecting emerging themes from media, patents, blogs
Scenario Generation via GANs	Creating plausible future worlds for planning
Agent-Based Modeling	Simulating stakeholder behavior in complex systems
Temporal Topic Modeling	Tracking how strategy-related themes evolve over time
Predictive Analytics + What-If	Forecasting outcomes under different strategic moves
Generative AI for Visioning	Ideating future business models, products, or risks

For example, using temporal topic modeling on academic and industry publications, a global health NGO may identify that *preventive genomics* is rising faster than *personalized medicine* in discourse across emerging markets, suggesting a pivot in its innovation focus.

Real-World Example: Future of Work Strategy

A multinational HR firm uses NLP and data mining to scan thousands of global job boards, white papers, and academic articles. Machine learning clusters emerging skillsets and work models—like hybrid scheduling algorithms, algorithmic management, and neurodiverse workplace design. These signals inform strategic investments in new workforce tools and shape the firm's 5-year innovation roadmap.

This approach exemplifies the power of **machine-augmented foresight**, where advanced analytics transform fragmented data into a forward-looking strategy. The benefits of this approach extend across multiple dimensions of strategic planning, as outlined below:

Table 3.21. Benefits of Machine-Augmented Foresight

Capability	Strategic Impact
Early signal detection	Captures weak or fast-moving trends
Synthetic scenario development	Enables stress-testing under future uncertainties
Strategic creativity at scale	Augments human ideation with machine-generated ideas
Dynamic feedback into planning	Keeps foresight alive, not static

Integrating Foresight into the Strategic Cycle

Strategic foresight must not be relegated to a single off-site session or annual report. With machine augmentation, foresight becomes a **living function**, embedded in:

* Ongoing strategic reviews
* Innovation lab workflows
* Product pipeline steering
* ESG and regulatory anticipation
* Talent and capacity planning

By integrating machine-augmented foresight tools with strategic analytics (e.g., XGBoost, SHAP, Digital Twins), organizations can **move from risk-aversion to risk-intelligence,** developing future-ready strategies that are imaginative, evidence-based, and continuously evolving.

From Peripheral to Core

Historically, foresight lived on the periphery of strategic planning, viewed as an abstract or speculative exercise. But in a volatile, AI-driven world, **foresight becomes a core competency**. It is not just about avoiding surprises but about surfacing ideas that wouldn't emerge through linear analysis.

Machine-augmented foresight ensures that leaders are not trapped by yesterday's logic or today's constraints. It turns open-ended questions into structured insights, future anxiety into proactive design, and uncertainty into advantage.

3.4.4 Strategic Fit as a Modeling Problem

For decades, strategic fit has been described as the golden thread of competitive advantage, the alignment between a firm's internal capabilities and its external environment. In classical strategic management, achieving fit meant ensuring that resources, activities, and positioning reinforced one another in ways that were difficult for rivals to imitate. But in today's fast-moving, data-rich, and AI-enabled world, the nature of fit is evolving.

Rather than treating fit as a conceptual ideal or static alignment, modern strategy must treat **strategic fit as a modeling problem,** a challenge that can be mapped, measured, simulated, and optimized. This approach does not discard the original wisdom of fit; it enhances it with precision, evidence, and adaptability.

By leveraging analytics, machine learning, and systems thinking, strategists can now **quantify degrees of fit**, test strategic configurations under varying conditions, and dynamically adjust to ensure continuous alignment. This makes fit not just a goal, but a process, alive, iterative, and responsive to both internal change and external turbulence.

In this section, we explore how to:

* Map and model strategic configuration using analytical tools.
* Use optimization methods to identify high-fit combinations of activities and environments.
* Evaluate fit over time, not just at a point in time.
* Apply simulation-based testing (via Digital Twins and Monte Carlo) to stress-test alignment under disruption.

We also address the paradox of fit: too much alignment can breed inertia. We show how **adaptive fit,** the ability to reconfigure strategically with minimal friction, becomes the new source of advantage in volatile environments.

Strategic fit is no longer just a narrative; it's a data-driven construct. And modeling it transforms how we design strategies that endure, evolve, and outperform.

3.4.4.1 Mapping the Strategic Configuration

To model strategic fit, we must first understand what we are aligning. Strategic fit emerges from the interplay between internal resources and capabilities (the "inside") and external opportunities and constraints (the "outside"). But in a data-rich world, this alignment can, and must, be **mapped systematically**.

This begins with constructing a **strategic configuration map**: a comprehensive framework that captures the interdependent components of an organization's strategy. Think of this as a multidimensional blueprint of how the organization is structured to compete, including:

* **Key resources** (e.g., intellectual property, talent, financial capital)
* **Core capabilities** (e.g., R&D speed, customer engagement, supply chain agility)
* **Value propositions** (e.g., low cost, differentiation, sustainability)
* **Strategic activities** (e.g., sourcing, distribution, innovation, after-sales)
* **Market positions** (e.g., geographic presence, segmentation focus, pricing model)

Each of these components represents a **node** in the configuration. The real insight, however, comes from **mapping the connections between these nodes**—the reinforcing loops, synergies, and tensions that define the firm's architecture.

Using Network Analysis to Model Strategic Systems

Modern strategy modeling borrows from systems thinking and network science. By modeling the firm as a **network of interconnected strategic choices**, we can identify:

* **Clusters** of tightly coupled activities that create a self-reinforcing advantage
* **Bottlenecks** or weak links where misalignment occurs
* **Redundant or underutilized capabilities** that contribute little to value
* **Leverage points** where small changes can produce outsized effects

This enables a shift from intuitive storytelling ("We are a customer-centric firm") to measurable architecture ("Customer insight capabilities reinforce both product design and loyalty programs, generating a virtuous cycle of retention").

From Static Fit to Dynamic Mapping

A critical innovation in this approach is that the strategic map is **not static**. It evolves over time as internal and external conditions shift. With real-time data flows, firms can **dynamically update** their configuration models to reflect:

* Shifts in consumer preferences
* Technological disruption
* Supply chain instability
* Regulatory or geopolitical change
* Competitor innovation

For instance, a retail firm can model how a new e-commerce platform interacts with logistics investments, labor capabilities, and brand positioning, and identify whether this reconfiguration enhances or erodes strategic fit.

Tools for Strategic Configuration Mapping

The modeling of strategic fit is enhanced through digital tools such as:

* **Bayesian Networks** to capture probabilistic relationships between strategic variables.
* **Causal Loop Diagrams** to visualize reinforcing and balancing feedback loops.
* **Digital Twin Models** to simulate the configuration in real time and under different future scenarios.
* **Strategy Mapping Platforms** (e.g., Strategyzer, iThink, AnyLogic) to visualize and test strategic coherence.

These tools help strategists move from qualitative intuition to quantitative simulation—transforming strategy development from PowerPoint slides to living, testable systems.

Strategic Example

Consider a multinational pharmaceutical company exploring a digital health expansion. Its existing strategic configuration includes robust clinical research, physical distribution channels, and a brand built on efficacy. Introducing AI-powered health monitoring requires new capabilities (data analytics, digital UX), new partners (tech platforms), and a new value proposition (personalized health). Mapping this shift in configuration allows the company to identify capability gaps, resource reallocations, and risks to core brand coherence before any capital is committed.

In short, **mapping the strategic configuration** allows leaders to visualize and interrogate their firm's architecture of advantage. It is the first, indispensable step toward modeling fit, not as a fixed outcome, but as a dynamic system of strategic design and adaptation.

3.4.4.2 Modeling Fit Across Layers of Strategy

Strategic fit is not a one-dimensional alignment; it is a **multilayered construct** that operates across and between different levels of an organization's strategy. To fully leverage advanced analytics and simulation tools, strategists must model how fit is achieved, or lost, across these interconnected layers:

1. **Corporate-Level Fit** – alignment between the firm's overall portfolio and its strategic objectives.
2. **Business-Level Fit** – coherence between a business unit's competitive strategy and its market environment.
3. **Functional-Level Fit** – integration of internal capabilities and activities to support strategic positioning.
4. **Temporal Fit** – consistency of choices over time, enabling adaptive learning and strategic continuity.

Each layer adds complexity and potential fragility to the overall architecture of fit. Modeling these layers requires breaking down traditional silos and adopting a systems view that captures both horizontal and vertical integration.

1. Corporate-Level Fit: Portfolio Coherence

At the corporate level, fit is about synergy and focus. Do the different businesses in the portfolio reinforce a shared strategic intent? Or do they compete for attention and resources without common leverage?

Modeling Tools:

* Correlation matrices to assess shared capabilities (e.g., brand strength, digital platforms)
* Value-at-risk simulations for portfolio-level exposure
* SHAP-informed analysis to isolate the most value-contributing business units

Strategic Example:
A conglomerate may discover through SHAP modeling that its fintech and e-commerce units derive shared benefits from a unified data infrastructure, while its unrelated real estate unit weakens portfolio coherence. The result: a divestment or spin-off to strengthen strategic fit.

2. Business-Level Fit: Positioning and Capability Match

This layer concerns how each business unit aligns its value proposition with both external market demands and internal strengths.

Strategic fit emerges when:

* Market positioning (e.g., cost leadership, differentiation) is supported by the right resource base

* Key activities reinforce rather than contradict one another
* Predictive analytics validate that customer needs are being met efficiently

Modeling Tools:

* VRIO + SHAP scorecards to test alignment of resources and outcomes
* Partial Dependence Plots (PDPs) to visualize marginal gains from strategic levers
* Sensitivity analysis to identify over- or under-invested capabilities

Strategic Example:
A mid-sized airline seeks to position itself as a premium regional carrier. PDP modeling shows that customer satisfaction increases only marginally beyond a certain threshold of in-flight services—implying that strategic fit requires investment in reliability and time efficiency rather than luxury upgrades.

3. Functional-Level Fit: Executional Coherence

Fit at the functional level concerns whether individual departments (marketing, operations, HR, IT) are aligned with each other and with the broader strategy. Misalignment here can sabotage even the most well-designed strategic plan.

Key Questions:

* Are IT investments enabling or slowing innovation?
* Do marketing efforts reinforce brand strategy or fragment it?
* Is talent development aligned with strategic priorities?

Modeling Tools:

- Causal loop diagrams showing functional feedback and conflicts
- Monte Carlo simulation to model execution risks
- NLP-based diagnostics to analyze strategic alignment in internal communications

Strategic Example:
A global telecom firm discovers that its HR policy—focused on short-term performance incentives—undermines its innovation strategy, which requires long-term collaboration. A reconfiguration of incentive systems is modeled and tested through digital twin simulations.

4. Temporal Fit: Consistency Over Time

Strategic fit must also endure—or adapt—over time. Misalignment often arises not from initial poor design but from failure to adjust as the environment shifts.

Indicators of Temporal Fit:

* Consistency in long-term investment patterns
* Stability in brand promise despite product evolution
* Strategic renewal through data-informed sensing

Modeling Tools:

* Longitudinal panel regressions tracking strategic outcome alignment
* Time series analysis of capability-performance lags
* Scenario planning integrated into digital twin updates

Strategic Example:

An electric vehicle startup originally positioned as an innovator sees its market leadership decline over five years. Time-series modeling reveals underinvestment in fast-charging infrastructure—a strategic misfit with its promise of convenience. Realignment strategies are simulated and costed through a Monte Carlo-informed investment dashboard.

Strategic Fit Is a System—Not a Snapshot

Modeling fit across layers moves strategy away from static templates toward **living systems** that can be stress-tested, refined, and evolved. Advanced tools—from SHAP and PDPs to digital twins—allow strategists to interrogate whether the organization is not just well-positioned, but well-integrated across all fronts.

This layered approach helps expose **invisible contradictions**, surface **latent synergies**, and support **cross-level learning**. In a STEM-powered strategy framework, such modeling becomes essential to crafting adaptive, resilient, and high-performing organizations

3.4.4.3 Simulating Strategic Fit Under Pressure

Strategic fit is often celebrated in ideal conditions, when environments are stable, resources are abundant, and goals are clear. But in reality, strategy is forged in turbulence. Whether due to economic shocks, technological disruption, regulatory shifts, or supply chain crises, pressure tests the integrity of strategic fit like nothing else. This is where simulation becomes a powerful diagnostic and planning tool.

To ensure that fit is not merely cosmetic but **structural and resilient**, organizations must simulate how their strategic architecture behaves under stress. The goal is not just to predict success, but to **anticipate failure**, **understand fragility**, and **build adaptation capacity**. By using dynamic simulation tools such as **Monte Carlo modeling**, **Digital Twins**, and **system dynamics**, strategists can observe how fit holds—or fractures—under volatility.

Why Simulate Strategic Fit?

Strategic fit under normal conditions can mask vulnerabilities. Only when the system is perturbed—by uncertainty, complexity, or constraint—do the true dynamics reveal themselves. Simulation exposes these dynamics by modeling cause-and-effect interactions, ripple effects, tipping points, and resilience thresholds.

Key reasons to simulate fit:

* Test how resource alignment behaves under resource cuts or shocks
* Evaluate interdependencies across business units and functions
* Identify the sequence in which misalignments propagate
* Assess which components of the strategy are robust, fragile, or over-dependent

Simulation is not an afterthought; it's a foresight mechanism that allows organizations to move from static fit to **adaptive fit**. Table 3.22 presents Simulation Tools for Strategic Fit.

Table 3.22. Simulation Tools for Strategic Fit

Simulation Tool	Application to Fit Under Pressure
Monte Carlo Simulation	Quantifies how fit-related KPIs (e.g., ROI, NPS, cycle time) respond to thousands of uncertain input scenarios
Digital Twin Simulation	Models real-time operations and simulates strategic changes, revealing system-wide impacts under stress
System Dynamics Modeling	Captures feedback loops (e.g., investment-learning-performance) and simulates strategic behavior over time
Agent-Based Models	Tests how local-level behavior (e.g., customer churn, employee morale) scales into global misfit or emergent alignment
Scenario Stress-Testing	Applies extreme hypothetical conditions (e.g., -30% revenue, +50% material costs) to test strategy resilience

These tools can be integrated into "strategy labs" where hypotheses about fit are tested like experiments, and decision-makers are able to engage with model outcomes interactively.

Case Example: Simulating Strategic Fit in a Retail Transformation

Context:
A global apparel retailer undergoes digital transformation, shifting from physical stores to e-commerce. Leadership assumes existing strengths (logistics, branding, pricing) will transfer seamlessly.

Simulation Setup:

* **Digital Twin:** A replica of the firm's supply chain, store operations, marketing mix, and customer engagement systems.
* **Monte Carlo Inputs:** Variability in online customer acquisition cost, digital returns, and delivery times.
* **Scenarios:** Sudden spike in online demand, drop in brick-and-mortar footfall, labor shortage in fulfillment centers.

Findings:

* Functional-level misfit: Logistics optimized for store distribution created bottlenecks in last-mile delivery.
* Temporal misfit: HR policies on performance evaluation lagged behind new team-based digital workflows.
* Business-level misfit: Brand value eroded in some markets due to inconsistencies in digital experience quality.

Strategic Action:
Leadership redesigned fulfillment centers, retrained frontline managers in agile teams, and rewrote service-level agreements with IT, restoring alignment and avoiding a $120M projected revenue loss.

Strategic Fragility Signals Uncovered Through Simulation

Simulation doesn't just predict outcomes, it reveals warning signs:

* **Overreliance on Single Capabilities**: A Fit that depends too heavily on one asset (e.g., a superstar product or team) becomes brittle.
* **Feedback Time Lags**: When performance data is delayed, adaptation is slow, leading to strategic drift.
* **Hidden Constraints**: Functional capacity (e.g., IT bandwidth or legal compliance cycles) may silently sabotage otherwise aligned strategy.

Simulating these pressure points lets leaders proactively redesign for **strategic slack**, **redundancy**, and **buffer zones** that protect in extreme conditions.

Embedding Simulation into Strategic Governance

Organizations can institutionalize simulation by embedding it into:

* **Annual Strategic Reviews**: Simulate external shock scenarios alongside traditional financial planning.
* **Pre-Investment Evaluation**: Use digital twins to test the strategic fit of new ventures or acquisitions.
* **Board Risk Committees**: Present Monte Carlo risk ranges and fit sensitivity analyses.
* **Scenario Planning Retreats**: Use role-playing and simulations to rehearse cross-functional responses to misalignment.

Simulation as Strategic Insurance

Strategic fit is not a guarantee; it is a **hypothesis** about how different parts of an organization reinforce one another to achieve performance. Simulation allows leaders to test that hypothesis under duress.

In the STEM-enhanced strategy toolkit, simulation becomes a **strategic insurance policy**—allowing firms not only to withstand disruption, but to adapt and emerge stronger.

3.4.4.4 Using Fit Models to Identify Points of Leverage and Innovation

Strategic fit is often thought of as an optimization problem: how to align internal capabilities with external opportunities. But fit is also a platform for **strategic innovation**. Once organizations understand where their resources, processes, and structures interact with environmental demands, they can move from alignment to **amplification,** uncovering leverage points where small changes generate outsized impacts. Fit models, especially when powered by analytics and simulations, can reveal these hidden zones of strategic leverage.

Rather than viewing fit as a fixed blueprint, leaders can treat it as a dynamic map, a living model that surfaces not just risks of misalignment but also areas of **creative potential**. These leverage points can drive innovation, resource redeployment, business model evolution, and even industry disruption.

From Fit to Leverage: The Logic of Amplification

Strategic leverage occurs when a relatively minor input or intervention leads to disproportionately large results. Fit models make these leverage points visible by highlighting:

* **Resource Intersections** – Where capabilities (e.g., data analytics + logistics) combine to create synergies

* **Capability Bottlenecks** – Where small upgrades (e.g., digital tools, AI-enabled decisions) relieve major constraints
* **Cultural Convergence Zones** – Where team values align with customer expectations, enabling rapid adoption of innovations
* **Temporal Advantage** – Where firms can act before competitors due to already-aligned internal structures

Fit models don't just ask: *Are we aligned?* They also ask: *Where is our system most responsive to innovation?*

These models help organizations identify leverage points—areas where small, strategic adjustments can yield outsized impact. Rather than treating misalignment as failure, fit models frame it as an opportunity to innovate around friction points.

To uncover these leverage points, strategy teams can apply a range of analytical methods. For example, **sensitivity analysis using partial dependence plots (PDPs)** quantifies how changes in key inputs affect strategic KPIs, helping prioritize high-impact variables. **SHAP-based contribution mapping** goes deeper, revealing which internal features, such as leadership diversity or digital maturity, most influence success predictions.

When comparing current versus transformed configurations, **scenario delta analysis** highlights where small tweaks unlock significant gains. **System dynamics feedback loops** expose self-reinforcing cycles (e.g., R&D → learning curve → product quality → market share) that can be accelerated. Finally, **fit disruption mapping** reframes misalignment as a signal, pointing to gaps or contradictions that could become sources of innovation. These methods are summarized below:

Table 3.23. Methods to Identify Leverage Points in Fit Models

Method	Strategic Application
Sensitivity Analysis (via PDPs)	Quantifies how changes in key inputs affect strategic KPIs—highlighting high-impact variables
SHAP-Based Contribution Mapping	Identifies which internal features (e.g., leadership diversity, digital maturity) most affect success predictions
Scenario Delta Analysis	Compares performance under current vs. transformed configurations, showing where small tweaks unlock big gains
System Dynamics Feedback Loops	Reveals self-reinforcing cycles (e.g., R&D → learning curve → product quality → market share) that can be accelerated
Fit Disruption Mapping	Looks at misalignment not as a failure—but as an opportunity to innovate around gaps or contradictions

These tools transform fit models from evaluative scorecards into **strategic innovation engines**.

Strategic Example: Leverage Point Discovery in a Financial Services Firm

Context:
A regional bank conducts a fit assessment to evaluate its ability to expand into mobile-first digital banking. The model includes talent readiness, cybersecurity infrastructure, app development cycles, customer preferences, and regulatory constraints.

Insights from Fit Model:

* High digital adoption among customers, but low mobile engagement due to app friction
* Talented dev team with limited agile training
* Robust data security but fragmented customer analytics

Leverage Points Identified:

1. **Agile Training Investment** – Boosting team capacity by reskilling rather than hiring
2. **Unified Analytics Platform** – Enabling hyper-personalized mobile UX
3. **Micro-Innovation Teams** – Creating rapid feedback loops with early adopter customers

Outcome:
A 90-day sprint produced a new feature set that doubled mobile app engagement, improved NPS scores by 15%, and created a reusable innovation model for other service lines.

Using Fit Models to Innovate at the Edge

Leverage points often reside not in the core of operations, but at the **edges**—customer touchpoints, fringe capabilities, and emerging partnerships. Fit models can identify these edge zones by:

* Mapping customer journey elements that are only partially supported by internal processes
* Visualizing interdependencies across business units where silos obscure opportunity
* Highlighting underutilized assets that could unlock new markets (e.g., data as a service, supply chain as a platform)

These edge innovations are often low-cost and low-risk—yet open high-value strategic pathways.

Innovation through Tension: Productive Misfits

Ironically, some misalignments uncovered by fit models are not problems to fix, but **creative tensions** to explore. When legacy systems clash with new technologies, or when team mindsets diverge from industry norms, the friction can spark strategic invention.

Example:
A hospitality company discovers a fit gap between its standardized global branding and rising demand for localized, community-rooted experiences. Rather than enforcing conformity, it embraces the misfit—launching a new "local edition" line of properties co-designed with community stakeholders. The result: both brand stretch and customer affinity increase.

Embedding Leverage Discovery into Strategic Processes

To make this repeatable, organizations can embed leverage identification into their planning cycles:

* **Quarterly Strategic Fit Reviews** – Include an explicit step for "fit-to-leverage" mapping
* **Innovation Charters** – Task cross-functional teams to test one identified leverage point per quarter
* **Executive Dashboards** – Integrate real-time analytics on system responsiveness to highlight emerging opportunities
* **Crisis Playbooks** – Use fit models to identify fast-acting pivots when disruptions strike

From Fit Assessment to Strategic Innovation Architecture

Strategic fit, when seen dynamically, is not a constraint—it's a **springboard**. Fit models become more than measurement tools; they become **discovery platforms** for uncovering new value, designing novel configurations, and scaling innovation through evidence-based leverage.

As we complete this section on strategic fit, we transition to the final segment of Chapter 3.4—where we synthesize these approaches and explore how strategy, analytics, and modeling come together to power a new kind of strategist: one who doesn't just align—but **architects strategic evolution** through intelligent systems.

Classic strategy frameworks like VRIO, SWOT, environmental scanning, and strategic fit have served as foundational pillars in the discipline of strategic management. They provide timeless insight into how organizations compete, adapt, and evolve. Yet in an age of algorithmic complexity, data abundance, and global uncertainty, these tools—while still relevant—require reinterpretation. They must evolve not just in form, but in **function**.

What this chapter has shown is that **strategic analytics doesn't replace classic thinking—it reactivates it.** Tools like SHAP, predictive modeling, NLP, and scenario simulation don't discard foundational models; they breathe new life into them, transforming static frameworks into dynamic engines of insight and action.

Reinvented frameworks do more than diagnose; they **simulate, experiment, explain, and iterate**:

* **VRIO becomes a living value engine** when paired with interpretability tools like SHAP and scorecards that track the real-time contribution of strategic resources.
* **SWOT transforms from a brainstorming template** to a predictive diagnostic that learns from data to surface latent strengths and emerging threats.
* **Environmental scanning evolves into strategic sensemaking,** where machine learning and NLP uncover pattern signals in oceans of noise.
* **Strategic fit becomes an adaptive modeling exercise,** one that doesn't just evaluate alignment but reveals points of leverage, innovation, and resilience.

Across all these reinventions, the core message is clear: **analytics is not an add-on—it is an amplifier.** It strengthens the logic of classic tools, enhances their accuracy, expands their relevance, and deepens their impact.

But adopting these analytics-enhanced frameworks requires more than technical upgrades. It calls for a **shift in mindset**: strategists must see themselves not only as thinkers and planners, but as **designers of intelligent systems**—able to test, model, and optimize strategic decisions in real time.

This is the essence of STEM-based strategy: the fusion of theory, data, and experimentation to support **decisions that are not only smart, but explainable, agile, and future-ready**.

As we now move into Section 3.5, we address a critical responsibility that comes with this power: the **ethical use of advanced tools in strategic modeling**. In a world where AI can shape decisions at scale, we must confront not only what we *can* model, but what we *should* model, and how we ensure accountability, transparency, and long-term responsibility in the decisions we help shape.

3.5 Ethical Intelligence in Strategic Modeling

As strategy becomes increasingly driven by data, powered by algorithms, and enacted through intelligent systems, a fundamental shift is underway, not just in how decisions are made, but in who (or what) makes them, how they are justified, and whom they affect. This evolution presents a double-edged sword: **unprecedented decision power and escalating ethical responsibility**.

While the strategic analytics toolkit equips leaders with tools of precision, prediction, and simulation, it also introduces new vulnerabilities, **bias in models, opacity in reasoning, overreliance on automation, unintended externalities**, and even algorithmic harm. A model that optimizes for profit may simultaneously exacerbate inequality. A simulation that increases efficiency may unknowingly undermine sustainability. A predictive system that identifies strategic winners may marginalize stakeholders or overlook societal impact.

In this environment, **ethical intelligence is not optional; it is strategic.** Just as leaders seek competitive intelligence to navigate markets, they must now cultivate ethical intelligence to navigate the algorithmic frontier. This means going beyond compliance checklists and toward **embedding moral reasoning, fairness principles, and stakeholder foresight directly into the modeling process.**

This section explores the **moral architecture of STEM-based strategic modeling.** We examine how organizations can:

* Detect and mitigate bias in AI-driven strategy tools
* Build transparency into complex models to foster trust and accountability
* Weigh long-term consequences alongside short-term optimization
* Establish governance systems that align analytics with organizational values

Just as the strategist must master data science, they must also become a **steward of ethical foresight,** ensuring that strategy serves not only organizational goals, but also the broader ecosystem in which those strategies operate.

The future of intelligent strategy will not belong solely to the most advanced models, but to the **most responsible, transparent, and human-centered ones**. In the following subsections, we outline the core principles, risks, and frameworks that will guide this transformation.

3.5.1 Bias, Transparency, and Responsibility in Model Use

Strategic models do not exist in a vacuum. They are shaped by the data they consume, the assumptions they encode, and the decisions they ultimately inform. When used wisely, they enable firms to optimize growth, predict disruptions, and allocate resources with scientific rigor. But when used carelessly, or worse, blindly, they can embed systemic bias, erode trust, and produce outcomes that are efficient but ethically flawed.

At the core of ethical intelligence in strategic modeling lies a simple yet powerful idea: **every model is a moral actor.** It influences what gets prioritized, who benefits, and who bears the cost. Therefore, every strategist must ask: **Is this model fair? Is it explainable? And are we accountable for its decisions?**

Bias in Strategic Models: The Silent Risk

Bias in modeling can enter at multiple stages: data collection, feature selection, algorithm design, or interpretation. For instance:

* A model trained on historical hiring data may reinforce gender or racial biases if past decisions were biased.
* A predictive model for loan approvals may inadvertently favor regions or demographics that already have privileged access to capital.
* An innovation scoring algorithm might overlook ideas from emerging markets if the training data skews toward established economies.

In strategic contexts, such biases don't just lead to unfairness; they distort foresight. If the model overlooks potential in underserved segments or misrepresents market risks due to biased inputs, then even the best optimization will drive a poor strategy.

Ethical intelligence requires continuous bias auditing, not only checking for statistical imbalance, but also questioning the framing of the problem: *What outcomes are we optimizing? Who defines success? And what isn't being measured?*

Transparency: Making Strategy Models Understandable

Transparency transforms strategic models from opaque machines into communicative partners. It ensures that decision-makers understand **why** a model recommends a specific course of action—not just what it recommends.

SHAP values (as discussed in Section 3.3.3) play a critical role here, offering localized, interpretable insights. But transparency also includes:

* **Model documentation** that clearly states assumptions, data sources, and performance limitations.
* **Visual explainability tools** like partial dependence plots and scenario trees enable board-level conversations.
* **Stakeholder briefings** that translate technical outputs into human narratives, with space for debate and feedback.

Transparency isn't just about compliance; it's a **strategic asset**. It builds trust among executives, confidence among regulators, and engagement among employees and customers.

Responsibility: Who Owns the Decision?

As models become more autonomous and embedded in strategic workflows, the question of responsibility becomes urgent: **When a model leads to a bad outcome, who is accountable, the analyst, the algorithm, the executive, or the institution?**

Ethical modeling practices call for:

* **Human-in-the-loop systems** where critical decisions are reviewed, not rubber-stamped.

* **Clear accountability chains** that define roles and escalation protocols.
* **Ethical review boards** or cross-functional governance committees that audit high-stakes models before deployment.

Responsibility also means **anticipating harm**. What happens if a demand forecasting model causes overproduction and environmental waste? What if an expansion strategy powered by AI leads to job loss in vulnerable regions? Strategic ethics demands we look not only at first-order gains, but second- and third-order consequences.

Bias unchecked is a strategy misaligned. Transparency unpracticed is trust forfeited. Responsibility ignored is risk unbounded. In the STEM era, **ethical intelligence is not a constraint; it is the compass that guides analytics toward human-centered strategic excellence.**

3.5.2 Long-Term Consequences of Automated Decisions

In a world increasingly driven by algorithms, the immediacy and precision of automated strategic decisions can be alluring. Models forecast demand, optimize pricing, recommend expansions, allocate talent, and even shape corporate narratives. But in this pursuit of speed and efficiency, a crucial question often goes unasked: **What happens when short-term optimization creates long-term distortion?**

Strategic decisions do not operate in isolation; they ripple through time. And when they are guided by automated models that lack contextual foresight or ethical grounding, these ripples can evolve into unintended waves: cultural erosion, ecosystem degradation, stakeholder backlash, and even systemic collapse.

In this section, we explore the **long-term risks** that arise when algorithms replace deliberation, when efficiency eclipses empathy, and when predictive logic becomes the primary lens through which organizations interpret their world.

1. Strategic Lock-In: When Models Freeze Innovation

Automation often reinforces the status quo. A model trained on past success may repeatedly recommend the same high-performing products, markets, or customer profiles. While this may yield short-term gains, it can also produce **strategic lock-in,** where the organization becomes trapped in yesterday's logic.

Example: A retail algorithm consistently directs advertising budget to existing high-margin products. Over time, investment in experimental products drops, eroding the firm's innovation pipeline. Five years later, the company struggles to compete in emerging categories it failed to incubate.

Lesson: Models should not just optimize the present—they must leave space for strategic emergence and serendipity. Otherwise, they risk stifling innovation in pursuit of certainty.

2. Structural Drift: When Automation Distances Leaders from Reality

The more strategy is delegated to models, the greater the distance between leadership and lived reality. When dashboards replace direct engagement, and recommendation engines shape every initiative, executives may lose sight of cultural nuances, frontline constraints, and qualitative shifts.

Example: An airline uses demand prediction algorithms to fine-tune pricing and staffing. But over time, it stops gathering direct customer feedback. The algorithm, blind to passenger sentiment, fails to detect a growing frustration with delayed flights—until it explodes on social media and triggers reputational damage.

Lesson: Strategic automation must be coupled with **strategic sensing**—regular human inquiry, feedback loops, and qualitative exploration that ground data in reality.

3. Ethical Drift: When Efficiency Outpaces Values

Automated models often optimize for a measurable outcome: profitability, efficiency, growth. But these outcomes may conflict with deeper organizational values such as fairness, sustainability, or inclusion. When not checked, models can lead to **ethical drift**—where behavior gradually departs from stated principles.

Example: A talent retention model predicts which employees are likely to leave and recommends preemptive incentives. However, the model consistently underrepresents employees from marginalized groups due to biased data. Over time, disparities in retention benefits widen, undermining the company's DEI commitments.

Lesson: Optimization must be bounded by **ethical constraints**. Strategic modeling should not only ask "what works" but "what's right."

4. Ecosystem Distortion: When Strategic Models Reshape Markets

Automated strategies do not just affect the company—they influence competitors, customers, partners, and even regulatory frameworks. At scale, strategic automation can produce **ecosystem-level effects**: pricing races to the bottom, homogenization of offerings, or collapse of fragile supply chains.

Example: An e-commerce platform uses dynamic pricing algorithms that constantly undercut competitors. Over time, small vendors exit the market, reducing diversity and resilience in the ecosystem. When a supply chain disruption hits, the lack of redundancy triggers cascading failures.

Lesson: Responsible strategy modeling requires **second-order thinking,** anticipating how automated actions reshape the broader system and designing for resilience, not dominance alone.

5. Reputational Fragility: When Automation Makes Decisions Without Empathy

Strategic decisions shaped by automated systems can sometimes produce outcomes that are tone-deaf, insensitive, or even offensive, especially when they fail to consider social, cultural, or emotional dimensions.

Example: A global hotel chain uses an AI system to optimize room rates during a natural disaster, spiking prices during evacuations. Though the system is technically correct, public backlash is swift and severe, damaging the brand's trust and goodwill.

Lesson: In moments of crisis or ambiguity, **human judgment must override machine logic**. Strategic models should include "break-glass protocols" for ethical overrides and empathetic discretion.

The Imperative of Strategic Foresight

The long-term consequences of automated decisions cannot be an afterthought. Just as we simulate financial risk and market volatility, we must simulate ethical erosion, reputational fragility, innovation decay, and ecosystem fragility. **Foresight becomes a moral responsibility.**

This means:

* **Scenario planning** that includes social, cultural, and ethical variables, not just economic ones.
* **Cross-disciplinary review** involving ethicists, anthropologists, and systems thinkers, not only data scientists.
* **Strategic memory systems** that track past model decisions and consequences for institutional learning.

As we accelerate into the algorithmic age, the essence of strategy remains unchanged: to navigate complexity with vision, agility, and wisdom. What changes are in the toolkit and the stakes? Automated models can amplify excellence or accelerate failure. The difference lies in whether we wield them with ethical intelligence and long-term perspective.

3.5.3 Strategic Accountability with Advanced Tools

In a world where algorithms shape high-stakes decisions, from hiring to pricing, market entry to resource allocation, **strategic accountability** becomes not just a regulatory checkbox, but a defining feature of responsible leadership. As advanced tools such as XGBoost, SHAP,

Digital Twins, and Bayesian Optimization become embedded in strategic processes, organizations must answer a critical question: **Who is accountable for the consequences of machine-assisted decisions?**

Strategic accountability is about creating structures that ensure **transparency, responsibility, traceability, and ethical governance** throughout the analytical decision-making process. It means designing decision systems where leaders can not only trust the tools but also defend their use, understand their limits, and intervene when needed.

This section unpacks how accountability must evolve alongside intelligent systems, and how advanced tools can be reconfigured not only for performance but also for principled use.

1. Transparent Decision Logic: Models Must Explain Themselves

Accountability begins with transparency. Leaders must be able to **understand and explain** the logic behind strategic recommendations, particularly when those decisions affect stakeholders, markets, or society at large.

Advanced interpretability tools like **SHAP** provide localized and global explanations for predictions, allowing strategists to defend decisions at the board level, in regulatory audits, or in public discourse.

Example: A financial services firm is investigated for bias in loan approvals. Instead of vague defenses, it presents a SHAP-based audit trail demonstrating how credit decisions were influenced by legitimate financial indicators rather than protected attributes. This transparency averts regulatory penalties and enhances public trust.

Accountability Practice: Implement explainability dashboards that accompany every major model output with justifications, SHAP plots, and risk flags.

2. Ethical Failsafes: Designing Human-in-the-Loop Protocols

Advanced models should not run on autopilot. Strategic accountability demands the presence of **human-in-the-loop mechanisms,** points where human oversight is required before high-impact decisions are executed.

Example: A multinational firm uses a digital twin to simulate facility closures. Before any action is taken, a cross-functional ethics board must review the outcomes to ensure compliance with labor agreements and community impact guidelines. Automation proposes; humans ratify.

Accountability Practice: Define escalation thresholds and ethical intervention points in the decision pipeline. Ensure that humans review not just outputs but assumptions and trade-offs encoded in the model.

3. Traceable Strategy: Keeping a Strategic Memory of Model Decisions

Many organizations lack a system for tracking how and why specific strategic decisions were made using models. This erodes accountability over time, especially when leadership changes or when decisions have delayed consequences.

Strategic accountability requires the creation of **strategic memory systems,** records of which model was used, under what parameters, with what data, and who signed off.

Example: A healthcare organization facing a malpractice lawsuit is able to show, through its strategic memory archive, that a resource allocation model was appropriately reviewed, approved, and monitored, demonstrating due diligence in uncertain conditions.

Accountability Practice: Maintain model governance logs that record version history, assumptions, performance metrics, and decision contexts.

4. Inclusive Model Governance: Expanding Stakeholder Voice

Strategic decisions shaped by advanced tools affect many stakeholders. Accountability requires that those stakeholders, especially underrepresented or vulnerable groups, have a voice in how models are designed, trained, and evaluated.

Example: A public transportation authority introduces a predictive maintenance model. Before deployment, it conducts community forums with riders, drivers, and disability advocates to identify potential blind spots and unintended consequences. Their input directly influences model constraints and performance metrics.

Accountability Practice: Institute participatory design processes where model objectives and constraints are co-developed with internal and external stakeholders.

5. Auditable Strategy: Preparing for External Scrutiny

With growing regulatory attention on AI and automated decision-making (e.g., the EU AI Act, U.S. Algorithmic Accountability Act), organizations must design their strategic modeling systems to withstand **external audits**.

This means more than technical compliance; it requires **proactive auditability**: clear documentation, explainable models, ethical justifications, and outcome monitoring.

Example: A large logistics firm builds a Monte Carlo simulation to assess the risk of cross-border disruptions. Anticipating scrutiny from international trade regulators, it includes a built-in audit layer that details every assumption, probability distribution, and risk parameter used. When disruption occurs, it can show that it was planned for it—and why.

Accountability Practice: Conduct regular strategic model audits involving legal, compliance, and external advisors—not just data scientists.

Embedding Accountability into the Strategic Fabric

Strategic accountability is not an afterthought; it must be baked into the very design of modern decision systems. This involves:

* **Policy frameworks** for how, when, and why advanced tools are used
* **Cross-disciplinary governance** to ensure balance between performance and principles
* **Cultural change** where data fluency is matched by ethical fluency

Advanced tools are not infallible, but they can be made **answerable**. When paired with responsible governance, they offer not just power, but **legitimacy**. They signal to employees, investors, regulators, and the public that strategy is not only smart but also just.

As strategic decision-making increasingly relies on intelligent systems, the question is no longer whether we *can* use advanced tools, but *how* we use them responsibly. In this section, we explored the ethical imperatives that arise when predictive models, simulations, and optimization engines become central to strategic choices.

We saw that:

* **Bias is not a side effect; it is a strategic risk.** Unexamined models can reinforce structural inequalities or produce misleading results. Ethical strategy demands proactive bias detection and correction.
* **Transparency is not optional; it is foundational.** Tools like SHAP and PDPs transform opaque models into accountable systems, enabling leaders to understand and justify decisions in a fast-moving, high-stakes world.
* **Accountability must evolve, just as strategy does.** From maintaining audit trails to implementing human-in-the-loop protocols, organizations must build governance frameworks that ensure decisions made by machines remain traceable, defensible, and just.

Navigating Strategic Frontiers with Ethical Intelligence

Ethical intelligence in strategic modeling is not about rejecting technology. It is about mastering it, embedding judgment, empathy, and responsibility into the digital architectures of strategy. When ethics and analytics work in tandem, strategy becomes not only more effective but also more credible, inclusive, and future-proof.

In the next chapter, we move from analytical modeling to the broader question of **innovation as a STEM process,** where strategic creativity meets systematic experimentation. The future of competitive advantage begins there.

In an age where data-driven models increasingly guide strategic decisions, the most powerful algorithms are not necessarily the most ethical or the most trustworthy. The true frontier of STEM-enabled strategy is not just predictive accuracy or optimization speed, but responsible intelligence. Ethical intelligence is not a soft constraint; it is a strategic necessity. Firms that fail to embed fairness, transparency, and accountability into their modeling practices risk more than poor decisions; they risk violating trust, undermining legitimacy, and losing their social license to operate.

Embedding ethical intelligence into strategic modeling requires a deliberate integration of value-based thinking, technical safeguards, and human judgment. This means proactively confronting the ethical trade-offs that arise when modeling future outcomes, especially when those outcomes affect customers, employees, markets, and the environment.

The table below outlines common ethical challenges, the strategic risks they pose, and responsible practices and tools that mitigate them. It serves as both a cautionary checklist and a forward-looking blueprint for ethical strategy design:

Table 3.24. Common Ethical Challenges

Ethical Challenge	Strategic Risk	Responsible Practice	Supporting Tool/Approach
Algorithmic Bias	Discrimination, reputational harm	Bias audits, diverse data sources	SHAP, fairness constraints in modeling
Black-box Models	Lack of trust, regulatory scrutiny	Model interpretability and explainability	SHAP, PDPs
Over-automation	Displacement of human judgment	Human-in-the-loop decision systems	Augmented analytics
Data Privacy Violations	Legal penalties, customer alienation	Data minimization, differential privacy techniques	Secure modeling pipelines
Irresponsible Risk Modeling	Overconfidence in simulations, ethical blind spots	Stress testing, transparency about assumptions	Monte Carlo Simulation, Digital Twins
Lack of Accountability	Decision denial, governance failure	Audit trails, ethical guidelines, and role clarity	Strategic model governance frameworks

These ethical touchpoints are not afterthoughts; they shape the way strategy is framed, interpreted, and implemented. As the role of analytics expands across domains, it is essential to design not just smart models, but fair ones. Not just strategic systems, but accountable systems.

Ultimately, ethical intelligence is what transforms powerful analytics into sustainable strategy. It aligns technological capabilities with human values and reinforces the legitimacy of strategic decision-making in a data-saturated world. As we now turn to the final section of this chapter, we explore how these advanced tools and principles can be combined into a coherent architecture for STEM-based strategy.

3.6. Chapter 3 Wrap-Up: Strategic Analytics and Decision Intelligence

As strategic challenges grow more complex and time-sensitive, the need for **intelligent strategy** has never been more urgent. This chapter has taken you deep into the **core engine of modern strategic thinking**: analytics. Not just as a reporting function, but as a generative, anticipatory, and ethically grounded force for navigating complexity.

We started by examining the roots of **strategic failure**, revealing how misjudged assumptions, poor feedback loops, and data illiteracy can silently undermine even the boldest vision. To counter these risks, we introduced the idea of **intelligence as a strategic capability,** not just for top leadership, but embedded across the organization.

Section 3.2 presented the **Analytics Continuum**, moving from raw data to actionable strategic foresight. You explored how descriptive, diagnostic, predictive, and prescriptive analytics each add value in different phases of the strategy process, and why leaders must be fluent in navigating this continuum, not just delegating it.

In Section 3.3, we introduced the **Strategic Analytics Toolkit,** a suite of STEM-powered tools including **XGBoost, Bayesian Optimization, SHAP, PDPs, Monte Carlo Simulation, and Digital Twin Modeling**. You saw how each tool helps leaders confront different dimensions of strategic uncertainty, whether modeling competitive scenarios, explaining algorithmic predictions, or running simulation-based stress tests. A key theme throughout: tools must not only predict well, they must also explain well, adapt fast, and align ethically.

Section 3.4 reimagined classic strategy frameworks, like VRIO, SWOT, environmental scanning, and strategic fit, through the lens of intelligent modeling. We demonstrated how models like **SHAP** can unlock hidden sources of value, how NLP can enhance sensemaking, and how predictive layering can transform static frameworks into dynamic learning systems.

Finally, Section 3.5 reminded us that **strategy without ethics is fragility in disguise**. Algorithmic decisions impact real lives, and without deliberate attention to **bias, transparency, and governance**, even the most accurate model can lead to unintended harm.

Ethical intelligence, like analytical intelligence, is a muscle that must be built and embedded into both process and culture.

As you move into the next chapters, remember: the most powerful strategist is not the one who knows the most models, but the one who can select the right model, adapt it to the moment, explain its implications, and account for its risks. That's not just a smart strategy. That's an intelligent strategy.

3.6.1. Discussion Questions

1. Why do strategic decisions often fail even when supported by data? What is typically missing?
2. In what ways does SHAP improve the interpretability of complex machine learning models in a strategic context?
3. How does the analytics continuum (from descriptive to prescriptive) help align analytical efforts with real-time strategic needs?
4. What distinguishes Bayesian Optimization from traditional forecasting or sensitivity analysis in strategy modeling?
5. When might a digital twin be a better strategic tool than a static business model or SWOT analysis? Why?
6. What ethical dilemmas might arise from using AI-driven tools in strategy formulation? How can these be mitigated?
7. How can Monte Carlo simulations support strategic resilience planning, especially in volatile industries?
8. Why is tool interpretability essential for strategic accountability at the leadership level?
9. How does reengineering classic strategy frameworks (like VRIO or SWOT) with machine learning improve their usefulness in practice?
10. How can companies embed both analytical and ethical intelligence into day-to-day decision processes, not just long-term strategic planning?

3.6.2. Micro-Case: Zara's Strategic Sprint

From Gut Feeling to Explainable Modeling – A Strategy Pivot at Zara

Zara, now promoted to **Director of Strategic Analytics**, is leading a high-impact decision at her firm: whether to pivot into the **circular economy fashion market** using a new AI-driven supply chain forecasting model. The decision involves selecting one of three major product lines to pilot a "design-for-reuse" initiative. Time is tight, and the board expects a recommendation within 21 days, with full transparency on why.

Zara's team has built a predictive model using **XGBoost** to simulate profitability, adoption, and environmental impact across product lines. The model performs well, but the board is concerned about the *why* behind the outputs.

To bridge this gap, Zara uses **SHAP values** to explain the drivers behind each product line's predicted performance. The results reveal that environmental impact metrics, originally assumed to be secondary, are in fact highly predictive of customer engagement in the premium line. This contradicts earlier assumptions based on traditional SWOT analysis.

Zara then integrates **Partial Dependence Plots (PDPs)** to visualize how small shifts in design cost and return logistics affect forecasted success. She also simulates the risks using **Monte Carlo modeling**, mapping best-case and worst-case scenarios under different policy regulations, using sustainability tax data.

Before presenting her final recommendation, she creates a **Digital Twin** prototype of the redesigned supply chain for the selected product line. This virtual model is used to simulate production, returns, emissions, and delivery logistics in real time under different global operating conditions.

Her pitch to the board is no longer based on instinct; it's a transparent, testable, and explainable case for launching the circular model with the premium line first.

Zara's sprint illustrates how analytics not only supports strategy, but also transforms it. She didn't just use tools; she chose the right ones, interpreted their outputs, accounted for risks, and translated complexity into actionable foresight.

Discussion Questions

1. How did SHAP and PDP enhance transparency in Zara's model?
2. What role did the Digital Twin play in stakeholder alignment?
3. How could Zara institutionalize this decision-making model across other functions?
4. How would you balance stakeholder concerns with model precision in a case like this?

3.6.3. Interactive Exercise: Strategy Under Uncertainty — The Analytics Playbook Challenge

Scenario:

You've been hired as a **Strategic Intelligence Lead** for **EcoFabric**, a fast-scaling global apparel company exploring market entry into three regions: **Brazil**, **South Korea**, and **Sweden**. Each market has differing levels of climate policy enforcement, consumer sustainability preferences, and economic volatility.

Your role is to help the executive team decide which market to enter using **analytics-driven strategy tools** introduced in this chapter. You will be working with the following datasets:

* National environmental regulations (numeric + textual)
* Real-time consumer sentiment data from social media
* Economic indicators (inflation, employment, retail sentiment)
* Internal simulations of supply chain CO_2 output

Your Task:

Using at least **three tools** from the Strategic Analytics Toolkit in Section 3.3 (e.g., SHAP, PDP, Bayesian Optimization, Monte Carlo Simulation, Digital Twin), create a **one-page analytics-backed strategic entry recommendation**.

Your deliverable should include:

1. **Risk Modeling** – Identify 2–3 key uncertainties (e.g., policy shifts, consumer backlash) and show how your tool(s) simulate outcomes.
2. **Interpretability Strategy** – Use SHAP or PDP to explain what variables drive key results in your model.
3. **Decision Justification** – Clearly state your recommended market and why, supported by analytics outputs.
4. **Ethical Lens** – Flag any fairness, transparency, or bias issues you encountered in model assumptions or data.

This exercise simulates how future leaders move from **complex data to confident, explainable decisions,** while upholding ethical accountability.

3.6.4. Recommended HBR Case Studies

❖ The Digital Transformation of Freeport McMoRan: The Strategic Use of Agile, AI, and Data Analytics

- **Overview**: Authored by William E. Youngdahl and Kannan Ramaswamy and published in November 2023 by the Thunderbird School of Global Management, this 11-page case explores how Freeport McMoRan, a global leader in copper mining, embraced digital transformation to stay competitive in a rapidly evolving industry. As copper demand surged due to the global shift toward renewable energy, Freeport faced the challenge of increasing efficiency without opening new mines, which can take decades. The company responded by investing in artificial intelligence, data analytics, and agile project management to optimize copper extraction from existing operations. The transformation began at its Baghdad, Arizona, site and expanded across U.S. operations under the America Concentrator program. However, as competitors like BHP and Rio Tinto adopted similar technologies, Freeport had to consider how to maintain a sustainable competitive advantage, especially through its human capital and organizational agility.

- **Citation**: Youngdahl, W. E., & Ramaswamy, K. (2023). *The Digital Transformation of Freeport McMoRan: The Strategic Use of Agile, AI, and Data Analytics* (Case No. TB0698). Thunderbird School of Global Management.

❖ Giving Data A Voice: The Rise of TalkingData

- **Overview**: Authored by Siew-Kien Sia, Jun Jie Yang, and Wee-Kiat Lim, and published on May 27, 2019, this case explores the rise of TalkingData, a Chinese data analytics startup, from its founding in 2012 through 2017. It highlights how the company built its capabilities in big data analytics and artificial intelligence (AI) to become one of China's leading data analytics providers. The case focuses on the strategic decisions, technological infrastructure, and market positioning that enabled TalkingData to thrive in a rapidly evolving digital landscape. It also examines the challenges the company faced as it scaled its operations and sought to expand its influence in both established and emerging markets. This case is particularly useful in courses on digital innovation, entrepreneurship, and business analytics, as it illustrates how firms can leverage data as a strategic asset.
- **Citation**: ia, S.-K., Yang, J. J., & Lim, W.-K. (2019). *Giving Data A Voice: The Rise of TalkingData* (Case No. NTU245-PDF-ENG). ABCC at Nanyang Technological University.

❖ HSBC: Leveraging Data Analytics and AI to Enhance Customer Life Cycle Management

- **Overview**: Published on February 8, 2024, and authored by Joon Nak Choi, Ohchan Kwon, and Joseph Santana Fernandez, this 27-page case focuses on how HSBC Hong Kong is using data analytics and generative AI to address a strategic challenge: retaining and attracting millennial customers in a highly competitive digital banking environment. Although HSBC holds a dominant market position in Hong Kong, it faces a long-term risk as younger customers increasingly prefer digital-first banking experiences offered by online-only competitors. The case centers on two executives, Jiahao Teo, Managing Director and Global Head of Data, Analytics & CRM, and Amy Hui, Head of Personal Banking and Customer Lifecycle Management, who must evaluate how to best deploy HSBC's advanced data infrastructure and AI capabilities to support a new Customer Life Cycle Management (CLCM) strategy. The case explores the strategic, technological, and organizational implications of using AI and analytics to personalize customer engagement, improve retention, and future-proof the bank's market leadership.
- **Citation**: Choi, J. N., Kwon, O., & Santana Fernandez, J. (2024). *HSBC: Leveraging Data Analytics and AI to Enhance Customer Life Cycle*

Management (Case No. ST138). The Hong Kong University of Science and Technology.

3.6.5. Strategy-in-Film: Learn by Watching

In modern strategy, data is no longer optional; it is foundational. The films below portray how organizations leverage information, interpret signals, and make critical decisions in high-stakes environments. As you engage with strategic analytics and decision intelligence in this chapter, use these narratives to reflect on the power and pitfalls of data-driven decision-making.

Movie / Documentary	**Strategic Theme Highlighted**
Moneyball	Predictive modeling in sports strategy, data vs intuition
The Imitation Game	Algorithmic thinking under pressure, cryptographic strategy
Margin Call	Risk modeling and ethical blind spots in financial strategy
Coded Bias	Algorithmic bias, ethical oversight, and AI interpretability
Ex Machina	AI autonomy, black-box decision systems, and unintended consequences
Her	Predictive adaptation, human-AI interaction in strategic product development
Oppenheimer	Strategic risk, simulation modeling, and ethical intelligence in scientific breakthroughs
The Social Dilemma	Algorithmic decision-making and strategic tradeoffs in tech platforms
21	Probabilistic reasoning, decision modeling under uncertainty
Hidden Figures	Data modeling, decision intelligence, and the importance of interpretable analytics

References

Adeleye, I. O. (2024). The impact of artificial intelligence on design: Enhancing creativity and efficiency. *Journal of Engineering and Applied Sciences*, *3*(1), 1-13.

Amason, A., & Ward, A. (2020). *Strategic management: From theory to practice*. Routledge.

Anwar, C. M. (2019). The Algorithmic Compass: Navigating Ethical Decision-Making in the Age of AI-Driven Management. *Management Review*, *61*(4), 15-42.

Bertsimas, D., & Kallus, N. (2020). From predictive to prescriptive analytics. *Management Science*, *66*(3), 1025-1044.

Brandmeier, R. A., & Rupp, F. (2024). Foundations of Systematic Innovation. In *Structured Innovation: Levers to Successful Innovation Beyond Benchmarking* (pp. 1-25). Cham: Springer Nature Switzerland.

Brynjolfsson, E., & McElheran, K. (2016). The rapid adoption of data-driven decision-making. *American Economic Review*, *106*(5), 133-139.

Burkart, N., & Huber, M. F. (2021). A survey on the explainability of supervised machine learning. *Journal of Artificial Intelligence Research*, *70*, 245-317.

Büber, H., & Seven, E. (2025). Strategic Decision-Making in the AI Era: An Integrated Approach Classical, Adaptive, Resource-Based, and Processual Views. *International Journal of Management and Administration*, *9*(17), 67-97.

Celis Bueno, C., Chow, P. S., & Popowicz, A. (2025). Not "what", but "where is creativity?": towards a relational-materialist approach to generative AI. *Ai & Society*, *40*(2), 339-351.

Chandrasekera, T., Hosseini, Z., & Perera, U. (2025). Can artificial intelligence support creativity in early design processes?. *International journal of Architectural Computing*, *23*(1), 122-136.

Colantoni, A., Villarini, M., Monarca, D., Carlini, M., Mosconi, E. M., Bocci, E., & Hamedani, S. R. (2021). Economic analysis and risk assessment of biomass gasification CHP systems of different sizes through Monte Carlo simulation. *Energy Reports*, *7*, 1954-1961.

De Cremer, D., Bianzino, N. M., & Falk, B. (2023). How generative AI could disrupt creative work. *Harvard Business Review*, *13*, 13.

Dyer, J. H., Godfrey, P. C., Jensen, R. J., & Bryce, D. J. (2021). *Strategic management: Concepts and cases*. John Wiley & Sons.

Ferlie, E., & Ongaro, E. (2022). *Strategic management in public services organizations: Concepts, schools, and contemporary issues*. Routledge.

Gitelman, L. D., Kozhevnikov, M. V., & Chebotareva, G. S. (2021). *Strategic intelligence of an organization amid uncertainty* (pp. 294-305). WIT Press: Ashurst, Southampton, UK.

Guo, D., & Mantravadi, S. (2025). The role of digital twins in lean supply chain management: review and research directions. *International Journal of Production Research*, *63*(5), 1851-1872.

Haddar, N., Tmar, M., & Gargouri, F. (2016). A data-centric approach to manage business processes. *Computing*, *98*(4), 375-406.

Haider, S. A., & Tehseen, S. (2022). Role of decision intelligence in strategic business planning. *Decision intelligence analytics and the implementation of strategic business management*, 125-133.

Hambrick, D. C. (1982). Environmental scanning and organizational strategy. *Strategic Management Journal*, *3*(2), 159-174.

Henry, A. (2021). *Understanding strategic management*. Oxford University Press.

Hodorog, A., Petri, I., & Rezgui, Y. (2022). Machine learning and Natural Language Processing of social media data for event detection in smart cities. *Sustainable Cities and Society*, *85*, 104026.

Iansiti, M., & Lakhani, K. R. (2020). *Competing in the age of AI: Strategy and leadership when algorithms and networks run the world*. Harvard Business Press.

Jeyanthi, P. M., Choudhury, T., Hack-Polay, D., Singh, T. P., & Abujar, S. (2022). *Decision Intelligence Analytics and the Implementation of Strategic Business Management*. Springer International Publishing.

Kaggwa, S., Eleogu, T. F., Okonkwo, F., Farayola, O. A., Uwaoma, P. U., & Akinoso, A. (2024). AI in decision making: transforming business strategies. *International Journal of Research and Scientific Innovation*, *10*(12), 423-444.

Karkaria, V., Goeckner, A., Zha, R., Chen, J., Zhang, J., Zhu, Q., ... & Chen, W. (2024). Towards a digital twin framework in additive manufacturing: Machine learning and Bayesian optimization for time series process optimization. *Journal of Manufacturing Systems*, *75*, 322-332.

Kim, J., & Maher, M. L. (2023). The effect of AI-based inspiration on human design ideation. *International Journal of Design Creativity and Innovation*, *11*(2), 81-98.

Köseoglu, M. A. (2025). Leveraging Entrepreneurship for Environmental Sustainability: A Machine Learning Approach to SDG Achievement. *Business Strategy and the Environment*.

Köseoglu, M. A. (2025). How does the entrepreneurship ecosystem foster tourism competitiveness?. *Tourism Economics*, 13548166251341021.

Kreuzer, T., Papapetrou, P., & Zdravkovic, J. (2024). Artificial intelligence in digital twins—A systematic literature review. *Data & Knowledge Engineering*, 102304.

Kroese, D. P., Brereton, T., Taimre, T., & Botev, Z. I. (2014). Why the Monte Carlo method is so important today. *Wiley Interdisciplinary Reviews: Computational Statistics*, *6*(6), 386-392.

Lasserre, P., & Monteiro, F. (2022). *Global strategic management*. Bloomsbury Publishing.

Lichtenthaler, U. (2020). A conceptual framework for combining agile and structured innovation processes. *Research-Technology Management*, *63*(5), 42-48.

Liu, Z., Jiang, P., De Bock, K. W., Wang, J., Zhang, L., & Niu, X. (2024). Extreme gradient boosting trees with efficient Bayesian optimization for profit-driven customer churn prediction. *Technological Forecasting and Social Change*, *198*, 122945.

Madanchian, M., & Taherdoost, H. (2025). Ethical theories, governance models, and strategic frameworks for responsible AI adoption and organizational success. *Frontiers in Artificial Intelligence*, *8*, 1619029.

Moesmann, M., & Pedersen, T. B. (2025). Data-driven prescriptive analytics applications: A comprehensive survey. *Information Systems*, 102576.

Naik, A. S., Yeniaras, E., Hellstern, G., Prasad, G., & Vishwakarma, S. K. L. P. (2025). From portfolio optimization to quantum blockchain and security: A systematic review of quantum computing in finance. *Financial Innovation*, *11*(1), 1-67.

Niu, Y., Ying, L., Yang, J., Bao, M., & Sivaparthipan, C. B. (2021). Organizational business intelligence and decision making using big data analytics. *Information Processing & Management*, *58*(6), 102725.

Pratt, L., Bisson, C., & Warin, T. (2023). Bringing advanced technology to strategic decision-making: The Decision Intelligence/Data Science (DI/DS) Integration framework. *Futures*, *152*, 103217.

Rêgo, B. S., Jayantilal, S., Ferreira, J. J., & Carayannis, E. G. (2021). Digital transformation and strategic management: A systematic review of the literature. *Journal of the Knowledge Economy*, 1-28.

Sadeghi, V. J., Garcia Perez, A., Vrontis, D., & Bedford, D. (2024). Digital resilience, new business models and international entrepreneurship in the era of knowledge-economy. *Journal of Enterprise Information Management*, *37*(5), 1401-1417.

Sewpersadh, N. S. (2023). Disruptive business value models in the digital era. *Journal of Innovation and Entrepreneurship*, *12*(1), 2.

Sharma, A. K., Sharma, D. M., Purohit, N., Rout, S. K., & Sharma, S. A. (2022). Analytics techniques: descriptive analytics, predictive analytics, and prescriptive analytics. *Decision Intelligence Analytics and the Implementation of Strategic Business Management*, 1-14.

Sheel, N., Verma, L. P., Kumar, S., & Singh, T. P. (2022). Artificial Intelligence and Analytics for Better Decision-Making and Strategy Management. *Decision Intelligence Analytics and the Implementation of Strategic Business Management*, 31-43.

Smyth, C., Dennehy, D., Fosso Wamba, S., Scott, M., & Harfouche, A. (2024). Artificial intelligence and prescriptive analytics for supply chain resilience: a systematic literature review and research agenda. *International Journal of Production Research*, *62*(23), 8537-8561.

Soykoth, M. W., Sim, W., & Frederick, S. (2024). Research trends in market intelligence: A review through a data-driven quantitative approach. *Journal of Marketing Analytics*, 1-27.

Susnjak, T. (2024). Beyond predictive learning analytics modelling and onto explainable artificial intelligence with prescriptive analytics and ChatGPT. *International Journal of Artificial Intelligence in Education*, *34*(2), 452-482.

Surma, J. (2011). *Business intelligence: making decisions through data analytics*. Business Expert Press.

Tao, F., Zhang, H., & Zhang, C. (2024). Advancements and challenges of digital twins in industry. *Nature Computational Science*, *4*(3), 169-177.

Trantas, A., & Pileggi, P. (2024). Digital Twin and Foundation Models: A New Frontier. In *ICAART (3)* (pp. 988-994).

Vudugula, S., Chebrolu, S. K., Bhuiyan, M., & Rozony, F. Z. (2023). Integrating artificial intelligence in strategic business decision-making: A systematic review of predictive models. *International Journal of Scientific Interdisciplinary Research*, *4*(1), 01-26.

Warren, K. (2008). *Strategic management dynamics*. John Wiley & Sons.

Chapter 4

Innovation as a STEM Process

"Innovation isn't inspiration—it's engineering. To lead the future, think like an inventor, build like a scientist, and scale like a strategist."

by MAK

Learning Objectives

Innovation isn't just a spark—it's the engineering of new worlds. Like the beaver, true innovators build, adapt, and transform the environment for generations to come.

By the end of this chapter, readers will be able to:

- Redefine innovation as a structured and systematized process, leveraging STEM tools such as TRIZ, generative design, and AI-assisted modeling.
- Distinguish between incremental, radical, and disruptive innovations, and evaluate their strategic fit using forecasting and simulation tools.
- Apply structured innovation methodologies, including TRIZ, Stage-Gate models, and Technology Roadmapping, to align innovation efforts with strategic objectives.
- Analyze the strategic trade-offs between first-mover and fast-follower approaches, supported by data modeling and real-time decision systems.
- Contrast traditional Blue Ocean Strategy with AI-enabled market space identification and design logic.
- Utilize AI-driven creativity tools to enhance product design, marketing, and personalization in innovation pipelines.
- Conduct technology forecasting and horizon scanning to anticipate emerging disruptions, detect weak signals, and inform strategic innovation investments.
- Integrate scenario planning and predictive analytics to model innovation risk, resilience, and opportunity under conditions of uncertainty.
- Develop a personal innovation leadership toolkit that combines creativity, ethical foresight, and system-level thinking for leading innovation in STEM-intensive environments.

Innovation is no longer an act of inspiration; it is an engineered discipline.

In today's complex, data-rich, and rapidly evolving world, organizations can no longer afford to treat innovation as a flash of genius or a side project of R&D. The most competitive firms have redefined innovation as a systematic, STEM-powered process, one that blends creativity with analytical rigor, and vision with verifiable outcomes. They don't simply ideate; they model. They don't just experiment, they simulate. They don't wait for the future; they forecast and build toward it.

This chapter explores the transformation of innovation from intuition-driven creativity into an architected capability fueled by science, technology, engineering, and mathematics. We will examine how frameworks like TRIZ and Stage-Gate, combined with AI tools, roadmapping, and predictive analytics, enable organizations to generate, test, and scale innovations with precision and strategic alignment. You'll see how structured innovation methods reduce risk, enhance speed, and connect directly to long-term goals, whether it's creating sustainable products, disrupting traditional markets, or designing adaptive experiences through generative AI.

But innovation is not just a corporate function. It's a strategic competency for every future leader. In a STEM world, students, managers, and decision-makers must develop the ability to engineer innovation, thinking beyond brainstorming, designing systems that learn and adapt, and aligning creativity with competitive advantage.

This chapter will prepare you to do exactly that. You'll gain a toolkit of methods, frameworks, and forecasting techniques that make innovation not only possible, but predictable, measurable, and strategically essential.

Welcome to innovation, engineered for impact.

Opening Case: Engineering Innovation at Ping An

By the early 2010s, Ping An was a traditional financial services giant in China, deeply entrenched in insurance, banking, and investment management. But like many legacy firms, it faced a pivotal challenge: its core markets were saturated, new digital entrants were disrupting value chains, and consumer expectations were shifting toward real-time, tech-enabled experiences. Traditional product development cycles, slow, linear, and compliance-heavy, were no match for the speed of innovation demanded in a digitally driven economy. The question wasn't just how Ping An could innovate. It was how to make innovation predictable, scalable, and central to its strategy.

Ping An chose not to treat innovation as a siloed R&D function; it embedded innovation as an engineered system. The company developed a cross-functional innovation platform powered by artificial intelligence, big data, and design logic. Using structured tools such as technology roadmapping, TRIZ-based problem solving, and algorithmic forecasting, Ping An began to systematically create new products, services, and business models.

Their innovation engine didn't just iterate, it anticipated. The company used machine learning to identify underserved markets and unmet financial needs, then deployed rapid prototyping and digital twins to model user experiences before launch. For instance, its Smart Healthcare platform was not just a tech upgrade; it was the result of scenario planning, predictive analytics, and medical knowledge graph modeling, bringing AI-powered diagnosis and telemedicine to millions across China.

Perhaps most importantly, Ping An built a culture of "innovation discipline." Instead of relying on creative spontaneity, it scaled innovation through structured systems, stage-gate funnels, AI-assisted ideation tools, and cross-industry foresight teams. Every initiative passed through gates linked to data, user validation, and long-term strategic alignment.

Strategic Insight

Ping An's story challenges the traditional myth of innovation as unpredictable genius. It shows that the future of innovation lies in engineering, not just creativity. By embedding STEM logic into every stage of innovation, from ideation to scaling, Ping An transformed itself into one of the most technologically advanced and strategically agile financial groups in the world.

Play the Strategist

* How did Ping An shift from intuition-led innovation to a systematized, scalable model?
* What role did AI, data, and scenario modeling play in reducing innovation risk and time to market?
* How could other legacy firms replicate Ping An's approach without losing flexibility or speed?

> Innovation doesn't live in labs anymore, it lives in systems, sensors, simulations, and strategy.

4.1. Why Innovation Must Be Engineered

For much of modern business history, innovation was viewed as an art, unpredictable, spontaneous, and dependent on creative flair. But in today's high-speed, high-stakes world, that view is no longer sustainable. Innovation is now a strategic imperative, and it demands structure.

In STEM-driven organizations, innovation is no longer confined to idea boards and brainstorming retreats. It is engineered through repeatable systems, modeled with simulations, and validated through data. Leading firms don't wait for inspiration; they build platforms that continuously sense opportunities, solve contradictions, and test ideas in real time. Innovation has evolved from an act of inspiration into a disciplined process of exploration, design, and execution.

This section redefines innovation for the STEM strategist. We explore how modern organizations treat innovation as a managed system, one that integrates design thinking, TRIZ logic, simulation models, and AI-driven insight. We show how creative intuition becomes more powerful, not less, when enhanced by structured methods and technological tools.

Whether you're developing the next product breakthrough, reinventing a service model, or transforming an entire industry, sustainable innovation starts with one principle: **it must be engineered.**

4.1.1 Innovation Is No Longer Accidental

The myth of the lone genius, locked away in a lab, scribbling formulas or experiencing sudden "aha" moments, has shaped popular perceptions of innovation for decades. While appealing, this narrative no longer aligns with reality. Contemporary challenges require organizations to innovate faster, more reliably, and more strategically, relying not on serendipity but on structured systems and deliberate methods.

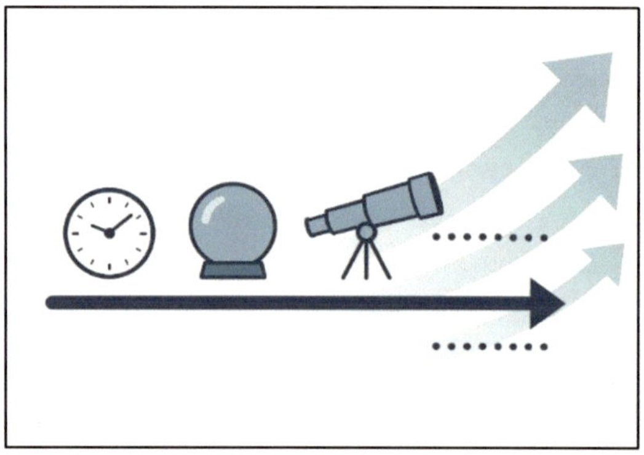

Organizations now face complexities that demand swift, systematic responses. Consider global firms contending with challenges such as sustainability regulations, rapid technology shifts, and evolving customer behaviors. Relying purely on intuition or isolated creativity risks delays, costly mistakes, and strategic misalignment. Firms must proactively engineer innovation pipelines, embedding disciplined methodologies that can reliably produce impactful outcomes.

Moreover, modern innovation often demands cross-functional collaboration involving teams distributed across the globe. To coordinate diverse talents and resources effectively, innovation processes must be well-engineered and transparent. Rather than depending on spontaneous insights or uncoordinated experiments, organizations use structured processes to align efforts, measure outcomes, and scale successes consistently.

In this new paradigm, organizations actively cultivate environments that support continuous idea generation and iterative experimentation. Structured methods empower organizations to track progress, rapidly discard ineffective paths, and double down on promising directions. Innovation, therefore, moves from a mysterious act of creativity to an organizational discipline, predictable, accountable, and strategically vital.

4.1.2 Innovation as a Repeatable Process

Today's successful organizations understand that innovation should not be sporadic or opportunistic, but repeatable and scalable. They deploy structured frameworks such as **Stage-Gate** and agile methodologies, allowing multiple ideas to pass systematically through rigorous checkpoints. These processes ensure alignment with strategic priorities, customer needs, and resource constraints, eliminating guesswork and subjective biases from critical decisions.

A well-designed innovation funnel integrates real-time analytics and predictive modeling at every stage. Organizations leverage digital tools like dashboards, simulation software, and feedback loops to provide clarity and transparency. This approach allows leaders to quantify innovation risks, anticipate market receptivity, and ensure strategic fit early in the development cycle. As a result, the innovation process becomes increasingly efficient, focused, and outcome-driven.

Furthermore, innovation portfolios must balance exploratory projects, those aiming for disruptive breakthroughs, with exploitative initiatives that refine and enhance existing products and services. Leading firms systematically monitor this balance using analytics, enabling them to optimize resource allocation across high-risk, high-reward opportunities and incremental improvements. Such rigorous management ensures the organization remains innovative yet financially disciplined.

This disciplined repeatability is central to building long-term innovation capabilities. Organizations no longer wait passively for inspiration; they actively cultivate pipelines that continuously generate, evaluate, and scale ideas. The outcome is an innovation culture that systematically learns from failures, replicates successes, and continuously improves its predictive and strategic accuracy over time.

4.1.3 Structured Innovation in a STEM World

Structured innovation methodologies, enhanced by STEM-based tools, are transforming how organizations conceptualize and manage creativity. Frameworks like TRIZ, design thinking, and technology roadmapping allow companies to systematically address contradictions, prioritize market opportunities, and accelerate product cycles through data-informed decision-making.

In practice, companies use data-driven ideation platforms powered by artificial intelligence and machine learning to identify unmet customer needs and innovation gaps. NLP tools parse millions of consumer interactions to pinpoint latent pain points. AI-generated simulations and digital twin models rapidly prototype and validate ideas, significantly shortening the ideation-to-market timeframe. Thus, structured innovation methodologies transform intuition into actionable insight.

Additionally, STEM-enhanced innovation methods encourage cross-disciplinary collaboration. For example, engineering teams, marketers, and data scientists collaborate through structured frameworks that translate customer insights into technical requirements. Design thinking workshops integrate user empathy, iterative prototyping, and customer validation into disciplined routines, systematically guiding innovation efforts toward viable, impactful outcomes.

When coupled with digital analytics, structured methods also facilitate transparency and accountability. Decision-makers can continuously monitor innovation performance, adjusting direction swiftly based on real-time data rather than subjective perceptions. This systematic transparency reduces friction, enhances stakeholder trust, and drives innovation strategies aligned precisely with organizational priorities.

4.1.4 The Strategic Payoff of Engineering Innovation

The strategic advantage of engineered innovation goes far beyond increased idea generation; it delivers targeted, sustainable competitive differentiation. Organizations that structure innovation processes systematically ensure alignment with long-term objectives, accelerating outcomes with precision and clarity.

Structured innovation enables better resource allocation, as projects undergo rigorous, data-informed assessment at each stage. Rather than funding innovation initiatives indiscriminately, leaders prioritize investments according to clear strategic criteria such as market potential, technical feasibility, and sustainability benchmarks. This disciplined approach reduces waste, optimizes return on innovation investment, and mitigates risk.

Structured innovation further integrates with long-term sustainability and ESG goals, allowing firms to proactively address emerging global imperatives. Organizations employing disciplined, predictive innovation pipelines can efficiently incorporate regulatory shifts and societal expectations into their product development and strategy. Such strategic foresight creates future-ready firms resilient against disruption and adept at capturing emerging opportunities.

Moreover, digital and analytical feedback loops embedded within structured innovation processes create cycles of continuous learning. Firms capture and analyze vast amounts of innovation performance data, enabling iterative improvement of decision-making algorithms and models. Consequently, engineered innovation processes become not only more efficient but increasingly intelligent, anticipating market shifts, identifying risks earlier, and consistently refining strategic outcomes.

4.1.5 Building Innovation Capability in Emerging Leaders

Innovation leadership is rapidly evolving from a loosely defined managerial attribute to a structured, core competency expected from future strategists. Emerging leaders must be adept at orchestrating structured innovation processes, balancing creativity with discipline, and aligning organizational resources with strategic goals.

Developing innovation capability requires cultivating a comprehensive toolkit, blending creative methodologies with analytical tools. Emerging leaders must master frameworks such as TRIZ for inventive problem-solving, technology roadmapping for strategic alignment, and agile methodologies for rapid experimentation. Fluency in predictive modeling, digital twins, and simulation technologies also becomes essential, enabling leaders to rigorously test and validate ideas before committing significant resources.

Moreover, tomorrow's innovation leaders must possess strong systems thinking capabilities. Rather than viewing innovation as isolated projects or inventions, they see interconnected ecosystems of technologies, markets, and stakeholders. Leaders skilled in systems thinking can strategically guide innovations that deliver cross-functional benefits and align coherently with broader organizational objectives, thus amplifying overall strategic value.

Finally, ethical foresight and strategic sensitivity become critical competencies in innovation leadership. Leaders must ensure innovation initiatives not only deliver financial returns but also align ethically with organizational values and societal expectations. Understanding the broader implications of innovation decisions and embedding transparency, accountability, and sustainability throughout the innovation process will distinguish the strategic innovators of the future.

Real-World Examples of Engineered Innovation

* **Vicarious Surgical** utilized anatomical modeling and digital twins to virtually prototype robotic surgery systems before physical trials, dramatically accelerating product development while ensuring regulatory compliance.
* **Climeworks**, through systematic modeling and scenario planning, identified optimal deployment regions for carbon capture plants, aligning technical feasibility with strategic climate-impact goals.
* **Decathlon** systematically applies TRIZ-based contradiction analysis to develop innovative sporting gear, systematically addressing consumer pain points, increasing product differentiation, and capturing greater market share.

When innovation is structured using STEM principles, such as simulation models, predictive analytics, and systems thinking, creativity becomes scalable, strategic, and sustainable. Organizations leveraging engineered innovation methods achieve higher efficiency, reduced risk, and consistent strategic alignment.

4.2 Structured Innovation Methods in Strategic Contexts

Innovation without structure is chaos. While creativity provides essential fuel for innovation, systematic frameworks translate creativity into sustainable strategic advantage. Modern organizations utilize structured innovation methods such as TRIZ, Stage-Gate processes, and Technology Roadmapping to strategically align innovation efforts with organizational objectives.

Structured innovation has evolved into an essential strategic discipline. These methods not only guide idea generation and implementation systematically but also enable precise strategic decisions, such as choosing between being a first mover or a fast follower in competitive markets. Enhanced by STEM-driven analytics and digital simulations, structured innovation methods have become powerful strategic tools, providing clarity, predictability, and competitive agility.

4.2.1 TRIZ: The Science of Inventive Problem Solving

TRIZ, the Theory of Inventive Problem Solving, systematically addresses contradictions inherent in innovation challenges. Originally developed by analyzing thousands of patents, TRIZ guides organizations to proven inventive principles and structured solutions. It moves innovation from chance to methodical logic, enabling consistent strategic alignment.

Organizations use TRIZ to systematically solve problems where contradictions hinder innovation. For example, automotive manufacturers leverage TRIZ to reduce vehicle weight without sacrificing safety, two traditionally conflicting goals. TRIZ methods help teams identify existing innovative patterns, applying structured logic to resolve such contradictions systematically.

In strategic contexts, TRIZ supports decisions related to first-mover versus fast-follower strategies. Firms pursuing first-mover strategies rely on TRIZ to systematically break through challenging contradictions, creating pioneering technologies. Conversely, fast followers deploy TRIZ to quickly and methodically resolve contradictions discovered by early entrants, often improving upon original innovations while reducing development risks and costs.

Combined with modern analytics and patent-mining AI tools, TRIZ further empowers strategic precision. Firms systematically scan innovation landscapes, rapidly identifying strategic opportunities for both pioneering and incremental innovations. Thus, TRIZ becomes a key strategic asset for systematically deciding when and how to enter competitive markets.

4.2.2 Stage-Gate and Innovation Funnels

The Stage-Gate method provides structured processes for systematically managing innovations from conception to commercialization. Through clearly defined decision gates, organizations rigorously evaluate potential projects based on strategic alignment, market readiness, and technical feasibility. Such disciplined assessments transform innovation management into a predictable, transparent, and strategically precise activity.

Strategically, Stage-Gate structures the choice between first-mover and fast-follower strategies. Companies pursuing first-mover advantage systematically apply stringent Stage-Gate criteria to rapidly screen and advance pioneering innovations, leveraging predictive analytics and simulation models to forecast market acceptance and mitigate pioneering risks.

Conversely, fast-follower firms utilize Stage-Gate methodologies to systematically analyze existing market innovations, rigorously identifying improvements and optimization opportunities. Real-time market analytics and competitor benchmarking, integrated within Stage-Gate gates, enable fast followers to swiftly and efficiently refine ideas, often achieving superior market success without bearing first-mover uncertainties.

Digital dashboards, simulation tools, and real-time analytics augment Stage-Gate frameworks, providing systematic precision in innovation execution. Leaders dynamically model project outcomes, assess innovation pipeline performance, and rapidly adjust strategies, ensuring rigorous strategic coherence regardless of whether their approach is pioneering or incrementally refining.

4.2.3 Technology Roadmapping for Strategic Planning

Technology Roadmapping systematically visualizes the integration of technological developments with long-term strategic objectives. Roadmaps dynamically align innovations, market opportunities, and internal capabilities over strategic time horizons. As strategic navigation tools, roadmaps enable organizations to proactively guide innovation pathways toward clearly defined objectives.

Critically, roadmapping informs first-mover versus fast-follower decisions with analytical precision. Firms aiming for first-mover status use roadmaps to systematically plot technological trajectories and identify early breakthrough opportunities. Predictive modeling within roadmaps enables proactive management of risks inherent in pioneering market entry.

On the other hand, fast followers leverage dynamic technology roadmaps differently. They systematically track technological trajectories initiated by first movers, clearly identifying emerging standards, customer expectations, and optimal timing for strategic entry. Integrated analytics provide precise foresight, allowing fast followers to strategically time market entry and product enhancements, often outmaneuvering early movers.

Advanced data analytics, machine learning, and real-time market intelligence continuously enhance roadmap accuracy. Organizations systematically monitor market signals, competitor actions, and technological advances, dynamically adjusting their roadmaps with predictive agility, precision, and clarity. Consequently, technology roadmapping becomes an indispensable strategic tool, precisely guiding organizations whether they pioneer new markets or strategically follow.

4.2.4 Integrating Structured Methods with Data and Analytics

The strategic value of structured innovation methods multiplies exponentially when integrated with STEM-based analytics and predictive tools. Companies now combine frameworks such as TRIZ, Stage-Gate, and roadmapping with AI-powered data analytics, NLP-driven horizon scanning, and digital twins, systematically optimizing strategic decisions with data-driven precision.

Structured methods, when enhanced by analytics, systematically inform first-mover and fast-follower choices with unprecedented accuracy. Advanced analytics precisely predict market dynamics, technological feasibility, and competitive responses. Organizations systematically model scenarios, evaluate strategic trade-offs, and dynamically decide whether early market entry or subsequent refinement will yield maximum strategic advantage.

For instance, predictive analytics embedded within Stage-Gate models enable precise, systematic forecasting of pioneering innovation risks—informing rigorous decisions about the feasibility of first-mover strategies. Similarly, NLP-driven competitor benchmarking systematically identifies strategic vulnerabilities of early movers, guiding fast followers to precise, market-dominating refinements.

Digital twins and real-time analytics further enhance systematic decision-making. Firms rapidly simulate innovation scenarios, systematically optimizing strategic timing, resource allocation, and market positioning. This integration systematically transforms structured innovation methods into comprehensive, analytically precise decision engines, capable of consistently executing both first-mover and fast-follower strategies with strategic clarity (Table 4.1).

Table 4.1. First-Mover vs. Fast-Follower: Strategic Fit of Structured Innovation

Strategy	Ideal Conditions	Role of Structured Methods	Risks
First-Mover	High uncertainty, potential for IP dominance	TRIZ to solve complex design challenges early; Roadmapping to guide entry timing	Market may not be ready; high R&D cost
Fast-Follower	Existing market established by pioneers	Stage-Gate to quickly iterate on others' solutions; AI tools to optimize marketing	Risk of being perceived as a copycat

Structured Innovation and First-mover vs. Fast-follower in Action

* **First-mover example:**

  Tesla leveraged technology roadmapping and predictive modeling systematically to establish pioneering market dominance in electric vehicles. Strategic simulations and analytical roadmaps enabled systematic anticipation of infrastructure challenges, regulatory shifts, and consumer adoption patterns, thereby mitigating first-mover risks.

* **Fast-follower example:**

  Samsung utilized structured Stage-Gate processes and systematic competitor benchmarking to quickly follow Apple's smartphone innovations. By systematically analyzing Apple's initial market responses and precisely forecasting consumer preferences through predictive analytics, Samsung introduced competitive products rapidly and efficiently, gaining significant market share without pioneering risks.

Structured innovation methods, integrated with advanced analytics and digital tools, provide systematic precision for executing first-mover and fast-follower strategies alike. Organizations employing these structured frameworks systematically assess strategic trade-offs, dynamically anticipate market trajectories, and precisely align innovation efforts with long-term strategic objectives.

For example, **TRIZ** helps teams resolve contradictions systematically using a proven set of inventive principles, ideal for engineering and R&D environments seeking breakthrough ideas. In contrast, the **Stage-Gate** process structures the innovation funnel through stage reviews and Go/Kill gates, reducing risk and ensuring cross-functional alignment during product development or commercialization.

Meanwhile, **technology roadmapping** offers a time-based visual framework that links technological evolution with strategic milestones. This method is particularly valuable in sectors like electronics, pharma, and automotive, where long-term investment alignment is critical.

These structured methods are compared below:

Table 4.2: Comparing Structured Innovation Methods

Method	Core Purpose	Key Tools/Techniques	Strategic Advantage	Typical Use Cases
TRIZ	Solve contradictions systematically	Contradiction matrix, 40 inventive principles	Encourages breakthrough ideas using proven patterns	Engineering design, manufacturing, R&D
Stage-Gate	Manage innovation funnel & project risk	Stage reviews, Go/Kill gates, project scorecards	Reduces uncertainty & aligns teams	Product development, tech commercialization
Tech Roadmapping	Align tech trajectory with strategy	Time-based visual maps, strategic milestones	Enables long-term planning & investment alignment	Electronics, pharma, automotive

4.3 AI-Driven Creativity and Design

Artificial intelligence is fundamentally redefining creativity. Traditionally, creativity was seen as uniquely human, a domain of imagination, intuition, and originality. Yet in the modern strategic landscape, creativity increasingly blends human insight with algorithmic precision. Organizations now harness AI not just to support but to actively co-create innovation, accelerating ideation, prototyping, and implementation at unprecedented speed and scale.

This section explores the revolutionary impact of AI-driven creativity and design in strategic contexts. We examine how generative design, machine learning creativity tools, and adaptive personalization engines transform innovation processes. Additionally, we contrast traditional Blue Ocean Strategy methods, based on intuitive identification of uncontested market spaces, with AI-powered predictive analytics that systematically discover these "blue oceans" algorithmically.

4.3.1 Generative Design: From Constraints to Concepts

Generative design uses AI and computational modeling to automatically create numerous solutions based on predefined constraints such as material strength, cost efficiency, environmental impact, or performance specifications. Rather than relying on human designers alone, organizations use AI to systematically explore vast design possibilities, uncovering highly optimized yet often unexpected solutions.

This methodology revolutionizes sectors like aerospace, automotive, consumer products, and architecture. For example, engineers at General Motors utilized generative design software to develop seat brackets that were simultaneously lighter and stronger, structures resembling organic forms previously unimaginable through traditional design approaches. AI systematically explored thousands of structural variations, rapidly converging on optimal designs through iterative simulation and evaluation.

Generative design dramatically enhances strategic agility by allowing organizations to systematically prototype and validate multiple innovation paths simultaneously. By setting clear strategic constraints, cost, sustainability, and manufacturability, leaders guide generative systems to produce designs precisely aligned with long-term goals, drastically reducing development timelines and costs.

Furthermore, generative design reshapes the role of human designers. Rather than directly crafting every design detail, designers become curators, defining objectives, interpreting AI-generated alternatives, and strategically guiding iterative refinements. Thus, generative design transforms creative processes from linear trial-and-error into dynamic, systematically optimized innovation workflows.

4.3.2 Machine Creativity and Generative AI

AI-driven creativity extends beyond product design to content creation, brand messaging, user experiences, and even strategic positioning. Generative AI tools such as GPT, DALL·E, Midjourney, and RunwayML systematically produce diverse creative outputs, texts, visuals, and music, rapidly accelerating the ideation and prototyping stages of innovation.

Strategically, machine creativity tools allow organizations to swiftly generate and evaluate multiple creative alternatives, systematically optimizing strategic communication and positioning. For example, marketing teams use generative AI to rapidly prototype personalized ad campaigns, systematically assessing different messaging strategies and refining them based on predictive analytics of consumer responses.

AI-driven creativity also fosters strategic differentiation through innovation in experience design. Companies systematically leverage generative AI to design intuitive, adaptive user interfaces and personalized content experiences. For instance, streaming services like Netflix use generative algorithms to dynamically curate content, systematically improving user engagement and retention through precisely tailored recommendations.

Machine creativity thus redefines strategic innovation. Leaders become orchestrators of intelligent systems, systematically guiding creative processes through clearly defined constraints, strategic objectives, and analytical feedback loops, resulting in innovations that consistently resonate with target markets and strategic goals.

4.3.3 Personalization Engines and Design Adaptivity

AI-driven personalization represents a new frontier in strategic innovation. Adaptive systems powered by AI continuously analyze user behaviors, preferences, and contextual factors to dynamically personalize products, services, and experiences in real-time. Rather than static, one-size-fits-all offerings, innovation becomes an ongoing dialogue uniquely tailored to individual users.

Strategically, personalization engines create competitive differentiation by systematically delivering superior customer experiences. For example, fashion companies such as The Yes (now integrated into Pinterest) systematically use AI algorithms to personalize clothing recommendations, dynamically adapting product displays and marketing content based on precise, real-time user data. This systematic personalization boosts conversion rates, customer loyalty, and brand strength.

AI-driven adaptivity also enhances strategic flexibility, allowing firms to rapidly respond to changing consumer preferences or competitive threats. Organizations systematically leverage real-time analytics from adaptive systems to proactively adjust offerings and experiences, consistently meeting evolving market demands with strategic agility.

The transformative potential of AI-driven adaptivity extends beyond marketing. In healthcare, adaptive patient monitoring systems personalize medical interventions systematically, dramatically improving patient outcomes through real-time predictive analytics. Thus, AI-driven adaptivity systematically embeds strategic responsiveness within innovation processes, enhancing both competitive differentiation and organizational resilience.

4.3.4 Strategic Implications of AI-Driven Design

AI-driven creativity profoundly reshapes strategic innovation management. Organizations employing generative AI, personalization engines, and adaptive design systematically accelerate innovation cycles, enhance market responsiveness, and strategically differentiate their offerings. Yet AI-driven design also demands nuanced strategic oversight, particularly regarding originality, intellectual property, and ethical considerations.

Here, we contrast traditional Blue Ocean Strategy (BOS), which intuitively identifies uncontested market spaces, with AI-powered predictive analytics and machine learning. Traditional BOS typically involves human strategists intuitively spotting unmet market needs or underserved niches. However, the accuracy and scalability of traditional intuition-based BOS are limited and subjective.

In contrast, AI-enhanced Blue Ocean identification systematically and algorithmically discovers uncontested markets. AI-driven predictive analytics, leveraging large datasets of customer behaviors, industry trends, and competitive intelligence, systematically pinpoint emerging or overlooked opportunities with precise quantitative validation. Firms thus identify and occupy "blue oceans" algorithmically, dramatically enhancing the accuracy, scale, and speed of strategic market entry.

For example, AI-driven market analytics systematically guided Amazon's early investments in cloud computing, pinpointing previously unidentified customer needs and strategically validating them through predictive modeling. Similarly, fintech startups algorithmically identify underserved financial segments by analyzing extensive consumer transaction data, systematically creating innovative offerings that traditional strategies might overlook.

Strategically, AI-enhanced Blue Ocean identification provides systematic foresight, enabling firms to proactively seize market leadership positions with greater precision. Rather than depending on subjective intuition alone, firms now use AI-driven tools to rigorously validate strategic opportunities, ensuring strategic alignment, reduced market-entry risk, and sustainable competitive advantage.

Traditionally, Blue Ocean Strategy (BOS) relied on human insight and qualitative tools like the strategic canvas. While insightful, this approach was often slow and difficult to scale. In contrast, AI-augmented BOS introduces predictive modeling, machine learning, and real-time data mining to uncover untapped market spaces with speed and analytical rigor.

The comparison below highlights how AI transforms the Blue Ocean discovery process:

Table 4.3. Blue Ocean Identification—Traditional vs. AI-Augmented

Feature	Traditional BOS	AI-Augmented BOS
Method	Human intuition, strategic canvas	Predictive modeling, machine learning
Speed	Slow, qualitative	Fast, data-driven
Data Sources	Market research, customer interviews	Web data, transaction logs, sentiment analysis
Outcome	Insightful but subjective	Scalable, testable predictions of new niches

AI-Driven Creativity in Action

* **The Yes** (Pinterest) utilized AI-driven personalization algorithms to systematically curate fashion choices, dramatically increasing customer engagement through highly tailored, real-time recommendations.

* **General Motors** leveraged generative design systematically to create automotive components, achieving structures lighter and stronger than traditional methods, strategically enhancing vehicle performance and efficiency.
* **Endlesss.fm** implemented AI-driven generative music software, systematically enabling artists to collaboratively produce original tracks, accelerating creative innovation and user engagement.

AI-driven creativity systematically integrates human intuition with computational precision, transforming innovation processes into data-driven, dynamic strategic activities. Organizations employing generative AI, adaptive design, and algorithmic market identification systematically achieve superior innovation outcomes, competitive differentiation, and strategic agility.

For instance, **ChatGPT** enables rapid ideation and content generation, supporting everything from product descriptions to personalized customer interactions. Visual creativity is enhanced through tools like **DALL·E** and **Midjourney**, which generate brand visuals and advertising concepts tailored to new markets. In multimedia storytelling, **RunwayML** empowers teams to prototype marketing videos and pitch decks with cinematic quality. Meanwhile, **generative design** tools optimize physical product forms—such as lightweight automotive parts—by exploring thousands of design permutations algorithmically.

These tools and their strategic functions are summarized below:

Table 4.4. AI Tools in Creative Strategy

AI Tool	Type of Output	Strategic Function	Example Use
ChatGPT	Text (content, dialogue, ideation)	Creative ideation, personalization	Generating product descriptions
DALL·E / Midjourney	Images, visual design	Branding, ad visuals	Ad concepts for new markets
RunwayML	Video, animation	Storytelling, pitch decks	Prototypes for marketing videos
Generative Design	3D models, optimized forms	Product innovation	Lightweight automotive parts

4.4 Technology Forecasting in Strategy

In today's rapidly evolving technological landscape, an effective innovation strategy depends significantly on accurate technology forecasting. Forecasting involves systematically identifying, analyzing, and anticipating emerging technological trends and disruptions, enabling organizations to proactively align their innovation efforts with future opportunities and risks. Strategic leaders who master forecasting gain decisive competitive advantages by entering markets early, influencing industry standards, and capturing value ahead of rivals.

This section examines how structured forecasting methodologies, including Technology Roadmapping, horizon scanning, and scenario planning, enable strategic foresight and organizational resilience. Critically, we integrate Clayton Christensen's Disruptive Innovation Theory, highlighting case studies where failure to effectively forecast technological disruptions resulted in strategic setbacks. By understanding these insights, organizations can strategically navigate technological uncertainties with precision and agility.

4.4.1 Why Forecasting Is Strategic

Technology forecasting provides a systematic means to anticipate and respond proactively to rapid changes in technology, customer preferences, and competitive dynamics. Unlike passive trend observation, forecasting is an active strategic discipline, enabling leaders to systematically analyze potential future scenarios, allocate resources efficiently, and make informed strategic investments.

Companies adept at forecasting leverage strategic foresight to establish competitive asymmetries. By systematically identifying technological developments before competitors, they strategically position themselves as industry leaders, shaping market perceptions and influencing regulatory frameworks. Conversely, firms neglecting robust forecasting often find themselves reacting belatedly, forced into costly, defensive strategies when disruptions inevitably arise.

Structured forecasting methods reduce uncertainty by systematically mapping technological trajectories and clarifying strategic implications. Predictive analytics, machine learning, patent analysis, and real-time monitoring of competitive landscapes significantly enhance forecasting accuracy, allowing firms to make timely and informed strategic decisions rather than guessing blindly or reacting impulsively.

Forecasting also provides a structured platform for continuous learning and adaptation. Organizations systematically track forecasting accuracy, iteratively refining predictive models based on real-world outcomes. Thus, technology forecasting transforms strategic decision-making from speculative guesswork into disciplined, data-driven precision, consistently positioning firms ahead of disruptive trends.

4.4.2 Technology Roadmapping

Technology roadmapping visually aligns strategic objectives with technological development timelines, market evolution, and internal capabilities. As dynamic strategic guides, roadmaps systematically clarify innovation priorities, resource allocation, and timing decisions, ensuring cohesive, forward-looking strategic execution.

Strategically, roadmaps offer a disciplined means of anticipating and managing technological disruptions. They clearly highlight emerging threats and opportunities, enabling organizations to proactively develop capabilities or shift strategic direction well in advance. Predictive modeling integrated into roadmaps further enhances strategic foresight, allowing leaders to quantify risks, forecast adoption rates, and systematically identify optimal market-entry points.

Companies like Intel and Samsung have leveraged sophisticated technology roadmaps for decades, systematically guiding strategic investments in semiconductors and consumer electronics. Their disciplined forecasting has consistently provided strategic clarity, enabling proactive management of technological shifts, regulatory developments, and competitive threats—positioning them as long-term industry leaders.

Thus, technology roadmapping systematically transforms strategic planning into a proactive, evidence-based discipline, one that ensures precise alignment of technological innovation with overarching corporate goals and market realities.

4.4.3 Emerging Tech Scanning and Weak Signals

Effective technology forecasting depends on systematically detecting early, subtle indicators, or weak signals, of emerging technologies and disruptions. Organizations engage in continuous horizon scanning, analyzing academic research, patent filings, startup investments, regulatory announcements, and customer behavior patterns to identify nascent trends before they reach mainstream awareness.

Advanced analytics, such as natural language processing (NLP), patent analytics, and AI-driven foresight platforms, significantly enhance organizations' capability to detect these weak signals systematically. These tools rapidly scan vast datasets, algorithmically identifying patterns and convergences that human analysts might overlook. Firms systematically leveraging such technologies gain strategic foresight, enabling proactive adjustments before competitors even recognize emerging opportunities or threats.

For example, pharmaceutical companies systematically use NLP tools to detect early signals in medical literature, identifying novel therapeutic approaches or emerging public health issues. Similarly, tech firms utilize predictive analytics to spot early-stage technology investments, systematically forecasting future market disruptions and strategic inflection points.

Organizations that systematically scan and interpret weak signals can strategically navigate uncertainties with greater precision, mitigating disruptive risks and strategically positioning themselves ahead of competitors.

4.4.4 Scenario Planning for Technological Uncertainty

Scenario planning systematically addresses forecasting uncertainties by modeling multiple plausible future scenarios. Rather than attempting a single precise prediction, organizations create several strategically divergent scenarios, each systematically exploring different assumptions about technological progress, market dynamics, and regulatory environments.

For example, an automotive company planning for autonomous vehicle technology systematically models multiple future scenarios: rapid regulatory approval and consumer acceptance, technological setbacks and delayed adoption, or fragmented regional developments. Each scenario provides strategic insights, systematically preparing the organization for diverse potential outcomes.

Strategically, scenario planning enables systematic contingency planning. Organizations proactively develop strategic responses for multiple plausible futures, enhancing agility and resilience. Scenario-based simulations further inform strategic investments, guiding organizations toward flexible strategies robust against a variety of possible technological disruptions.

Moreover, predictive analytics and digital twins significantly enhance scenario accuracy. Firms systematically validate scenarios through computational modeling, dynamically adjusting assumptions and refining predictions as new data emerges. Consequently, scenario planning systematically equips organizations to strategically navigate technological uncertainty with analytical rigor and strategic agility.

4.4.5 Forecasting with STEM-Enhanced Tools and the Cost of Failure

Clayton Christensen's Disruptive Innovation Theory highlights strategic failures resulting from inadequate technology forecasting. Christensen demonstrated that many industry leaders fail precisely because they overlook or underestimate emerging technologies that initially appear inferior or insignificant, only to rapidly gain market traction and disrupt entire industries.

A classic example is Kodak's failure to systematically forecast the impact of digital photography. Initially viewing digital technology as inferior to traditional film, Kodak systematically underinvested in digital forecasting and innovation. Consequently, Kodak suffered severe disruption, unable to respond swiftly as digital cameras rapidly improved and gained widespread adoption.

Similarly, Blockbuster famously underestimated streaming technology and failed to forecast consumer preferences shifting toward digital content delivery. Their strategic oversight in technology forecasting allowed Netflix, originally a niche player, to disrupt the home-entertainment market systematically. Netflix strategically employed rigorous predictive analytics to anticipate market shifts, precisely timing their transition into streaming and systematically dominating a previously uncontested digital "blue ocean."

These cases underscore the strategic imperative of systematic, STEM-enhanced forecasting. Organizations must proactively deploy predictive analytics, scenario planning, and horizon scanning to avoid strategic blind spots. Firms that systematically forecast technological trajectories, anticipate disruptive threats, and respond proactively can avoid costly strategic miscalculations, securing sustainable competitive advantages in increasingly turbulent markets.

Forecasting Success and Failure

* **Failure Example (Kodak)**: Kodak systematically underestimated the strategic importance of digital photography, ignoring predictive signals, and experiencing profound disruption due to inadequate technological forecasting.
* **Failure Example (Blockbuster)**: Blockbuster failed to anticipate streaming services' disruptive potential, resulting in strategic collapse due to neglected predictive analytics and inadequate technology forecasting.
* **Success Example (Netflix)**: Netflix strategically utilized sophisticated predictive analytics and scenario modeling to forecast the disruptive potential of streaming technology, proactively securing a market-leading position.

Systematic forecasting using STEM-enhanced predictive analytics, roadmapping, and scenario planning provides strategic clarity, resilience, and competitive advantage. Christensen's Disruptive Innovation Theory underscores forecasting's critical role; organizations systematically neglecting predictive foresight risk severe disruption, while those effectively forecasting systematically achieve sustainable strategic success.

To operationalize this foresight, organizations deploy a range of forecasting tools—each offering distinct methodologies and strategic benefits. For instance, technology roadmapping visually aligns technological evolution with market trends, enabling long-term planning and resource alignment in sectors like semiconductors and biotech. Horizon scanning systematically detects weak signals, such as shifts in regulation or consumer sentiment, serving as an early warning system for disruption.

Scenario planning models plausible futures, helping organizations build strategic flexibility in uncertain domains like autonomous vehicles or AI regulation. Meanwhile, predictive analytics applies data modeling to forecast trends with tactical precision, supporting decisions around customer demand or technology adoption.

These tools and their strategic value are summarized below:

Table 4.4. Forecasting Tools and Strategic Value

Forecasting Tool	Methodology	Key Benefit	Example Application
Technology Roadmapping	Visual alignment of tech and market trends	Long-term planning & resource alignment	Semiconductors, biotech
Horizon Scanning	Systematic detection of weak signals	Early warning system for disruption	Regulatory shifts, consumer behavior
Scenario Planning	Plausible future modeling	Strategic flexibility	Autonomous vehicles, AI regulation
Predictive Analytics	Data modeling of future trends	Tactical accuracy	Customer demand, tech adoption

4.5. Chapter 4 Wrap-Up: Engineering Innovation for Strategic Impact

Innovation is no longer a creative luxury; it's a strategic necessity, and it must be engineered with precision.

In this chapter, we reframed innovation as a structured, STEM-enabled process that is central to strategy execution. We explored how leading organizations move beyond inspiration and instead deploy disciplined innovation methods such as TRIZ, Stage-Gate, and design thinking, augmented by AI, forecasting, and predictive analytics. These methods make innovation measurable, repeatable, and aligned with broader strategic goals.

We examined how data-driven approaches support both first-mover and fast-follower strategies, and how AI-enhanced tools enable firms to identify "blue oceans" not just through intuition, but algorithmically. Through the lens of technology forecasting and scenario planning, we emphasized how strategic foresight, especially when grounded in systems and analytics, can protect firms from disruption and position them to lead future markets.

Using integration points such as Disruptive Innovation Theory and Blue Ocean Strategy, we illustrated how structured innovation creates not just products, but adaptive systems of advantage. Ultimately, innovation strategy today is not about isolated breakthroughs, but engineered processes that learn, evolve, and scale.

The future of innovation is fast, uncertain, and tech-driven. This chapter equips you to navigate that future, not with guesswork, but with systematic insight.

4.5.1. Discussion Questions

1. How does structured innovation (e.g., TRIZ, Stage-Gate, Roadmapping) reduce the risks associated with strategic innovation?
2. In what ways can first-mover and fast-follower strategies both benefit from structured innovation tools? Which do you think is more suitable in today's tech-driven markets—and why?
3. How does AI change the way organizations generate, test, and scale creative ideas? Provide examples from industries you're familiar with.
4. Compare traditional Blue Ocean Strategy with AI-enhanced Blue Ocean discovery. What are the strategic trade-offs between human intuition and algorithmic insight?
5. What role does predictive analytics play in innovation forecasting? How can it help organizations avoid being disrupted?
6. What lessons can be drawn from companies that failed to adapt to disruptive innovation (e.g., Kodak, Blockbuster)? How could structured foresight have helped them?
7. In AI-driven innovation, what ethical challenges might arise during product design or deployment? How should strategy teams anticipate and address them?
8. How can structured innovation methods like TRIZ be adapted for non-technical industries, such as education or government?
9. What are the risks of over-relying on algorithmic creativity or forecasting tools? How can leaders balance human judgment with machine intelligence?
10. Imagine you are leading a cross-functional innovation sprint. Which tools from this chapter would you prioritize—and in what order? Justify your approach based on strategic goals.

4.5.2. Micro-Case: Zara's Innovation Forecast

Zara is now the Director of Strategic Innovation at a global logistics and retail analytics firm. Her team has just won a contract to help a major international retailer reimagine its last-mile delivery model. But there's a twist: the client is based in Southeast Asia, where urban density, informal economies, and real-time consumer behaviors complicate traditional logistics models.

Zara decides not to rely on existing Western logistics templates. Instead, she assembles a cross-functional team of AI engineers, behavioral economists, urban planners, and sustainability analysts. They begin with a **Blue Ocean analysis**, but instead of relying only on strategic canvases, they integrate real-time traffic, climate, and consumer purchasing data into machine learning models. The goal? Algorithmically identify unmet delivery needs and micro-markets the client could enter with custom, on-demand services.

To mitigate future disruption, Zara launches a **technology forecasting sprint**. Her team uses scenario modeling tools to test what happens if autonomous vehicles become the norm, or if a regional carbon tax shifts transportation costs. She also runs parallel **TRIZ workshops**, focusing on resolving contradictions between sustainability goals and delivery efficiency.

But the turning point comes when Zara's team uses **AI-driven simulations** to test how consumers would react to variable delivery pricing models. Using sentiment analysis from regional social media platforms, they forecast acceptance curves for new models, enabling the client to build a dynamic, adaptive pricing structure from day one.

By the end of the project, Zara's team has not only mapped a new strategic delivery system—they've built a simulation engine that the client can continue using to adjust logistics strategy in real time. Zara's approach shows how structured innovation, fused with predictive foresight, transforms complex problems into continuously evolving advantage.

Discussion Questions

1. How does structured innovation (e.g., TRIZ, Stage-Gate) enhance a firm's strategic agility?
2. In what ways does AI enable the discovery of "blue oceans" that might be invisible to traditional analysis?
3. What lessons from Disruptive Innovation Theory should firms internalize to avoid strategic failure?
4. How can scenario planning be used to prepare for sudden shifts in consumer behavior or regulation?
5. What innovation challenges in your industry could benefit from technology forecasting?

4.5.3. Interactive Exercise: Structured Innovation Sprint

You are part of a strategy team at a global consumer electronics company. The company wants to enter a new market segment: **AI-powered wearable devices** for health tracking, focused on early detection of chronic conditions (e.g., cardiovascular, diabetes). The product must launch within 9 months across three continents.

But the challenges are immense:

* Consumer health data is highly regulated and varies across regions.
* The market is crowded with competitors—some are already launching second-generation devices.
* The product must be personalized, ethically transparent, and capable of adapting to different use environments (urban/rural, developed/developing economies).

Your team must lead a **structured innovation sprint** to guide the product strategy.

Your Task:

Using **at least two structured innovation tools** from Chapter 4 (e.g., TRIZ, AI-enhanced Blue Ocean, Technology Forecasting, Roadmapping), outline your strategic approach to innovation. Your submission should address:

1. **Identify contradictions or market gaps** (e.g., affordability vs. precision, privacy vs. personalization)
2. **Select innovation tools** and describe how you will use them (e.g., AI for uncovering blue oceans, TRIZ to solve system design constraints)
3. **Forecast and map future risks or disruptions** (e.g., regulatory change, new AI models)
4. **Design a decision logic**: What data will guide your sprint priorities? What are your go/no-go milestones?

4.5.4. Recommended HBR Case Studies

❖ **Barcelona Supercomputing Center: A strategic partner for industrial innovation**

* **Overview**: Published on February 5, 2025, and authored by M. Julia Prats and Guillermo Yanez, this 27-page case focuses on the Barcelona Supercomputing Center (BSC), a leading European research institution in high-performance computing (HPC). The case explores how BSC collaborates with industrial partners to apply cutting-edge computing technologies to real-world challenges. The narrative centers on Laura Sanchez, the CIO of an automotive company, who must evaluate whether partnering with BSC could help her firm maintain a competitive edge. The case highlights themes such as open innovation, public-private partnerships, and the strategic use of technology and analytics to accelerate R&D, optimize operations, and reduce risk.
* **Citation**: Prats, M. J., & Yanez, G. (2025). *Barcelona Supercomputing Center: A strategic partner for industrial innovation* (Case No. I00125-PDF-ENG). IESE Business School.

❖ **Four Inter Catering Group: Combining Inheritance and Innovation**

* **Overview**: Published on July 26, 2024, and authored by Haifen Lin, Xuan Wang, Yu Sun, and Yingwei Ren, this 11-page case focuses on Four Inter Catering Group Co. Ltd., a company based in Chengdu, China, known for its dedication to Sichuan cuisine and cultural heritage. Founded in 2002, Four Inter developed a business model that balanced traditional culinary inheritance with modern innovation. By 2023, as the food and beverage industry began recovering from the COVID-19 pandemic, the company faced a strategic crossroads. Its regional business model was limiting growth, and CEO Xiaohong Xu had to decide whether to deepen brand awareness in existing markets or expand into new territories. The

case explores themes of growth strategy, business model innovation, and cultural branding, making it ideal for discussions on how traditional businesses can evolve in a modern, competitive landscape.

* **Citation**: Lin, H., Wang, X., Sun, Y., & Ren, Y. (2024). *Four Inter Catering Group: Combining Inheritance and Innovation* (Case No. W37086-PDF-ENG). Ivey Publishing.

❖ **NIDO: The Innovation Lab of the Belgian Government**

* **Overview**: Published on April 1, 2025, and authored by Wim Vanhaverbeke and Riyaad Ismail, this 10-page case explores the role of NIDO, the innovation lab of the Belgian federal government, in fostering public sector innovation. NIDO was created to encourage experimentation and develop innovative solutions within government services. It had launched several notable initiatives, including the Network Innovation Program, the Federal Innovation Award, and Gov Buys Innovation. By September 2023, however, NIDO faced significant challenges: a lack of a clearly defined mission, internal resistance within its parent organization (BOSA), and uncertainty about future funding. The case centers on the strategic decisions facing NIDO's leadership, including how to better communicate its value, whether to integrate more closely with other BOSA departments, and how to secure sustainable funding—either through direct government support or alternative models like resource-sharing.
* **Citation**: Vanhaverbeke, W., & Ismail, R. (2025). *NIDO: The Innovation Lab of the Belgian Government* (Case No. W38701-PDF-ENG). Ivey Publishing.

4.5.5. Strategy-in-Film: Innovation on Screen

Innovation is rarely accidental; it is the product of deliberate processes, disciplined experimentation, and strategic foresight. The films below illustrate how organizations identify opportunities, manage innovation cycles, and adapt to changing market dynamics. As you explore innovation through a STEM lens in this chapter, use these narratives to observe how systematic approaches to innovation can drive sustained strategic advantage.

Movie / Documentary	Strategic Theme Highlighted
The Founder	Structured innovation vs. operational scaling (McDonald's systematization)
Steve Jobs	Vision-driven innovation, forecasting user needs, and product-market timing
General Magic (Documentary)	First-mover risks, forecasting failure, and tech-driven disruption
Iron Man	TRIZ-style engineering and iterative prototyping under constraint
Moneyball	Predictive analytics in decision-making, challenging traditional models
Her	AI-enabled product design, emotional algorithms, and ethical foresight
Ex Machina	Generative AI, strategic containment, and the risks of autonomous systems
The Social Network	Fast-follower adaptation, data-driven scaling, and digital platform strategy
Tucker: The Man and His Dream	Disruption theory—innovation blocked by legacy industry structures
The Great Hack (Documentary)	Algorithmic targeting, AI ethics, and the power of data in shaping behavior
Black Mirror: "Smithereens" (Episode)	Human unpredictability in AI-structured environments: limits of forecasting

References

Adeleye, I. O. (2024). The Impact of Artificial Intelligence on Design: Enhancing Creativity and Efficiency. *Journal of Engineering and Applied Sciences*, *3*(1), 1-13.

Alake, O. R., Adegbuyi, O., Babajide, A. A., Salau, O. P., Onayemi, O. O., Joel, O. O., & Adesanya, O. D. (2024). Business model innovation and SMEs' resilience: technological roadmap in Nigeria. *Business: Theory and Practice*, *25*(2), 614-624.

An, J., & Mikhaylov, A. (2024). Technology-based forecasting approach for recognizing trade-off between time-to-market reduction and devising a scheduling process in open innovation management. *Journal of Open Innovation: Technology, Market, and Complexity*, *10*(1), 100207.

Andersen, P. D., Hansen, M., & Selin, C. (2021). Stakeholder inclusion in scenario planning—A review of European projects. *Technological Forecasting and Social Change*, *169*, 120802.

Bigliardi, B., Ferraro, G., Filippelli, S., & Galati, F. (2021). The past, present, and future of open innovation. *European Journal of Innovation Management*, *24*(4), 1130-1161.

Blind, K., Kenney, M., Leiponen, A., & Simcoe, T. (2023). Standards and innovation: A review and introduction to the special issue. *Research Policy*, *52*(8), 104830.

Brandmeier, R. A., & Rupp, F. (2024). Foundations of Systematic Innovation. In *Structured Innovation: Levers to Successful Innovation Beyond Benchmarking* (pp. 1-25). Cham: Springer Nature Switzerland.

Brownlie, D. T. (1992). The role of technology forecasting and planning: formulating business strategy. *Industrial Management & Data Systems*, *92*(2), 3-16.

Celis Bueno, C., Chow, P. S., & Popowicz, A. (2025). Not "what", but "where is creativity?": towards a relational-materialist approach to generative AI. *Ai & Society*, *40*(2), 339-351.

Chandrasekera, T., Hosseini, Z., & Perera, U. (2025). Can artificial intelligence support creativity in early design processes?. *International journal of architectural computing*, *23*(1), 122-136.

Chesbrough, H. (2024). Twenty years of open innovation. *MIT Sloan Management Review*, *65*(2), 1-3.

Ciarli, T., Kenney, M., Massini, S., & Piscitello, L. (2021). Digital technologies, innovation, and skills: Emerging trajectories and challenges. *Research Policy*, *50*(7), 104289.

Cordova-Pozo, K., & Rouwette, E. A. (2023). Types of scenario planning and their effectiveness: A review of reviews. *Futures*, *149*, 103153.

Creely, E., & Blannin, J. (2025). Creative partnerships with generative AI. Possibilities for education and beyond. *Thinking Skills and Creativity*, *56*, 101727.

Dahlander, L., Gann, D. M., & Wallin, M. W. (2021). How open is innovation? A retrospective and ideas forward. *Research Policy*, *50*(4), 104218.

De Cremer, D., Bianzino, N. M., & Falk, B. (2023). How generative AI could disrupt creative work. *Harvard Business Review*, *13*.

De Weck, O. L. (2022). *Technology roadmapping and development: A quantitative approach to the management of technology*. Springer Nature.

Doshi, A. R., & Hauser, O. P. (2024). Generative AI enhances individual creativity but reduces the collective diversity of novel content. *Science Advances*, *10*(28), eadn5290.

Dutta, S., Lanvin, B., Wunsch-Vincent, S., & León, L. R. (Eds.). (2022). *Global Innovation Index 2022: What is the future of innovation-driven growth?* (Vol. 2000). WIPO.

Ebadi, A., Auger, A., & Gauthier, Y. (2022). Detecting emerging technologies and their evolution using deep learning and weak signal analysis. *Journal of Informetrics*, *16*(4), 101344.

Farida, I., & Setiawan, D. (2022). Business strategies and competitive advantage: the role of performance and innovation. *Journal of Open Innovation: Technology, Market, and Complexity*, *8*(3), 163.

Favoreu, C., Maurel, C., & Queyroi, Y. (2024). Influence of public innovation laboratories on the development of public sector ambidexterity. *Public Management Review*, 1-27.

Garcia, M. B. (2024). The paradox of artificial creativity: Challenges and opportunities of generative AI artistry. *Creativity Research Journal*, 1-14.

Göbel, S., & Wendel, V. (2016). Personalization and adaptation. *Serious games: Foundations, concepts and practice*, 161-210.

Haase, J., & Hanel, P. H. (2023). Artificial muses: Generative artificial intelligence chatbots have risen to human-level creativity. *Journal of Creativity*, *33*(3), 100066.

Hervás-Oliver, J. L., Parrilli, M. D., Rodríguez-Pose, A., & Sempere-Ripoll, F. (2021). The drivers of SME innovation in the regions of the EU. *Research Policy*, *50*(9), 104316.

Kim, J., & Maher, M. L. (2023). The effect of AI-based inspiration on human design ideation. *International Journal of Design Creativity and Innovation*, *11*(2), 81-98.

Kuang, X., & Wager, S. (2024). Weak signal asymptotics for sequentially randomized experiments. *Management Science*, *70*(10), 7024-7041.

Lichtenthaler, U. (2020). A conceptual framework for combining agile and structured innovation processes. *Research-Technology Management*, *63*(5), 42-48.

Mäntysalo, R., Granqvist, K., Duman, O., & Mladenović, M. N. (2023). From forecasts to scenarios in strategic city-regional land-use and transportation planning. *Regional Studies*, *57*(4), 629-641.

Mendonça, S., Cardoso, G., & Caraça, J. (2012). The strategic strength of weak signal analysis. *Futures*, *44*(3), 218-228.

Mourlas, C., & Germanakos, P. (Eds.). (2008). *Intelligent user interfaces: adaptation and personalization systems and technologies: adaptation and personalization systems and technologies*. IGI Global.

Muralidharan, V. (2020). A structured innovation strategy. *IEEE Engineering Management Review*, *48*(4), 12-16.

Mühlroth, C., & Grottke, M. (2018). A systematic literature review of mining weak signals and trends for corporate foresight. *Journal of Business Economics*, *88*(5), 643-687.

Osorio, F., Cruz, F., Camargo, M., Dupont, L., & Peña, J. I. (2024). Exploring team roles for social innovation labs: Toward a competence-based role self-assessment approach. *Journal of Engineering and Technology Management*, *71*, 101799.

Petrov, V. (2019). *TRIZ. Theory of inventive problem solving*. Springer International Publishing.

Ramdurai, B., & Adhithya, P. (2023). The impact, advancements, and applications of generative AI. *International Journal of Computer Science and Engineering*, *10*(6), 1-8.

Samimian-Darash, L. (2022). *Uncertainty by design: preparing for the future with scenario technology*. Cornell University Press.

Santarsiero, F., Carlucci, D., & Schiuma, G. (2024). Driving digital transformation and business model innovation in tourism through innovation labs: An empirical study. *Journal of Engineering and Technology Management*, *74*, 101841.

Savransky, S. D. (2000). *Engineering of creativity: Introduction to TRIZ methodology of inventive problem solving*. CRC Press.

Schiuma, G., & Santarsiero, F. (2023). Innovation labs as organisational catalysts for innovation capacity development: A systematic literature review. *Technovation*, *123*, 102690.

Silva, A. C. D., & Emmendoerfer, M. L. (2023). Innovation labs in South American governments: Congruencies and peculiarities. *BAR-Brazilian Administration Review*, *20*(03), e220173.

Singh, S., & Vijay, T. S. (2024). Technology roadmapping for the e-commerce sector: A text-mining approach. *Journal of Retailing and Consumer Services*, *81*, 103977.

Tavana, M., Ghasrikhouzani, M., & Abtahi, A. R. (2022). A technology development framework for scenario planning and futures studies using causal modeling. *Technology Analysis & Strategic Management*, *34*(8), 859-875.

Terninko, J., Zusman, A., & Zlotin, B. (1998). *Systematic innovation: an introduction to TRIZ (theory of inventive problem solving)*. CRC Press.

Tjandrawinata, R. R., & Budi, H. S. (2025). The Structured Innovation System: Optimizing Patent Filing and Obtaining Global Protection. *Journal of World Science*, *4*(3), 317-328.

Vinchon, F., Lubart, T., Bartolotta, S., Gironnay, V., Botella, M., Bourgeois-Bougrine, S., ... & Gaggioli, A. (2023). Artificial intelligence & creativity: A manifesto for collaboration. *The Journal of Creative Behavior*, *57*(4), 472-484.

Zhang, H. (2022). Does combining different types of innovation always improve SME performance? An analysis of innovation complementarity. *Journal of Innovation & Knowledge*, *7*(3), 100192.

Zhao, D., Tang, Z., & He, D. (2024). A systematic literature review of weak signal identification and evolution for corporate foresight. *Kybernetes*, *53*(10), 3160-3188.

CHAPTER 5

Business Models and Digital Value Architecture

"A business model is no longer a blueprint on paper it's a living algorithm shaped by data designed for adaptation and tested in real time"

by MAK

Learning Outcomes

A business model is not a static plan, but a living web—designed, connected, and always sensing the next opportunity.

By the end of this chapter, you will be able to:

- Explain the foundational elements of modern business models and articulate how they interrelate in dynamic, digitally-driven environments.
- Apply the Business Model Canvas (BMC) enhanced with data overlays to map, analyze, simulate, and refine strategic business model designs.
- Differentiate between linear, circular, and platform-based business models, evaluating their suitability and performance under various market and competitive scenarios.
- Utilize STEM-powered analytical tools, such as predictive modeling, agent-based simulations, and scenario planning, to rigorously test business model assumptions and validate strategic hypotheses.
- Conduct strategic group mapping using clustering algorithms to visualize competitive dynamics and market positioning clearly and systematically.
- Design adaptive and resilient business models capable of evolving in response to real-time data signals, market volatility, and technological disruption.
- Demonstrate an understanding of platform and ecosystem strategies, identifying network effects, data-driven design principles, and strategic orchestration methods to leverage ecosystem opportunities.

- ❖ Evaluate the strategic impact of analytics and simulation methods on business model innovation, identifying key levers for enhanced competitive advantage.
- ❖ Reflect critically on your role as a future strategic innovator, equipped to architect and evolve digitally integrated business models that respond proactively to changing environments and emerging technologies.

In today's digitally interconnected environment, the traditional notion of a business model as a fixed, static plan has become obsolete. Instead, modern business models function as dynamic, data-infused architectures, capable of continuously sensing opportunities, responding rapidly to change, and systematically innovating to create, deliver, and capture value.

This chapter explores the evolution from conventional, linear business models to flexible, digitally integrated ecosystems powered by analytics, artificial intelligence, and real-time feedback. You will discover how the strategic management of value creation has transitioned from merely selling products or services to orchestrating intelligent, adaptive networks of partnerships, data streams, and user experiences.

We will delve into proven frameworks like the Business Model Canvas, augmented with digital overlays and simulation tools, to illustrate how modern strategists prototype, test, and refine their business models before committing to large-scale investments. You'll gain insights into how platform thinking and ecosystem orchestration have reshaped competitive boundaries, enabling companies to leverage network effects, modular design, and agile adaptation as sustainable strategic advantages.

By mastering these digital and analytical tools, you will learn how to architect business models that do not just withstand market volatility but thrive within it. This chapter empowers you, as future strategic innovators, to design resilient, data-driven business models that are as responsive, adaptive, and innovative as the digital world around us.

Welcome to the era of digital value architecture, where business modeling meets STEM intelligence to shape tomorrow's competitive landscapes.

Opening Case: Ping An's Digital Rebuild – From Insurer to Intelligent Ecosystem

In the early 2000s, Ping An was a traditional insurance company, operating in a sector known more for actuarial tables than artificial intelligence. But by 2025, it had transformed into one of the world's most advanced integrated digital ecosystems, offering not just insurance, but banking, health care, investment services, and even smart city infrastructure, all delivered through a seamless, data-powered platform.

At the heart of this radical transformation was a deliberate re-engineering of Ping An's business model. Rather than viewing digital tools as enhancements to legacy operations, Ping

An reconceived its entire value architecture as modular, interconnected, and predictive. Its executives didn't ask, "How do we improve insurance services?" They asked, "What new kinds of value can we generate when insurance, AI, and user behavior data converge?"

Using a layered platform model, Ping An developed internal AI capabilities that would eventually power both consumer-facing services and back-end efficiency engines. Products like **Ping An Good Doctor**, a telemedicine app that uses AI to triage symptoms and connect users with physicians, created exponential network effects, bringing in over 400 million users and generating valuable health data. Meanwhile, its **OneConnect** fintech platform offered cloud-based risk modeling, fraud detection, and lending tools, not just for Ping An's operations, but for competitors and partners across Asia.

Behind the scenes, data lakes and AI models served as the connective tissue across business units. With predictive analytics, Ping An could simulate customer lifetime value, model churn scenarios, or identify emerging health risks before claims even occurred. Every business line, from property insurance to investment advising, was rewired around digital insight, real-time interaction, and feedback loops.

Critically, Ping An didn't just digitize its operations; it embedded simulation logic, algorithmic foresight, and modular design into its very business model. When the COVID-19 pandemic disrupted physical infrastructure across China, Ping An's virtual platforms allowed it to scale telehealth, process loans remotely, and deliver predictive insurance products without interruption. Its business model wasn't just resilient, it was anticipatory.

Today, Ping An stands not merely as a diversified company but as a living example of a **STEM-powered business model** in motion, one that learns, evolves, and scales with data at its core. Its story illustrates that the future of strategy doesn't belong to the biggest firms, but to those that can reimagine their business models as intelligent, adaptive systems.

Play the Strategist

1. How did Ping An transition from a linear business model to a platform-based architecture, and what role did data play in that shift?
2. Which elements of the Business Model Canvas would require the most redesign in a digital transformation like Ping An's? Why?
3. How did Ping An integrate predictive analytics and simulation tools into its business model, and what strategic advantages did this create?
4. What makes Ping An's business model resilient and adaptive in a volatile environment? How can these principles be applied to other industries?

5.1 Mapping Business Models in the Digital Era

The digital revolution has profoundly transformed the strategic landscape. In the past, firms could thrive by simply excelling in product innovation, operational efficiency, or market positioning. Today, competitive advantage increasingly depends not on what you sell, but how your entire organization creates, delivers, and captures value. This is the essence of **business model thinking**.

In a digital world powered by connectivity, data analytics, and real-time responsiveness, traditional, static business models have given way to dynamic, adaptive architectures. Successful strategists no longer design fixed blueprints; instead, they engineer flexible business systems capable of continual learning and real-time evolution. This section provides the foundational knowledge necessary to understand, visualize, and strategically manage these new digital business model architectures.

5.1.1 From Product Thinking to Business Model Thinking

Historically, competitive strategy focused heavily on product superiority, better features, higher quality, lower cost, or greater innovation. Companies like Sony, Ford, and Kodak exemplified product-centric strategies, achieving success through meticulous engineering, design breakthroughs, and operational excellence. But as digital technologies reshaped consumer expectations and market boundaries, product thinking alone proved insufficient.

Today, digital-native firms like Netflix, Amazon, and Spotify demonstrate that sustainable advantage often resides not within product features, but within innovative business models. Netflix's success did not originate merely from streaming technology itself but from fundamentally rethinking how customers access and consume entertainment, through subscription models, personalized recommendations, and predictive content investments driven by real-time user analytics.

This shift toward business model thinking moves the strategic focus from optimizing products and processes to orchestrating ecosystems of value. Strategic success now depends on crafting intelligent value webs, leveraging data-driven decision-making, and building platforms that can dynamically respond to market signals. Firms must continuously prototype and refine their business models, incorporating consumer insights, competitive dynamics, and technological advancements into the very fabric of their strategic architecture.

For strategists, this new reality requires fluency not only in traditional strategy tools but also in digital analytics, scenario modeling, and platform logic. The capacity to visualize and redesign the business model itself has become a crucial competitive skill.

5.1.2 Anatomy of a Modern Business Model

The **Business Model Canvas (BMC)** remains the most effective visual framework to systematically map the logic behind how companies create, deliver, and capture value. It comprises nine interconnected building blocks:

1. **Customer Segments:** Who are the primary target users or consumers?
2. **Value Propositions:** What core value is delivered to customers?
3. **Channels:** How are offerings communicated, distributed, and sold?
4. **Customer Relationships:** How are interactions managed and relationships fostered?
5. **Revenue Streams:** How does the firm earn money?
6. **Key Resources:** What assets are essential to delivering the value proposition?
7. **Key Activities:** Which processes and actions underpin value creation?
8. **Key Partnerships:** Who are critical suppliers, collaborators, or ecosystem partners?
9. **Cost Structure:** What are the main sources of expense or investment?

However, in digital contexts, each of these building blocks becomes dynamic, continuously informed by data and optimized through analytics. Customer segments can rapidly shift based on behavior analytics, value propositions can be personalized through AI, and channels can dynamically adapt through predictive modeling.

5.1.3 Digital Overlays: Augmenting the Canvas with Data

A modern business model is incomplete without data overlays, layers of real-time analytics, and predictive insights that bring clarity and dynamism to traditional frameworks. Rather than static strategic planning documents, digital overlays transform business model canvases into live decision-support systems.

For instance, Netflix continually analyzes viewer habits to dynamically refine its value proposition and optimize content investment. Similarly, Amazon overlays customer interaction data onto its channels block, using analytics to predict consumer purchasing behaviors and optimize supply chains. By embedding real-time dashboards, predictive models, and sentiment analysis directly into business model mapping, companies can iterate rapidly, pivot intelligently, and refine their strategic direction with data-driven precision.

Digital overlays also make it possible to simulate strategic decisions before execution. Firms use digital twins, virtual replicas of their business model, to test how changes in pricing, customer segmentation, or cost structures would impact financial performance, customer satisfaction, and operational efficiency. Such digital modeling transforms assumptions into validated hypotheses, allowing strategists to manage risk more effectively.

5.1.4 From Linear Models to Platform Models

While traditional business models often follow a linear logic, inputs, processes, outputs, platform business models thrive on networked interactions. Platforms, such as Airbnb or Alibaba, connect multiple user groups (producers, consumers, partners) in ecosystems that co-create value. Platform logic thus fundamentally alters strategic considerations:

* **Value Creation:** Platforms leverage network effects, exponentially increasing their value as more users engage.
* **Value Delivery:** Platforms orchestrate rather than directly deliver value, enabling interactions between multiple ecosystem players.
* **Revenue Streams:** Platforms often rely on transaction fees, subscriptions, or data monetization rather than traditional sales.

This shift requires strategists to manage ecosystems as complex adaptive systems. Questions such as "How do we enhance network effects?", "How do we design ecosystem incentives?" and "How do we foster trust among diverse participants?" become central to strategy formulation.

To navigate this complexity, organizations must understand how different business models, linear, platform-based, and circular, operate across strategic dimensions. Each model brings distinct logic, value creation mechanisms, revenue structures, and analytical toolsets. For example, linear models focus on internal efficiency and process optimization, while platform models thrive on external network effects and real-time feedback loops. In contrast, circular models prioritize sustainability, resource regeneration, and lifecycle optimization.

> "Strategy in the digital age is less about what you make, and more about how you orchestrate and dynamically evolve value creation."

The comparison below illustrates how these models differ across key strategic dimensions:

Table 5.1. Linear, Platform, and Circular Models

Dimension	Linear Model	Platform Model	Circular Model
Logic	Input → Process → Output	Ecosystem-driven interactions	Resource regeneration & reuse
Value Creation	Internal operations & efficiency	External network effects	Sustainability & lifecycle optimization
Revenue Model	Direct sales, transactional	Subscription, transaction fees, and data monetization	Extended lifecycle revenues, service models
Analytics & Tools	Process optimization, ERP systems	Network analytics, real-time dashboards, user feedback loops	Lifecycle analysis, sustainability metrics
Strategic Advantage	Operational efficiency, cost minimization	Scalability, adaptability, market expansion	Sustainability, regulatory alignment, brand reputation
Typical Examples	Manufacturing, traditional retail	Uber, Alibaba, Airbnb	Patagonia, IKEA, Philips

Examples

* **Peloton:** Combines hardware sales with subscription-based fitness content, data analytics guiding personalized content development.
* **Ant Group:** Digital financial platform leveraging predictive analytics, ecosystem partnerships, and AI-driven risk management.
* **LEGO Ideas:** Platform leveraging user-generated content and product co-creation, redefining traditional consumer roles.

5.1.5 Strategic Group Mapping with Clustering Algorithms

To make informed strategic decisions in highly competitive markets, it is essential to understand how firms are positioned relative to one another. Strategic group mapping offers a visual technique to analyze the competitive landscape by identifying clusters of firms with similar strategic characteristics, such as pricing, distribution model, value propositions, or target markets.

In a digitally advanced strategy environment, this process can be elevated by using clustering algorithms, such as **K-means**, **hierarchical clustering**, or **DBSCAN,** to detect naturally occurring groupings in multidimensional data. These groups represent strategic "neighborhoods," enabling firms to benchmark, spot white space opportunities, or identify threats from emerging competitors.

For example, a strategist using customer value proposition and revenue model data across competitors could apply K-means clustering to identify three distinct groups: traditional product-sellers, hybrid platform-product models, and full ecosystem orchestrators. A resulting strategic group map would not only clarify competitive positions but also reveal mobility barriers and attractive niches.

Clustering algorithms also complement strategic modeling. When combined with data overlays on the Business Model Canvas, firms can dynamically reposition themselves based on evolving market clusters, optimizing both differentiation and alignment with consumer demand.

Key steps in clustering-based group mapping:

1. **Select strategic dimensions** (e.g., cost structure, innovation level, delivery model).
2. **Normalize and process data** using statistical tools or Python/R packages.
3. **Apply clustering algorithms** to identify group boundaries.
4. **Visualize group positions** using 2D or 3D mapping for interpretation and action.

Clustering adds empirical rigor to competitive strategy. Instead of relying on intuition or static segmentation, strategists can continuously monitor how firm clusters evolve, detect shifts in the competitive field, and respond with precision.

5.2 Business Model Innovation and Analytics

In today's rapidly shifting business environment, sustainable competitive advantage rarely originates solely from product innovation or incremental operational improvements. Instead, firms increasingly rely on systematic business model innovation, transforming not just what they sell, but how they create, deliver, and capture value. This shift has elevated the strategic significance of analytics and data-driven modeling to unprecedented levels, making innovation processes not only systematic but predictively powerful.

> In the age of big data, strategic mapping is no longer drawn by hand—it's discovered through algorithms.

This section explores the role of analytics, simulations, and evidence-based experimentation in innovating and refining business models. We'll demonstrate why data-driven business model innovation matters strategically more than traditional product-centric approaches, and how firms leverage STEM-enhanced analytical tools to systematically innovate, validate, and adapt their business models.

5.2.1 Why Business Model Innovation Matters More Than Product Innovation

Historically, competitive differentiation focused heavily on superior product features, pricing, or brand strength. However, digital disruption and rapid technological advancement have substantially reduced the competitive lifespan of product-based advantages. Firms like Nokia, Kodak, and Blockbuster exemplify how once-dominant products can quickly become obsolete when strategic emphasis is placed solely on product innovation without business model evolution.

Today, industry leaders, such as Amazon, Spotify, and Tesla, demonstrate that enduring competitive advantage is more likely to result from innovative business models that continuously adapt to new opportunities and disruptions. Spotify disrupted traditional music distribution not simply by offering streaming, but by completely transforming the monetization model from pay-per-download to subscription-based streaming. Tesla's competitive differentiation relies not just on electric vehicles but on a vertically integrated model that combines software-driven vehicles, direct-to-consumer sales, and a digital ecosystem approach.

Business model innovation matters precisely because it reshapes competitive boundaries. It moves the strategic focus beyond incremental product improvements toward creating unique value ecosystems. Strategists must now continuously reevaluate business models, leveraging analytics and data-driven insights, to proactively adapt and shape their competitive positions.

5.2.2 Using Analytics to Identify Business Model Levers

Strategic analytics play a critical role in identifying, testing, and validating the components of a robust business model. Firms deploy advanced analytics, such as predictive modeling, customer sentiment analysis, and A/B testing, to pinpoint precisely where and how their business models can generate the most impactful strategic outcomes.

For example, Netflix systematically employs predictive analytics to identify which original shows to develop, minimizing risks associated with high-cost content production. Likewise, Airbnb analyzes vast amounts of user-generated data to refine host recommendations, pricing structures, and platform features dynamically. These analytics not only enhance the customer experience but also drive business model refinements that directly impact revenue, cost structure, and competitive positioning.

The strategic advantage of analytics is the capability to rapidly detect and adjust to market shifts, consumer preferences, and competitive threats. Analytics-driven companies proactively identify which segments, products, or channels will yield the highest returns, providing agility and precise control over their business models.

5.2.3 From Assumptions to Evidence: Hypothesis-Driven Modeling

The core of business model innovation lies in systematically testing and validating strategic assumptions. Traditional strategy relied heavily on intuition and experience; today's strategy emphasizes empirical validation through structured hypothesis-driven experiments.

Hypothesis-driven modeling starts by clearly defining strategic assumptions, such as customer willingness to pay, market adoption rates, or competitor reactions. Using approaches such as the Lean Canvas or Minimum Viable Products (MVPs), companies iteratively test these assumptions in real-world conditions, gathering data-driven evidence to inform strategic decisions.

AI-driven tools further enhance hypothesis testing. Companies leverage machine learning algorithms and simulations to run multiple innovation scenarios concurrently, evaluating potential outcomes before committing significant resources. Firms such as Google and Amazon continuously experiment with user experiences, pricing strategies, and product features, refining their business models based on real-time feedback and predictive analytics.

5.2.4 Simulation and Scenario Analysis for Strategic Resilience

Business model resilience requires understanding how models perform under diverse future scenarios, including market disruptions, technological shifts, and competitive surprises. Advanced simulation tools, such as Monte Carlo simulations, agent-based modeling, and dynamic scenario planning, enable organizations to systematically explore potential futures, anticipate risks, and proactively develop adaptive strategies.

Monte Carlo simulations, for example, allow companies to model thousands of possible outcomes based on varying assumptions of customer behavior, cost fluctuations, or competitor actions. Similarly, agent-based modeling enables firms to simulate complex interactions between customers, competitors, and markets, revealing hidden vulnerabilities or untapped opportunities.

Dynamic scenario planning helps firms prepare strategically for uncertain futures by systematically modeling scenarios ranging from economic shocks to new regulatory constraints. These methods provide robust, forward-looking insights that empower decision-makers to create business models designed not merely to survive disruptions, but to capitalize upon them strategically.

To support this, organizations are increasingly turning to strategic analytics, a suite of methods that combine data science, behavioral modeling, and experimentation to inform business model innovation. For example, predictive modeling helps forecast customer behavior and revenue patterns, enabling pricing optimization and churn reduction. Sentiment analysis taps into consumer perceptions, guiding product development and brand positioning through real-time social listening.

When uncertainty is high, Monte Carlo simulations model a wide range of possible outcomes, helping firms stress-test financial plans and strategic bets. Agent-based modeling simulates how individual actors—such as customers, competitors, or partners- interact within complex ecosystems, revealing emergent dynamics in platform strategies. For tactical experimentation, A/B testing validates product-market assumptions, while Lean Canvas and MVP testing refine strategic hypotheses through rapid iteration and customer feedback.

> Strategic resilience is not accidental, it is systematically engineered through rigorous analytics, empirical validation, and scenario modeling.

These tools and their strategic applications are summarized below:

Table 5.2: Strategic Analytics for Business Model Innovation

Analytical Method	Purpose & Strategic Use	Example Applications	Tools & Techniques
Predictive Modeling	Forecast customer behavior, revenue	Pricing optimization, customer churn	Machine learning, regression analysis
Sentiment Analysis	Understand consumer perceptions	Product development, brand positioning	NLP, social media analytics
Monte Carlo Simulation	Model risks and strategic uncertainties	Financial modeling, scenario planning	Excel, @Risk, Python/R simulation
Agent-based Modeling	Simulate market dynamics and interactions	Platform ecosystem modeling	NetLogo, AnyLogic
A/B Testing	Validate product-market assumptions	UI/UX, conversion optimization	Google Optimize, Optimizely
Lean Canvas/MVP Testing	Test and refine strategic assumptions	Product-market fit, business validation	Customer interviews, online surveys

Examples: Analytics-Driven Business Model Innovation

* **Zara** uses predictive analytics and real-time sales data to optimize inventory and dynamically adjust its fast-fashion business model.
* **Peloton** leverages user engagement data and predictive models to innovate its subscription and hardware integration.

* **Uber** applies A/B testing extensively to refine pricing strategies, enhance user experiences, and improve driver retention.

5.3 Platform Strategies and Ecosystem Modeling

In the digital economy, strategic success increasingly depends not only on individual products or services, but also on how effectively companies can orchestrate interactions among diverse ecosystem participants. Platform strategies represent a fundamental shift in how value is created and captured, emphasizing network effects, user engagement, and data-driven optimization.

Platforms such as Airbnb, Alibaba, and Uber have redefined competitive advantage by leveraging sophisticated data analytics, predictive modeling, and ecosystem design. This section explores the strategic architecture of platform business models, highlighting the critical role of STEM-based tools such as analytics, digital twins, game theory, and simulations in successfully managing and scaling platform ecosystems.

5.3.1 Understanding Platforms as Strategic Architectures

Unlike traditional linear business models, platforms succeed by facilitating value-generating interactions between independent participants, consumers, providers, suppliers, and third-party developers. Rather than producing products, platforms create ecosystems that allow multiple groups to co-create and exchange value dynamically. These interactions often produce **network effects**, where the platform's value grows exponentially as more users join and participate.

Airbnb exemplifies this strategy: its value arises from facilitating exchanges between hosts and guests. Each additional participant enhances the attractiveness and usefulness of the platform to other users. Platforms are thus uniquely positioned strategically, not through direct ownership of assets, but through the orchestration of interactions, trust-building, and ecosystem incentives.

Strategically, platforms require nuanced ecosystem management. Platform operators must systematically manage risks such as market imbalance (too many providers or users), loss of trust, or competitive entry. Success depends not just on technical excellence, but on strategic orchestration capabilities and continual, data-driven adaptation.

5.3.2 Data-Driven Design of Platform Models

Platforms thrive on real-time, actionable insights derived from data analytics. Successful platforms continuously optimize user engagement, customer satisfaction, and platform health metrics through advanced analytics. Techniques such as churn prediction, network analytics, and engagement modeling inform every strategic decision, from pricing to feature launches.

For example, Spotify utilizes deep analytics to personalize content recommendations and optimize user retention. Uber employs sophisticated real-time analytics to manage supply-demand balances, dynamically pricing rides, and minimizing customer wait times. Such platforms leverage vast streams of user-generated data to drive constant platform improvements and personalized experiences.

Analytics also enables predictive ecosystem modeling, allowing platforms to anticipate shifts in user behavior, competitor moves, or regulatory changes before they occur. Strategically, this capacity transforms platform management from reactive troubleshooting into proactive ecosystem shaping.

5.3.3 Ecosystem Strategy: Beyond the Platform Core

Successful platforms strategically leverage ecosystem partners, extending their reach and enriching their offerings. Platform strategy involves orchestrating interactions not just between consumers and producers, but across a broader network of third-party innovators, complementary businesses, and even competitors.

Apple's App Store demonstrates strategic ecosystem management, providing a robust infrastructure that encourages thousands of external developers to innovate on its platform. Amazon Web Services similarly positions itself strategically by offering cloud infrastructure that enables ecosystem participants, from startups to multinational corporations, to rapidly innovate, scale, and deliver new services.

However, managing an ecosystem strategically involves balancing openness and control. Platforms must provide sufficient openness to attract innovation, yet maintain governance standards to ensure quality, security, and user trust. Platforms systematically use analytics and modeling tools, such as ecosystem simulations, to navigate these trade-offs, dynamically adjusting governance policies based on data-driven insights.

5.3.4 Simulating Platform Dynamics with STEM Tools

Advanced STEM tools play a critical role in understanding and managing platform ecosystems. Firms employ techniques such as agent-based modeling, game theory, and digital twins to simulate complex interactions among ecosystem participants and anticipate how strategic choices will affect platform dynamics.

Agent-based modeling enables strategists to simulate interactions between thousands of independent actors (users, suppliers, competitors), identifying potential network bottlenecks, pricing sensitivities, or incentive structures that could maximize platform engagement. Game theory further aids strategic decision-making by modeling competitive behaviors and interactions, helping platforms anticipate competitor reactions and optimize strategic moves.

Digital twins, virtual replicas of platform ecosystems, allow firms to test innovations, policy changes, or crisis scenarios in a risk-free virtual environment before implementation. Platforms like Uber and Airbnb use digital twins extensively, modeling scenarios like demand surges, regulatory impacts, or competitor pricing strategies, ensuring strategic resilience.

To effectively manage these ecosystems, firms must also monitor key performance metrics that reflect platform health, user behavior, and systemic sustainability. These metrics guide strategic interventions and continuous optimization, as outlined below:

Table 5.3. Key Metrics for Platform Ecosystem Management

Ecosystem Metric	Strategic Relevance	Measurement Methods	Strategic Actions
Network Effects	Drives exponential growth & competitive advantage	User growth rates, referral metrics	Encourage user referrals, strategic partnerships
User Churn Rate	Indicates platform health & retention	Predictive churn modeling	Targeted retention strategies, dynamic pricing
Matchmaking Efficiency	Reflects quality of interactions & satisfaction	Wait time analysis, recommendation accuracy	Enhance AI recommendation engines, real-time adjustments
User Engagement	Directly impacts revenue & loyalty	Session analytics, engagement scores	Personalization strategies, gamification features
Ecosystem Health	Ensures long-term sustainability & growth	Supplier quality ratings, contributor satisfaction	Adjust platform incentives, refine governance policies

Examples: Platform Strategy in Action

> Platform success comes not from controlling assets but from mastering the strategic orchestration of dynamic interactions within data-rich ecosystems.

* **Alibaba:** Uses analytics-driven ecosystem modeling to balance the interests of buyers, sellers, and third-party service providers, continuously optimizing market dynamics.
* **Uber:** Employs real-time demand forecasting and surge pricing simulations, maintaining marketplace equilibrium and optimizing user experiences.

* **Apple:** Strategically manages ecosystem openness and security via analytics-informed app approval processes and developer incentives, fostering ecosystem innovation.

5.4 Business Model Resilience and Adaptive Architecture

In an era marked by uncertainty, volatility, and rapid technological shifts, strategic resilience has emerged as a vital competitive advantage. Business model resilience refers to the ability of an organization's strategic architecture not merely to withstand disruption, but to adapt, evolve, and thrive within it. Resilient business models are inherently flexible, continuously responsive, and proactively designed for dynamic market conditions.

This section delves into the principles and practices of adaptive business model architectures, highlighting how strategic use of simulation, analytics, modular design, and scenario analysis can systematically enhance resilience. Through STEM-powered tools and frameworks, organizations can build business models capable of anticipating, responding to, and capitalizing on disruptive changes.

5.4.1 What Makes a Business Model Resilient?

A resilient business model differs fundamentally from conventional static designs. While traditional business models often rely on stability, predictability, and controlled environments, resilient models incorporate flexibility, scenario readiness, and adaptive capacity into their core architecture.

Strategically resilient business models exhibit three defining characteristics:

* **Adaptability:** The ability to swiftly alter core elements such as customer segments, revenue models, or value propositions in response to changing market conditions.
* **Anticipatory Logic:** Systematic use of forecasting, scenario planning, and predictive analytics to anticipate disruptions before they occur.
* **Modular Structure:** Architectural flexibility allowing independent business units or value streams to pivot without destabilizing the broader organization.

Companies like Amazon, Ping An, and Tesla exemplify resilience by continually adapting to external disruptions, economic downturns, technological breakthroughs, regulatory shifts, and proactively restructuring their strategic architecture. Their strategic success emerges from a systematic, disciplined approach to managing uncertainty, rather than mere reaction or improvisation.

5.4.2 Simulation and Adaptive Strategy Design

To achieve adaptive resilience, strategic leaders rely heavily on advanced simulation techniques, such as Monte Carlo simulations, scenario modeling, and agent-based models, to systematically test strategic decisions before implementation. Simulations allow firms to explore thousands of potential futures, measure sensitivity to disruptions, and evaluate the robustness of business model choices.

For instance, Monte Carlo simulations help organizations assess how external shocks, such as economic recessions, regulatory shifts, or competitor entry, could impact key revenue streams, operational costs, or market positions. Such simulations produce probability-based insights, enabling decision-makers to quantify risks and systematically prioritize strategic adaptations.

Agent-based modeling further enhances resilience by simulating interactions within ecosystems or markets. Firms can model how customer segments, competitors, or suppliers might respond to strategic decisions under various scenarios. Thus, adaptive strategies become proactive rather than reactive, systematically designed for robustness under uncertainty.

5.4.3 Scenario Analysis and Strategic Flexibility

Scenario analysis is critical for strategic resilience, enabling organizations to proactively develop strategic responses for multiple plausible future outcomes. Scenario modeling systematically challenges organizational assumptions and uncovers hidden vulnerabilities, preparing the business model architecture for flexibility in uncertain environments.

Organizations such as Shell and IBM have long used scenario analysis as a strategic tool, systematically considering divergent futures, economic shifts, geopolitical developments, and disruptive technologies, to develop strategic agility. Today, scenario analysis is enhanced through STEM-based predictive analytics and digital twins, increasing accuracy, clarity, and strategic preparedness.

Digital twins, for example, simulate entire business models, allowing strategists to virtually test the impact of strategic decisions, such as entering new markets, launching new services, or adjusting pricing strategies. This systematic scenario-testing capability significantly reduces strategic risks and enhances adaptive flexibility.

5.4.4 Modular Business Model Architecture

At the structural level, modular business model design greatly enhances resilience by allowing strategic pivots without destabilizing overall operations. Modular design breaks the business model into discrete, interchangeable components, each with clearly defined strategic functions. If market conditions change, modules can be rapidly reconfigured, adapted, or replaced, minimizing disruption to the broader strategic system.

Apple's product ecosystem demonstrates modular resilience, with discrete modules, hardware, software, content, and services that can evolve independently and rapidly adapt to market dynamics. Similarly, Amazon's modular architecture separates logistics, cloud computing, and retail operations, enabling strategic flexibility in rapidly changing market environments.

Modularity also facilitates systematic experimentation, rapid scaling, and continuous innovation. Firms can test new strategic modules (new product segments, pricing strategies, or market-entry approaches) independently, integrating successful innovations seamlessly into the broader model.

To support this modular agility, organizations increasingly rely on advanced tools that simulate, test, and optimize business model components under uncertainty. These tools are outlined below:

Table 5.4: Tools for Enhancing Business Model Resilience

Tool/Technique	Strategic Role	Example Application	Strategic Impact
Monte Carlo Simulations	Risk quantification, scenario testing	Financial stress-testing, market volatility	Precise risk management
Agent-based Modeling	Ecosystem interaction & strategic response	Competitor reactions, customer dynamics	Anticipatory strategic agility
Scenario Analysis	Multiple futures modeling, strategic flexibility	Regulatory impact, tech disruption	Proactive decision-making
Digital Twins	Real-time simulation of business model impact	Product launch impacts, market entry	Reduced uncertainty & risk
Modular Design	Flexible adaptation of model components	Product line pivots, strategic reorientation	Rapid adaptability & innovation

Examples: Adaptive Resilience in Practice

- **Amazon:** Continuously adapts its modular business model by testing and scaling new modules such as AWS, Alexa, and Prime Video, leveraging advanced analytics and scenario modeling.
- **Tesla:** Employs simulation extensively to test market acceptance, production scenarios, and regulatory impacts before significant strategic commitments, enabling rapid strategic pivots.
- **Shell:** Uses robust scenario planning and predictive analytics systematically to adapt strategic decisions to changing geopolitical and market conditions, enhancing long-term resilience.

By embracing these STEM-driven techniques, organizations systematically embed resilience into their strategic DNA, enabling them not merely to survive disruptions but to strategically leverage them for competitive advantage.

5.5. Chapter 5 Wrap-Up: Business Models in a Digital Architecture

As we conclude Chapter 5, one truth is undeniable: business models are no longer static frameworks. In the STEM-driven era, they have become dynamic systems, designed, simulated, and adapted with the same precision as engineered prototypes. The most resilient organizations do not simply "have" a business model; they architect it, evolve it, and reimagine it continuously based on real-time data and predictive insights.

In this chapter, we moved beyond the traditional Business Model Canvas to explore how STEM tools such as simulations, predictive modeling, and scenario analysis empower firms to refine every dimension of their value architecture. We examined the rise of platform strategies and ecosystem modeling, where firms like Ping An, Uber, and Apple create and orchestrate complex, data-driven networks rather than just sell products or services.

Through algorithmic strategic group mapping and modular architectures, we explored how businesses can test assumptions, anticipate disruption, and design models capable of rapid adaptation. We introduced the concept of adaptive resilience, not as a reactive stance, but as a built-in strategic capability powered by simulation, scenario thinking, and modular design.

> True strategic resilience isn't reactive, it is proactively designed into the very architecture of your business model through simulation, modularity, and adaptive foresight.

This chapter also emphasized the power of strategic modeling, not only to improve decision accuracy but to unlock new markets, optimize pricing, and align with emerging ethical and regulatory standards. From platform ecosystems to business model transformation, STEM isn't just supporting strategy; it is strategy.

5.5.1. Discussion Questions

1. How does predictive simulation reshape how we evaluate business model success?
2. What distinguishes a platform business model from a traditional linear model?
3. In what ways can modular design improve a company's resilience during market shifts?
4. How do you define "ecosystem orchestration," and why is it central to platform strategy?
5. What metrics would you track to measure the health of a platform business?
6. How can digital twins be used to test new business model innovations?
7. What ethical or governance issues must be considered when using AI in value modeling?
8. How would you redesign a legacy business model using data overlays?
9. Why is adaptability more important than efficiency in volatile markets?
10. What tools from this chapter would you apply if launching your own venture?

5.5.2. Micro-Case: Zara's Modular Reinvention

Zara, now Chief Strategy Architect at a global fashion-tech firm, faces her toughest challenge yet: rebuilding the firm's business model after AI-driven disruptions have upended global retail. Supply chains are volatile, customer behavior is unpredictable, and platform-native competitors are gaining traction.

But Zara is undeterred. She begins by identifying weak links in the legacy value chain, especially the rigid production cycles and region-specific pricing models. Drawing on the Business Model Canvas with real-time data overlays, she redesigns the value proposition to prioritize predictive personalization and adaptive logistics.

Zara implements a **modular architecture** where each region operates semi-autonomously but syncs through a shared analytics platform. Using **digital twins**, her team simulates demand spikes, supply disruptions, and market-entry scenarios. Each unit is empowered to run micro-experiments with pricing, product bundling, and influencer partnerships, ensuring the entire model evolves in real time.

Her KPI dashboard tracks resilience factors: real-time sales elasticity, inventory velocity, customer re-engagement, and external shock absorption rates. As the system adapts, Zara fine-tunes platform partnerships for last-mile delivery and AI-based style prediction, balancing innovation with risk controls.

In just four quarters, Zara's re-architecture reduces cycle time by 40%, boosts customer lifetime value by 28%, and earns praise from investors for strategic adaptability. Her story reflects the future of strategic management, where STEM meets storytelling, simulation meets strategy, and adaptability is engineered into every decision.

Micro-Case Discussion Questions

1. Which part of Zara's redesigned business model best reflects adaptive resilience?
2. How could agent-based modeling help test her pricing strategies?
3. What governance risks might arise from decentralized regional experimentation?
4. How would you use KPIs to monitor innovation without sacrificing consistency?

5.5.3. Interactive Simulation: Platform Pivot or Product Focus?

You are the Chief Strategy Officer of a growing digital fitness startup that began with one product: a smart treadmill. Now you must decide whether to:

* Double down on the product with enhanced features and marketing (product focus), or
* Evolve the business model into a fitness data platform with third-party integrations (platform pivot).

Your Tools:

* Use strategic group mapping to visualize the market landscape.
* Run a business model simulation using scenario analysis (e.g., growth under partner integration vs. solo product development).

Your Task:
Draft a 1-page recommendation that includes:

* Visual of the strategic group map.
* Risks and benefits of each path.
* Key resilience strategies and metrics.

5.5.4. Recommended HBR Case Studies for Chapter 5

❖ CSL: Rebranding "The Biggest Company No One's Ever Heard Of"

- **Overview**: Published on October 13, 2023, and authored by Bertrand Guillotin, Sheri Lambert, and Phoebe Dickinson, this 16-page case focuses on CSL Limited, a major Australian biopharmaceutical company. Despite its global scale and scientific achievements, CSL struggled with brand recognition outside its industry. In October 2021, CSL's head of digital communications and branding was tasked with leading a strategic rebranding initiative. By August 2022, the company had successfully implemented a new brand strategy that improved its brand associations and net promoter score. The case explores how CSL aligned its corporate brand with its product brands, maintained trust, and positioned itself to attract top talent and outpace competitors. It raises key questions about brand positioning, strategic communication, and long-term brand management in a science-driven, global enterprise.
- **Citation**: Guillotin, B., Lambert, S., & Dickinson, P. (2023). *CSL: Rebranding "The Biggest Company No One's Ever Heard Of"* (Case No. W34093-PDF-ENG). Ivey Publishing.

❖ Uber: Competing Globally

- **Overview**: Authored by Alexander J. MacKay, Amram Migdal, and John Masko, and published on April 14, 2020, this 29-page case explores Uber's international expansion strategy and how it navigated regulatory challenges, local competition, and market dynamics in six major cities: New York City, Bogotá, Delhi, Shanghai, Accra, and London. The case provides a comparative look at how Uber adapted its business model to different legal, cultural, and economic environments. It also examines the company's nonmarket strategy, including its interactions with regulators and its efforts to shape public opinion and policy. This case is ideal for discussions on global strategy, disruptive innovation, market entry, and regulatory economics.

- **Citation**: MacKay, A. J., Migdal, A., & Masko, J. (2020). *Uber: Competing Globally* (Case No. 720404-PDF-ENG). Harvard Business School.

❖ KNOLSKAPE: Transforming Learning Dynamics

- **Overview**: Published on February 1, 2023, and authored by Srinivasan R, Satya Nandini A, and Srividya V, this 15-page case traces the evolution of KNOLSKAPE, an edtech company that began as a provider of simulation games for academia and transformed into a full-stack learning and development platform for corporations worldwide. The case is set during the 2019–2020 period, when the COVID-19 pandemic disrupted traditional learning models and accelerated the shift to online and hybrid learning. It explores how KNOLSKAPE transitioned from a pipeline business model to a platform-based model, integrating technology, content, and services to deliver personalized learning journeys. Students are invited to analyze KNOLSKAPE's platform architecture, including decisions around open vs. closed systems, proprietary vs. shared content, and the challenges of global expansion. The case also touches on strategic issues like scalability, innovation, and digital transformation in the learning industry.
- **Citation**: Srinivasan, R., Nandini, S. A., & Srividya, V. (2023). *KNOLSKAPE: Transforming Learning Dynamics* (Case No. IMB939-PDF-ENG). Indian Institute of Management Bangalore.

❖ Netflix (A): Business model innovation (Cartoon case)

- **Overview**: This graphic (cartoon-style) case, published by INSEAD and authored by Felipe Monteiro, presents the early story of Netflix, focusing on its business model innovation from its founding in 1997 through its rapid growth to over 4 million subscribers by 2007. The case uses a visual storytelling format to illustrate how Reed Hastings and his team disrupted the traditional video rental industry with a mail-based DVD subscription model, challenging incumbents like Blockbuster. It highlights the strategic decisions that enabled Netflix to scale, innovate, and eventually pivot toward streaming. This format is designed to engage students visually and is ideal for discussions on disruption, business model evolution, and strategic innovation.
- **Citation**: Monteiro, F. (2014). *Netflix (A): Business model innovation (Cartoon case)* (Case No. IM1417). INSEAD.

5.5.5. Strategy-in-Film: Learn by Watching

Business models are not fixed; they evolve in response to shifting technologies, customer needs, and competitive landscapes. The films below depict how organizations reimagine value creation, leverage digital platforms, and transform their operating models. As you explore business model innovation in this chapter, use these narratives to reflect on how strategic adaptability can unlock new forms of growth and competitive advantage.

Movie / Documentary	Strategic Theme Highlighted
The Founder	Business model design, franchising, scalability
Startup.com	Platform scaling, ecosystem collapse
The Great Hack	Platform monetization, data ethics
The Social Dilemma	Strategic risk in platform ecosystems
General Magic	Innovation ahead of market readiness
The Inventor: Out for Blood in Silicon Valley	Ethics in tech-based business models
Money Heist	Simulation, contingency planning
Inside Bill's Brain	Modular philanthropy, platform thinking
WeCrashed	Platform hype vs. value reality
Zara: The True Story of the World's Richest Man	Agile architecture and supply chain innovation

Reference

Amit, R., & Zott, C. (2020). *Business model innovation strategy: Transformational concepts and tools for entrepreneurial leaders*. John Wiley & Sons.

Andersen, T. C. K., Aagaard, A., & Magnusson, M. (2022). Exploring business model innovation in SMEs in a digital context: Organizing search behaviours, experimentation and decision-making. *Creativity and Innovation Management*, *31*(1), 19-34.

Audretsch, D. B., Belitski, M., Caiazza, R., Drapeau, M. D., Menter, M., & Wales, W. J. (2024). Resilience and digitally-advanced entrepreneurship. *Entrepreneurship & Regional Development*, *36*(1-2), 1-9.

Autio, E., Nambisan, S., Thomas, L. D., & Wright, M. (2018). Digital affordances, spatial affordances, and the genesis of entrepreneurial ecosystems. *Strategic Entrepreneurship Journal*, *12*(1), 72-95.

Bouhalleb, A., & Smida, A. (2020). Exploring the relationship between scenario planning and strategic flexibility and complexity. *European Journal of International Management*, *14*(3), 476-493.

Bresciani, S., Huarng, K. H., Malhotra, A., & Ferraris, A. (2021). Digital transformation as a springboard for product, process, and business model innovation. *Journal of Business Research*, *128*, 204-210.

Breuer, H., Fichter, K., Lüdeke-Freund, F., & Tiemann, I. (2018). Sustainability-oriented business model development: Principles, criteria, and tools. *International Journal of Entrepreneurial Venturing*, *10*(2), 256-286.

Bromiley, P., & Rau, D. (2014). Towards a practice-based view of strategy. *Strategic Management Journal*, *35*(8), 1249-1256.

Cenamor, J., Parida, V., & Wincent, J. (2019). How entrepreneurial SMEs compete through digital platforms: The roles of digital platform capability, network capability, and ambidexterity. *Journal of Business Research*, *100*, 196-206.

Chesbrough, H. (2020). To recover faster from Covid-19, open up: Managerial implications from an open innovation perspective. *Industrial Marketing Management*, *88*, 410-413.

Ciasullo, M. V., & Lim, W. M. (2022). Digital transformation and business model innovation: Advances, challenges, and opportunities. *International Journal of Quality and Innovation*, *6*(1), 1-6.

Ciampi, F., Demi, S., Magrini, A., Marzi, G., & Papa, A. (2021). Exploring the impact of big data analytics capabilities on business model innovation: The mediating role of entrepreneurial orientation. *Journal of Business Research*, *123*, 1-13.

Daymond, J., Knight, E., Rumyantseva, M., & Maguire, S. (2023). Managing ecosystem emergence and evolution: Strategies for ecosystem architects. *Strategic Management Journal*, *44*(4), O1-O27.

Foss, N. J., & Saebi, T. (2018). Business models and business model innovation: Between wicked and paradigmatic problems. *Long range planning*, *51*(1), 9-21.

Fu, X., Ghauri, P., Ogbonna, N., & Xing, X. (2023). Platform-based business model and entrepreneurs from Base of the Pyramid. *Technovation*, *119*, 102451.

Gómez, J., Krammer, S. M., Pérez-Aradros, B., & Salazar, I. (2024). Resilience to the pandemic: The role of female management, multi-unit structure, and business model innovation. *Journal of Business Research*, *172*, 114428.

Grego, M., Magnani, G., & Denicolai, S. (2024). Transform to adapt or resilient by design? How organizations can foster resilience through business model transformation. *Journal of Business Research*, *171*, 114359.

Iansiti, M., & Lakhani, K. R. (2020). *Competing in the age of AI: Strategy and leadership when algorithms and networks run the world*. Harvard Business Press.

Kretschmer, T., Leiponen, A., Schilling, M., & Vasudeva, G. (2022). Platform ecosystems as meta-organizations: Implications for platform strategies. *Strategic Management Journal*, *43*(3), 405-424.

Langley, D. J., Van Doorn, J., Ng, I. C., Stieglitz, S., Lazovik, A., & Boonstra, A. (2021). The Internet of Everything: Smart things and their impact on business models. *Journal of Business Research*, *122*, 853-863.

Lüdeke-Freund, F., Carroux, S., Joyce, A., Massa, L., & Breuer, H. (2018). The sustainable business model pattern taxonomy—45 patterns to support sustainability-oriented business model innovation. *Sustainable Production and Consumption*, *15*, 145-162.

Mirghaderi, S. A., Aboumasoudi, A. S., & Amindoust, A. (2023). Developing an open innovation model in the startup ecosystem industries based on the attitude of organizational resilience and blue ocean strategy. *Computers & Industrial Engineering*, *181*, 109301.

Mishra, S., & Tripathi, A. R. (2021). AI business model: an integrative business approach. *Journal of Innovation and Entrepreneurship*, *10*(1), 18.

Nambisan, S., Wright, M., & Feldman, M. (2019). The digital transformation of innovation and entrepreneurship: Progress, challenges, and key themes. *Research Policy*, *48*(8), 103773.

Napier, E., Liu, S. Y., & Liu, J. (2024). Adaptive strength: Unveiling a multilevel dynamic process model for organizational resilience. *Journal of Business Research*, *171*, 114334.

Pooya, G., Cheng, N., Marshall, A., Dencik, J., & Agrawal, N. (2021). Business success in the platform economy depends on the right ecosystem strategy and execution. *Strategy & Leadership*, *49*(5), 23-30.

Prasetyo, E. H. (2024). Digital platforms' strategies in Indonesia: Navigating between technology and informal economy. *Technology in Society*, *76*, 102414.

Ritala, P. (2012). Coopetition strategy–when is it successful? Empirical evidence on innovation and market performance. *British Journal of Management*, *23*(3), 307-324.

Settembre-Blundo, D., González-Sánchez, R., Medina-Salgado, S., & García-Muiña, F. E. (2021). Flexibility and resilience in corporate decision making: a new sustainability-based risk management system in uncertain times. *Global Journal of Flexible Systems Management*, *22*(Suppl 2), 107-132.

Tavoletti, E., Kazemargi, N., Cerruti, C., Grieco, C., & Appolloni, A. (2022). Business model innovation and digital transformation in global management consulting firms. *European Journal of Innovation Management*, *25*(6), 612-636.

Vaska, S., Massaro, M., Bagarotto, E. M., & Dal Mas, F. (2021). The digital transformation of business model innovation: A structured literature review. *Frontiers in Psychology*, *11*, 539363.

Wirtz, B. W., Göttel, V., & Daiser, P. (2016). Business model innovation: Development, concept, and future research directions. *Journal of Business Models*, *4*(1).

Zhang, H., Xiao, H., Wang, Y., Shareef, M. A., Akram, M. S., & Goraya, M. A. S. (2021). An integration of antecedents and outcomes of business model innovation: A meta-analytic review. *Journal of Business Research*, *131*, 803-814.

Zhang, X., Antonialli, F., Bonnardel, S. M., & Bareille, O. (2024). Where business model innovation comes from and where it goes: a bibliometric review. *Creativity and Innovation Management*, *33*(2), 109-126.

Zhang, Y., Ma, X., Pang, J., Xing, H., & Wang, J. (2023). The impact of digital transformation of manufacturing on corporate performance—The mediating effect of business model innovation and the moderating effect of innovation capability. *Research in International Business and Finance*, *64*, 101890.

Zhou, Q., Wang, S., Ma, X., & Xu, W. (2024). Digital technologies and corporate green innovation: opening the "black box" of resource orchestration mechanisms. *Sustainability Accounting, Management and Policy Journal*, *15*(4), 884-912.

Zhao, W., & Liu, W. (2025). Co-evolution of AI-driven business models, innovation, and business ecosystems in Chinese cross-border e-commerce firm: an exploratory study based on a single case. *Asia Pacific Business Review*, 1-31.

Chapter 6

Corporate Strategy In a STEM World: Navigating Growth, Diversification, & Ecosystems

"In the STEM driven era corporate strategy is no longer about empire building—it's about ecosystem shaping data powered foresight and agile orchestration across industries"

by MAK

A chipmunk doesn't build a forest, it curates its ecosystem, storing resources wisely, sensing change early, and adapting with precision. That's corporate strategy in a STEM world.

Learning Outcomes

By the end of this chapter, you will be able to:

- Define corporate strategy and distinguish it from business-level strategy, understanding its evolving role in multi-business organizations.
- Identify classical corporate strategies, including growth, diversification, vertical integration, and portfolio management, and recognize their limitations in a modern data-driven environment.
- Apply predictive analytics and AI tools to enhance corporate decision-making, particularly in diversification, mergers and acquisitions (M&A), and resource allocation.
- Use simulation models and system dynamics to evaluate corporate expansion strategies under uncertainty.
- Assess the role of ecosystem strategy as an alternative to traditional vertical integration, and understand how platform-based strategies redefine corporate boundaries.
- Diagnose corporate strategic coherence using STEM-enhanced tools, ensuring alignment across business units and strategic objectives.
- Critically evaluate corporate portfolio decisions with a focus on real-time data integration, dynamic modeling, and strategic fit in complex markets.

- ❖ Develop STEM-augmented corporate strategies that are adaptive, evidence-based, and aligned with long-term value creation in volatile environments.

Corporate strategy has always been about the big choices, decisions that determine what industries a company should compete in, how it should expand or contract, and how resources are allocated across a portfolio of businesses. Classic corporate strategy revolved around diversification, vertical integration, growth models, and portfolio optimization using tools like the BCG matrix or Ansoff's product-market expansion grid. Yet in today's hyper-connected, data-driven, platform-dominated world, these approaches are no longer enough.

The emergence of STEM has fundamentally changed how corporations think about growth, scope, and value creation. Successful corporations today are not only optimizing existing business units—they are continuously reshaping portfolios, designing dynamic ecosystems, and leveraging AI-driven insights to make complex multi-business decisions at speed. Firms like Amazon, Ping An, and Alphabet illustrate this transformation, evolving into intelligent conglomerates that use data, platforms, and real-time models to orchestrate growth across industries.

In this chapter, we reimagine corporate strategy through a STEM lens. We explore how predictive analytics, digital twins, AI-enhanced M&A screening, and system dynamics can modernize classic corporate decisions. We shift from static portfolio planning to dynamic resource allocation, from top-down diversification logic to data-augmented growth modeling, and from rigid vertical integration to ecosystem orchestration powered by real-time feedback loops.

Our objective is clear: to provide future strategists and corporate leaders with the analytical tools, modeling capabilities, and system-level thinking necessary to thrive in complex, multi-business environments. You will learn how to blend timeless principles with modern STEM-powered tools, making corporate strategy smarter, faster, and more adaptive to the demands of the 21st century.

By the end of this chapter, you will not only understand how corporate strategy is evolving, but also gain practical exposure to the tools and techniques that make it actionable in the STEM-driven economy.

Opening Case: Ping An's Data-Driven Expansion Blueprint

Imagine a company that started as a traditional insurance provider and transformed itself into a multi-business powerhouse, operating across financial services, healthcare, smart cities, and technology platforms. That company is Ping An, and its corporate strategy reveals how data, AI, and system modeling are redefining multi-business growth.

In the past, Ping An might have used classical corporate strategy tools like market share analysis or basic financial ratios to decide where to expand. But today, Ping An's corporate strategy engine runs on AI-driven diversification models, digital twins, and predictive

simulations. Before entering a new sector like telemedicine or smart city infrastructure, Ping An uses real-time market scanning, customer need forecasting, and scenario-based risk modeling to assess strategic fit and timing.
Rather than relying on boardroom intuition, Ping An uses STEM-enabled decision systems to:

* Predict synergy potential across business units,
* Simulate ecosystem effects before launching new ventures,
* Optimize capital allocation using real-options frameworks and Monte Carlo simulations.

Ping An didn't just grow bigger—they grew smarter. Their corporate strategy playbook is a masterclass in how data-powered corporate decisions create sustainable, cross-industry dominance.

Play the Strategist:

1. Imagine you are a corporate strategist at Ping An, tasked with deciding whether to enter the renewable energy sector.
2. How would you use AI, predictive analytics, and system simulations to evaluate the opportunity?

   How would you assess ecosystem synergies versus standalone market opportunities?

6.1 Foundations of Corporate Strategy

Corporate strategy answers a fundamental question: In what businesses should we compete? Unlike business-level strategy, which focuses on how to compete within a particular market or industry, corporate strategy determines the overall direction and scope of an organization. It defines the portfolio of industries and markets in which the company operates, how resources are allocated across different businesses, and how value is created through the coordination of diverse units.

In today's complex and fast-evolving environment, corporate strategy is no longer a one-time decision; it is a continuous process of portfolio design, strategic foresight, and systemic alignment. As companies expand into new domains, acquire emerging ventures, or participate in digital ecosystems, the role of corporate strategy becomes increasingly critical in ensuring that growth is coherent, diversification is synergistic, and resources are deployed with precision.

Modern corporate strategists must navigate a landscape shaped by technological convergence, platform economies, and data abundance. This requires moving beyond traditional frameworks and embracing STEM-powered tools, from AI-driven market scanning to digital twin simulations, to make informed, adaptive decisions. The goal is not just to choose the right industries, but to orchestrate a portfolio that evolves intelligently, responds to disruption, and creates sustainable corporate advantage.

In this section, we revisit the classical foundations of corporate strategy and explore how they are being redefined in a world where speed, complexity, and interdependence are the new strategic constants.

6.1.1 Classic Corporate Strategy: Growth, Stability, Retrenchment

Historically, corporate strategy has been categorized into three broad directions:

* Growth Strategies: These involve expanding into new industries, markets, or geographies through diversification, mergers and acquisitions (M&A), or vertical integration. Amazon exemplifies a growth strategy through diversification, expanding from e-commerce into cloud computing (AWS), entertainment (Prime Video), and physical retail (Whole Foods). Similarly, *Apple* has pursued vertical integration by designing its own chips (Apple Silicon), reducing reliance on external suppliers and enhancing product performance.

* Stability Strategies: These focus on maintaining the current portfolio and emphasizing incremental improvements, often adopted during periods of market maturity, regulatory uncertainty, or economic slowdown. Coca-Cola has often employed a stability strategy in its core beverage business, focusing on brand loyalty, operational efficiency, and gradual product innovation (e.g., zero-sugar variants) rather than aggressive expansion.

* Retrenchment Strategies: These involve reducing scope through divestitures, spin-offs, or market exits, typically used when businesses underperform, face strategic misalignment, or require repositioning. General Electric (GE) implemented a retrenchment strategy by divesting its financial services and media businesses to refocus on industrial technology and healthcare. Similarly, *IBM* spun off its managed infrastructure services unit (Kyndryl) to streamline operations and concentrate on cloud and AI.

These classic directions provided companies with a roadmap to scale, focus, or streamline their operations in alignment with external market conditions. They served as foundational choices for corporate leaders navigating economic cycles, competitive pressures, and organizational maturity.

However, in today's volatile, uncertain, complex, and ambiguous (VUCA) environment, these strategies must be reinterpreted through a dynamic and data-informed lens. Growth is no longer just about entering new markets; it's about identifying high-synergy adjacencies, leveraging digital platforms, and orchestrating ecosystems. Stability is not about standing still, but about resilience, operational agility, and continuous optimization. Retrenchment is no longer a sign of failure, but a strategic reset, often driven by real-time performance analytics and scenario-based simulations.

Moreover, the boundaries between these strategies are increasingly blurred. A firm may pursue growth in one business unit, stability in another, and retrenchment in a third, all within the same fiscal year. This requires a portfolio-level perspective, supported by STEM-enabled tools that allow for real-time monitoring, predictive foresight, and adaptive resource allocation.

Understanding these classical directions remains essential, but modern strategists must also recognize their limitations and evolve their application to match the speed, complexity, and interdependence of today's corporate landscape.

6.1.2 Corporate-Level vs. Business-Level Strategy

It is crucial to distinguish between corporate-level and business-level strategy, as they operate at different layers of decision-making and serve distinct strategic purposes:

- Corporate Strategy operates at the macro level, answering:
  "Which industries or markets should we compete in?"
- Business Strategy operates at the micro level, answering:
  "How should we compete within a specific market?"

This distinction is not merely academic—it shapes how organizations allocate resources, design structures, and pursue growth. For example, Ping An's corporate strategy involves deciding to operate in insurance, banking, telemedicine, and smart cities. These are high-level portfolio choices. Meanwhile, its business strategy in insurance focuses on AI-powered underwriting, customer personalization, and digital claims processing, tactics aimed at winning within that specific market.

Another example is Alphabet (Google's parent company). At the corporate level, Alphabet decides to operate in search, cloud computing, autonomous vehicles (Waymo), and life sciences (Verily). At the business level, Google Search competes through algorithmic relevance, ad targeting, and user experience optimization, while Waymo focuses on autonomous driving safety and regulatory partnerships.

This distinction matters because the mechanisms of value creation differ:

* Business-level strategy creates competitive advantage within industries, through differentiation, cost leadership, or niche focus.
* Corporate-level strategy creates synergies across industries, enabling the firm to achieve greater collective value than the sum of its parts.

The table below summarizes the key differences between the two levels of strategy, highlighting their distinct focus areas, goals, and STEM integration points:

Table 6.1. Corporate Strategy vs. Business Strategy

Dimension	Corporate Strategy	Business Strategy
Primary Question	Which industries or markets should we compete in?	How should we compete within a particular industry?
Focus Level	Portfolio of businesses and industries	Competitive positioning within a single market
Main Goal	Optimize scope, synergy, and resource allocation	Achieve competitive advantage and superior market performance
Decision Scope	Diversification, M&A, vertical integration, ecosystem design	Product-market fit, pricing, differentiation, cost leadership
Key Outcome	Portfolio-level value creation	Business-unit profitability and market share growth
STEM Integration	Data-driven portfolio modeling, AI in M&A, dynamic simulations	Predictive analytics, personalization, and AI-based pricing

6.1.3 Revisiting Scope: Diversification and Integration

Two major themes dominate corporate strategy scope decisions:

* Diversification: Expanding into new product lines, services, or industries to leverage existing capabilities or capture new growth opportunities.
* Vertical Integration: Moving upstream (backward integration) into supply chains or downstream (forward integration) into distribution channels to control value chains, reduce costs, or enhance customer experience.

Traditionally, these decisions were made using market attractiveness assessments, synergy calculations, and qualitative judgment. Executives relied on tools like the Ansoff Matrix, Porter's Five Forces, or SWOT analysis to evaluate whether to diversify or integrate. While these frameworks provided structure, they often lacked the granularity, speed, and predictive power needed in today's volatile markets.

In the STEM-powered environment, companies like Ping An, Amazon, and Alphabet are transforming how scope decisions are made. They deploy AI models, predictive simulations, and real-time market scanning to dynamically optimize their strategic moves.

For example, Amazon's diversification into cloud computing (AWS) was guided by internal infrastructure data and external demand forecasts. Alphabet's entry into autonomous vehicles (Waymo) and life sciences (Verily) was supported by AI-driven scenario modeling and ecosystem mapping. On the integration front, Tesla's control over battery production and retail channels is modeled using digital twins to simulate supply chain resilience and customer experience outcomes.

The table below contrasts traditional and STEM-driven approaches to scope decisions, highlighting how technology enhances strategic precision:

Table 6.2. Traditional vs. STEM-Based Scope Decision Tools

Aspect	Traditional Approach	STEM-Powered Approach
Decision Basis	Market reports, executive judgment	Real-time data, predictive analytics, and AI simulations
Diversification Assessment	Qualitative market attractiveness	Machine learning forecasting, adjacency mapping
Integration Evaluation	Cost-benefit analysis, strategic control assumptions	Digital twins, scenario modeling, supply chain stress testing
Risk Management	Static risk matrices	Monte Carlo simulations, dynamic risk modeling
Speed of Decision	Periodic reviews (quarterly/annual)	Continuous monitoring and adaptive modeling
Strategic Outcome	Incremental growth or control	Systemic value creation, ecosystem orchestration

These tools allow strategists to move from intuition to intelligence, enabling evidence-based decisions that are faster, more adaptive, and better aligned with long-term value creation.
In short, while the strategic logic of diversification and integration remains relevant, the execution has evolved. In a STEM-driven world, scope decisions are no longer static bets—they are continuously optimized, data-informed, and systemically modeled to reflect the complexity of modern business ecosystems.

Why This Matters Today

In a fast-moving, digital economy, corporate strategies are no longer static roadmaps but dynamic portfolios that evolve through continuous data feedback loops, AI-powered foresight, and ecosystem modeling. The traditional approach—where strategic decisions were made annually based on historical data and executive intuition—is being replaced by real-time, evidence-based orchestration.

Today's corporate strategists must operate in environments characterized by technological convergence, platform competition, and geopolitical volatility. In such settings, the ability to adapt scope, reallocate resources, and reconfigure business models quickly is not just a competitive advantage—it's a survival imperative.

The essence of corporate strategy remains: deciding where to play and how to configure the portfolio for long-term value creation. But the tools, pace, and complexity of decision-making have changed fundamentally. Strategic decisions now require:

* Predictive analytics to anticipate market shifts before they materialize.
* Digital twins to simulate the impact of expansion or integration moves.
* AI-driven M&A screening to identify high-fit acquisition targets.
* System dynamics modeling to understand ripple effects across business units.

Moreover, the rise of ecosystem-based competition means that firms must think beyond traditional industry boundaries. Instead of owning every part of the value chain, leading firms now orchestrate networks of partners, using platforms, APIs, and data-sharing protocols to scale rapidly and flexibly.

This transformation demands a new kind of strategist—one who is not only fluent in classical frameworks but also proficient in data science, systems thinking, and digital experimentation. In the next section, we will explore how STEM tools transform classic corporate growth strategies, enabling firms to make smarter, faster, and more resilient expansion decisions in an increasingly complex world.

6.2 STEM-Powered Growth Strategies

In the traditional era of corporate strategy, growth decisions were often based on historical financial data, qualitative market analyses, and executive intuition. Corporate leaders made pivotal decisions about entering new industries, expanding product lines, or acquiring firms based on static reports and periodic strategy reviews. While these methods provided a foundation for long-term planning, they often lacked the speed, precision, and adaptability required in today's fast-changing business environment.

In the STEM-driven world, these outdated approaches are being replaced by continuous data monitoring, predictive analytics, and advanced modeling techniques that enhance the accuracy, agility, and resilience of corporate growth decisions. Strategy is no longer a once-a-year exercise—it is a real-time, iterative process powered by machine learning, digital twins, and scenario-based simulations.

Modern firms operate in environments where customer preferences shift rapidly, technologies converge unpredictably, and competitive landscapes evolve overnight. In this context, growth strategies must be data-informed, systemically modeled, and dynamically adjusted. STEM tools enable strategists to:

* Forecast emerging market opportunities with greater precision.
* Simulate the impact of expansion moves under multiple future scenarios.
* Identify hidden adjacencies where existing capabilities can be leveraged for new growth.
* De-risk M&A decisions by modeling post-acquisition performance and cultural fit.

For example, Ping An uses AI-driven diversification models to evaluate entry into sectors like telemedicine and smart cities. Amazon leverages real-time customer behavior data and machine learning algorithms to guide product line expansion and geographic growth. Zara uses digital twins to simulate the operational impact of entering new lifestyle categories like home décor and cosmetics.

These capabilities allow firms to move faster than competitors, allocate capital more effectively, and respond to uncertainty with confidence. Growth is no longer just about scale—it's about strategic precision, ecosystem alignment, and long-term adaptability.

In the following sections, we explore how specific STEM tools—such as predictive analytics, AI-augmented M&A platforms, and simulation models—are transforming the way companies pursue growth. These tools empower strategists to make smarter, faster, and more resilient expansion decisions in an increasingly complex world.

6.2.1 Predictive Analytics for Diversification Decisions

Diversification remains a central growth strategy—whether it's related diversification (expanding into adjacent industries) or unrelated diversification (entering completely new industries). Traditionally, companies relied on market trend reports, financial projections, and managerial experience to assess diversification opportunities. These methods, while useful, were often retrospective, slow to update, and limited in scope.

In the STEM-powered era, leading firms like Ping An, Amazon, and Alphabet are transforming their diversification strategy using predictive analytics. These tools allow strategists to move from intuition-based decisions to data-driven foresight, enabling more accurate, timely, and resilient growth planning.

Specifically, predictive analytics enables firms to:

* Forecast industry growth trajectories using machine learning algorithms trained on macroeconomic indicators, consumer sentiment data, and technological adoption rates.
* Identify hidden adjacencies through network analysis, revealing sectors where existing capabilities, assets, or customer bases can be cross-leveraged.
* Simulate future demand scenarios, allowing strategy teams to evaluate long-term market potential and risk exposure before committing capital.

For example, Ping An uses AI to scan real-time data across healthcare, fintech, and smart city domains to identify high-synergy diversification paths. Amazon applies machine learning to customer behavior and logistics data to guide its expansion into areas like grocery retail (Whole Foods), healthcare (Amazon Clinic), and entertainment (Prime Video).
The shift from traditional to STEM-driven diversification is summarized in the table below:

Table 6.3. Traditional vs. STEM-Driven Growth Strategy Tools

Aspect	Traditional Tools	STEM-Driven Tools
Decision Basis	Historical data, static market reports	Real-time data streams, predictive analytics
Diversification Assessment	Qualitative market attractiveness	Machine learning forecasting, adjacency mapping
M&A Evaluation	Financial ratios, strategic fit assessment	AI-driven deal screening, synergy prediction models
Scenario Planning	Manual scenario crafting	Monte Carlo simulations, digital twins
Speed of Decision	Periodic reviews (annual/quarterly)	Continuous monitoring and adjustment

These STEM-driven tools allow companies to de-risk diversification moves, prioritize high-probability success paths, and respond to market shifts in real time. Rather than relying on static snapshots of the market, firms can now build dynamic, forward-looking models that evolve with the data—enabling smarter, faster, and more resilient diversification strategies.
In the next section, we explore how AI-augmented M&A platforms are transforming one of the most complex and high-stakes growth strategies: mergers and acquisitions.

6.2.2 AI-Augmented M&A Screening and Target Evaluation

Mergers and acquisitions (M&A) are among the most impactful—and riskiest—moves in corporate strategy. A successful acquisition can unlock new markets, accelerate innovation, and generate powerful synergies. But poorly executed deals often result in overestimated synergies, cultural clashes, and value destruction. Traditionally, M&A processes were labor-intensive, subjective, and slow, relying heavily on manual due diligence, financial ratio analysis, and executive judgment.

In the STEM-driven era, AI-powered M&A platforms are revolutionizing how firms identify, evaluate, and execute acquisitions. These tools bring speed, scale, and objectivity to a process that was once opaque and intuition-driven.

Modern AI systems enhance M&A strategy by:

* Screening thousands of potential acquisition targets in real time using customizable filters such as financial health, market reach, patent portfolios, and innovation velocity.
* Predicting post-merger performance trajectories by analyzing historical M&A outcomes, industry benchmarks, and integration complexity factors.
* Assessing cultural fit using natural language processing (NLP) applied to internal communications, employee reviews, and social media sentiment—helping to flag potential integration risks early.
* Automating due diligence workflows, enabling faster, more consistent evaluations of legal exposure, supply chain dependencies, and regulatory compliance.

For example, Alphabet uses AI to scan and rank startups in emerging tech sectors, while Ping An applies machine learning to assess the strategic fit of fintech and healthtech targets. These firms no longer rely solely on spreadsheets and boardroom intuition—they use data-driven foresight to optimize deal flow and reduce acquisition risk.

The shift from traditional to STEM-powered M&A is part of a broader transformation in corporate growth strategy. As shown in Table 6.3 (from the previous section), traditional M&A relied on financial ratios and strategic fit assessments, while STEM-driven approaches use AI screening engines and synergy prediction models to evaluate opportunities with greater precision and speed.

This transformation enables corporate strategists to:

* Accelerate decision-making without sacrificing analytical depth.
* Reduce cognitive and confirmation biases in target selection.

* Continuously update M&A pipelines based on real-time market signals and internal capability shifts.

In short, AI-augmented M&A tools are turning what was once a high-risk, high-stakes gamble into a systematic, evidence-based growth engine. As the pace of innovation accelerates and industries converge, these tools will become essential for firms seeking to scale intelligently, diversify strategically, and build resilient multi-business portfolios.
In the next section, we explore how simulation models—including Monte Carlo simulations and digital twins—help firms evaluate expansion strategies under uncertainty, enabling smarter decisions in volatile environments.

6.2.3 Simulation Models for Expansion Under Uncertainty

One of the greatest challenges in corporate strategy is making expansion decisions under uncertainty. External shocks—such as economic downturns, regulatory changes, supply chain disruptions, or technological breakthroughs—can derail even the most promising growth plans. In such volatile environments, relying solely on historical data or static forecasts is no longer sufficient.

To address this complexity, leading firms are turning to simulation models that allow them to test strategic moves in virtual environments before committing real-world resources. These tools enable decision-makers to anticipate risks, explore multiple futures, and build resilience into growth strategies.

Key simulation approaches include:

* Monte Carlo Simulations: These models run thousands of randomized scenarios based on variable inputs—such as market demand, cost fluctuations, or competitor actions—to estimate the probability distribution of outcomes. This helps firms understand the range of possible futures and make decisions that are robust across scenarios.
* Digital Twins: A digital twin is a virtual replica of a business unit, product, or strategic initiative. Strategy teams use digital twins to simulate expansion plans, test operational assumptions, and evaluate performance under different market conditions. For example, a company entering a new geography can simulate logistics, pricing, and customer behavior before launching.
* Scenario-Based Simulation Engines: These tools allow firms to model best-case, worst-case, and base-case scenarios, assigning quantified probabilities to each. This helps leaders prepare contingency plans and allocate resources more effectively.

For instance, Zara uses digital twins to simulate the impact of entering new lifestyle categories like home décor, testing how supply chain capacity, customer demand, and brand coherence would respond. Ping An models healthcare expansion scenarios using Monte Carlo simulations to assess regulatory risk and capital efficiency under different policy regimes. These STEM-based simulation tools allow organizations to move from guesswork to probabilistic strategy. Instead of asking "What will happen?", strategists ask, "What could happen—and how do we prepare for it?" This shift enables:

* Evidence-backed decision-making under uncertainty.
* Faster iteration and learning before capital is deployed.
* Greater resilience in the face of market volatility.

Simulation models are especially valuable in industries undergoing rapid technological change, where first-mover advantage must be balanced against execution risk. By integrating simulation into the strategic planning process, firms can stress-test their assumptions, optimize timing, and build optionality into their growth strategies.

In the pre-STEM era, corporate growth was largely a linear, intuition-driven process. Today, the integration of predictive analytics, AI decision support, and advanced simulations enables smarter, faster, and more defensible growth strategies. Corporate strategists no longer have to choose between speed and accuracy—STEM allows them to achieve both.

In the next section, we shift focus from growth to portfolio management—exploring how real-time data and system dynamics help firms not just expand, but intelligently orchestrate and rebalance their multi-business portfolios.

6.3 Managing Corporate Portfolios in Complex Markets

Managing a **corporate portfolio**—deciding which businesses to **grow**, **hold**, or **divest**—has always been a central challenge of corporate strategy. Traditional tools like the **BCG Growth-Share Matrix**, **GE-McKinsey Matrix**, and **cash flow analysis** provided useful starting points, but they were often based on **static data**, **simplified assumptions**, and **periodic reviews**. In today's volatile, data-rich business environment, these limitations have become more apparent.

Modern corporations operate in **fluid market ecosystems**, where portfolio performance can shift rapidly due to **technological disruption**, **consumer behavior changes**, **supply chain shocks**, and **geopolitical risks**. In such environments, portfolio strategy must evolve from a **snapshot-based exercise** to a **continuous, adaptive process**.

To manage this complexity effectively, organizations are integrating:

* **Real-time data streams** from internal operations, customer behavior, and external market signals.
* **AI-powered portfolio analytics** that detect emerging trends, performance anomalies, and synergy opportunities.
* **Dynamic simulation models** that allow strategists to test resource allocation scenarios and anticipate systemic ripple effects.

For example, **Ping An** uses real-time dashboards to monitor the performance of its insurance, healthcare, and fintech units, enabling dynamic capital reallocation based on market shifts. **Alphabet** continuously evaluates its diverse portfolio—from Google Search to Waymo to Verily—using AI to assess innovation potential, ecosystem contribution, and strategic fit.

This shift from static to dynamic portfolio management is part of a broader transformation in corporate strategy. As summarized earlier in **Table 6.3**, traditional tools relied on **periodic reviews** and **manual scenario crafting**, while STEM-driven approaches enable **continuous monitoring**, **predictive modeling**, and **automated decision support**.

In the sections that follow, we explore how firms are reimagining portfolio strategy through:

* **BCG Matrix 2.0**: Enhanced with real-time data and AI clustering.
* **Dynamic portfolio rebalancing**: Using live performance signals and scenario simulations.
* **System dynamics modeling**: To visualize interdependencies and avoid resource misallocation.

Together, these tools empower strategists to treat the corporate portfolio not as a collection of isolated units, but as a **living ecosystem**—one that must be **monitored, modeled, and optimized** continuously to create long-term value.

In the next section, we examine how firms are upgrading the classic BCG matrix using STEM tools to create more nuanced, real-time portfolio maps that reflect the complexity of modern markets.

6.3.1 Portfolio Strategy in a STEM World: BCG Matrix 2.0

The **BCG Growth-Share Matrix** has long been a staple of corporate portfolio strategy. It classifies business units into four categories—**Stars**, **Cash Cows**, **Question Marks**, and **Dogs**—based on two dimensions: **market growth rate** and **relative market share**. While this framework provided a simple and intuitive way to allocate resources, it was built for a more stable, slower-moving business environment.

In today's **data-saturated, rapidly evolving markets**, the traditional BCG matrix is no longer sufficient. Modern portfolio management tools have evolved into what we might call **"BCG Matrix 2.0"**—a dynamic, multi-dimensional, and AI-enhanced approach to portfolio strategy.

Key enhancements include:

- **Data-driven portfolio mapping**: Instead of relying on static growth and share metrics, firms now incorporate **real-time market signals**, **competitive intensity scores**, and **disruption risk indices**. This allows for more accurate and timely assessments of business unit performance and potential.
- **AI-enhanced clustering algorithms**: These tools automatically group business units based on **hidden patterns** in financial, operational, and market data. This reveals **non-obvious synergies**, **emerging stars**, and **latent risks** that traditional models might miss.
- **Dynamic portfolio dashboards**: Rather than labeling businesses once per year, organizations now use **continuously updating dashboards** that track performance across multiple dimensions—such as **innovation velocity**, **customer experience metrics**, and **ecosystem contribution**. These dashboards help identify **emerging winners** and **declining units** in real time.

For example, **Ping An** uses a multi-layered portfolio visualization system that goes beyond profitability and growth. It includes **AI-generated synergy scores**, **ecosystem alignment indicators**, and **real-time risk alerts**. This enables the firm to make **faster, more nuanced decisions** about where to invest, hold, or divest.

The table below illustrates how the traditional BCG matrix compares to its STEM-powered evolution:

This evolution reflects a broader shift in corporate strategy—from **static classification** to **dynamic orchestration**. In a STEM-powered world, portfolio strategy is no longer about placing businesses in boxes. It's about **understanding interdependencies**, **anticipating change**, and **allocating resources fluidly** to maximize long-term value.

In the next section, we explore how firms use **real-time data and simulation engines** to rebalance their portfolios dynamically, responding to shifting market conditions with speed and precision.

Table 6.4. Traditional BCG Matrix vs. BCG Matrix 2.0

Aspect	Traditional BCG Matrix	BCG Matrix 2.0 (STEM-Enhanced)
Core Dimensions	Market growth rate, relative market share	Multi-dimensional: growth, innovation, disruption risk, ecosystem fit
Data Source	Periodic financial reports, market research	Real-time data streams, AI-driven analytics
Update Frequency	Annual or semi-annual reviews	Continuous, real-time updates
Business Unit Categories	Stars, Cash Cows, Question Marks, Dogs	Dynamic clusters based on AI pattern recognition
Decision Support	Manual interpretation, static resource allocation	Automated alerts, scenario simulations, predictive modeling
Strategic Focus	Profitability and market share	Systemic value creation, innovation potential, ecosystem contribution

6.3.2 Dynamic Portfolio Rebalancing with Real-Time Data

In rapidly evolving industries, **annual or semi-annual portfolio reviews are too slow** to keep pace with shifting market dynamics. Strategic opportunities can emerge—and vanish—within weeks or even days. In this environment, firms must move beyond static planning cycles and adopt **dynamic portfolio rebalancing**, powered by **real-time data**, **AI analytics**, and **scenario-based simulations**.

Dynamic rebalancing enables organizations to:

* **Trigger automatic alerts** when a business unit's performance deviates from key benchmarks—such as revenue growth, innovation velocity, or customer churn—allowing for immediate strategic attention.
* **Continuously reallocate resources** by simulating the impact of shifting capital, talent, or technology investments between business units under varying market conditions.
* **Link external market signals**—such as **social media sentiment**, **regulatory developments**, or **technological breakthroughs**—to internal portfolio decisions, ensuring that strategy remains responsive and forward-looking.

For example, during the COVID-19 pandemic, firms with dynamic rebalancing engines were able to **rapidly increase funding for digital health ventures**, while **scaling back investments in physical retail or travel-related businesses**. These decisions were not based on quarterly reports, but on **live data feeds**, **predictive models**, and **real-time scenario testing**.

Companies like **Ping An** and **Amazon** use AI-powered dashboards that integrate performance metrics with external signals—such as competitor moves, policy changes, and consumer behavior trends. These systems recommend **portfolio adjustments in near real-time**, helping leaders stay ahead of disruption.

Dynamic rebalancing also supports **strategic experimentation**. Firms can allocate small amounts of capital to test new ventures, monitor early signals, and scale up only when data validates the opportunity. This approach reduces risk while preserving agility.

In essence, dynamic portfolio rebalancing transforms corporate strategy from a **reactive, calendar-driven process** into a **proactive, data-responsive system**. It empowers firms to:

* **Act faster than competitors** in reallocating resources.
* **Avoid overcommitting to declining units**.
* **Capitalize on emerging growth areas** before they become mainstream.

In the next section, we explore how **system dynamics modeling** helps firms visualize interdependencies across business units and anticipate the ripple effects of resource allocation decisions—ensuring that portfolio moves enhance, rather than undermine, overall corporate value.

6.3.3 System Dynamics for Resource Allocation

In complex, multi-business organizations, resource allocation is rarely a linear decision. Business units are often **interdependent**, sharing customers, technologies, talent, or infrastructure. A decision to invest in one unit can **accelerate growth elsewhere**, while underfunding another may create **bottlenecks or unintended ripple effects**. Traditional portfolio tools often overlook these systemic interactions.

To address this, leading firms are turning to **system dynamics modeling**—a method that treats the corporate portfolio as a **living system** with **feedback loops**, **delays**, and **nonlinear relationships**. This approach enables strategists to **visualize**

interdependencies, **simulate resource allocation scenarios**, and **anticipate long-term consequences** of strategic moves.

Key capabilities of system dynamics in portfolio strategy include:

* **Visualizing interdependencies**: System maps help identify how growth in one business unit (e.g., digital health) may drive demand in another (e.g., cloud infrastructure), or how resource constraints in one area may limit performance elsewhere.
* **Simulating allocation scenarios**: Strategy teams can test different capital allocation models such as increasing R&D in one unit while reducing marketing in another—and observe how these changes affect the overall system over time.
* **Modeling long-term ripple effects**: System dynamics helps firms avoid short-term optimization traps (e.g., overfunding a high-growth unit at the expense of ecosystem coherence) by revealing how today's decisions shape tomorrow's outcomes.

For example, **Ping An** uses **digital twin simulations** to model how reallocating capital to its healthcare ventures affects **customer retention**, **cross-selling opportunities**, and **ecosystem value** in its financial services units. This allows the firm to make **coordinated investment decisions** that enhance total portfolio performance—not just individual unit metrics.

System dynamics also supports **strategic coherence** by ensuring that resource flows align with the firm's overarching goals. It helps detect when units are **competing for the same resources**, **duplicating efforts**, or **drifting from the corporate mission**.

In a STEM-powered strategy environment, system dynamics modeling transforms resource allocation from a **spreadsheet exercise** into a **strategic simulation**—one that reflects the **complexity, interconnectivity, and dynamism** of modern corporate portfolios.

In the next section, we explore how firms use STEM tools to make strategic decisions about **vertical integration, outsourcing, and ecosystem participation**, redefining the boundaries of the firm in a platform-driven economy.

6.4 Vertical Integration, Outsourcing, and Platform Ecosystems

A central element of corporate strategy is deciding the optimal scope of operations—what to own and control internally versus what to outsource or manage through partnerships. These decisions define the boundaries of the firm, shape its cost structure, and influence its strategic agility.

In classical strategy, firms chose between vertical integration—controlling upstream suppliers or downstream distributors—and outsourcing, based on factors such as cost efficiencies, control advantages, or scale economies. These decisions were typically made using financial analysis, supply chain assessments, and qualitative trade-off evaluations.

However, in the STEM-driven era, these choices are no longer static or binary. Instead, they are becoming dynamic, data-driven, and modular, enabled by technologies such as digital twins, AI-powered trade-off modeling, and ecosystem simulation platforms. Increasingly, firms are moving beyond traditional value chains and embracing platform ecosystems flexible, scalable networks of partners, suppliers, and customers connected through digital infrastructure.

Vertical Integration Reimagined

Vertical integration still offers strategic benefits—such as greater control, quality assurance, and supply chain resilience—but it also introduces capital intensity and organizational rigidity. To navigate these trade-offs, firms now use digital twins to simulate end-to-end value chains, testing how changes in vertical scope affect cost structures, customer experience, and resilience to disruption.

For example, Tesla vertically integrates battery production and software development to ensure performance and innovation speed. It uses real-time manufacturing data and predictive analytics to optimize its vertically integrated operations. Similarly, Zara integrates design, production, and distribution, supported by AI-driven demand forecasting and inventory analytics to maintain its fast-fashion agility.

Outsourcing with Intelligence

Outsourcing remains a valuable strategy for accessing specialized capabilities, reducing fixed costs, and scaling quickly. But it also introduces risks—such as supplier dependency, intellectual property leakage, and quality inconsistency. Modern firms mitigate these risks using AI-based vendor scoring, blockchain for traceability, and scenario modeling to assess outsourcing resilience under stress conditions.

The Rise of Platform Ecosystems

Increasingly, firms are adopting a third model: ecosystem orchestration. Rather than owning every component of the value chain, companies act as platform architects, enabling others to build, innovate, and deliver value through shared infrastructure. This model promotes scalability, modularity, and rapid innovation, while reducing capital burden.

For instance, Ping An's OneConnect platform allows external banks and insurers to access its fintech infrastructure, turning Ping An into an ecosystem orchestrator. Amazon Web Services (AWS) enables thousands of businesses to build on Amazon's cloud infrastructure, while Apple's App Store creates a global ecosystem of developers and users.

These models are not mutually exclusive. In fact, many firms now blend integration, outsourcing, and orchestration, dynamically adjusting their boundaries based on real-time data, strategic priorities, and market conditions.

In the following subsections, we explore how STEM tools—such as digital twins, AI-based make-or-buy models, and ecosystem simulation engines—are transforming how firms configure their value creation systems in an increasingly interconnected world.

6.4.1 Strategic Decision-Making on Vertical Scope Using Digital Twins

Vertical integration decisions have historically been among the most complex in corporate strategy. They involve difficult trade-offs: **greater control and coordination** versus **higher capital intensity and reduced flexibility**. While integration can improve margins, ensure quality, and protect intellectual property, it can also lead to **organizational rigidity**, **slower innovation**, and **exposure to internal inefficiencies**. Conversely, outsourcing offers **agility and cost savings**, but increases **dependency on external partners** and **risks to supply chain continuity**.

In the **STEM-powered era**, firms are no longer forced to make these decisions based on static forecasts or intuition. Instead, they are leveraging **digital twins**—virtual replicas of physical systems, processes, or entire value chains—to simulate, test, and monitor vertical scope decisions in real time.

Digital twins allow corporate strategists to:

* **Simulate end-to-end value chains**, testing how changes in vertical scope (e.g., acquiring a supplier or launching a direct-to-consumer channel) impact **cost structures**, **customer experience**, **cycle times**, and **resilience to supply disruptions**.
* **Model stress scenarios**, such as **raw material shortages**, **logistics bottlenecks**, or **demand surges**, to predict how vertically integrated operations would perform under volatility—before committing to irreversible investments.

* **Continuously monitor integration decisions post-implementation** using **live operational data**, enabling firms to adjust scope dynamically as market conditions evolve.

For example, **Ping An** uses digital twins to model its **healthcare and fintech supply chains**, allowing for **real-time scope adjustments** based on performance feedback. If a particular vendor underperforms or a regulatory shift increases compliance costs, Ping An can simulate the impact of bringing that function in-house—or vice versa—before making a move.

Similarly, **automotive manufacturers** like **BMW** and **Ford** use digital twins to simulate the effects of vertically integrating battery production or software development. These simulations help them evaluate not only cost and efficiency, but also **innovation speed**, **supply chain resilience**, and **customer satisfaction**.

The use of digital twins in vertical scope decisions offers several strategic advantages:

* **De-risking capital-intensive moves** by testing them virtually first.
* **Enhancing agility** by enabling real-time reconfiguration of the value chain.
* **Improving cross-functional alignment** by visualizing how integration affects finance, operations, marketing, and customer experience simultaneously.

In short, digital twins transform vertical integration from a **high-stakes strategic gamble** into a **data-informed, continuously optimized decision process**. They empower firms to **balance control and flexibility**, **cost and innovation**, and **efficiency and resilience**—with far greater precision than ever before.

In the next section, we explore how firms are shifting from traditional integration models to **ecosystem orchestration**, using platforms and modular architectures to scale without owning every part of the value chain.

6.4.2 Ecosystem Orchestration vs. Vertical Integration

In a **platform-dominated economy**, many leading firms are shifting from **owning assets** to **orchestrating ecosystems**. Rather than building and controlling every part of the value chain, these firms focus on **enabling others**—suppliers, partners, developers, and even competitors to co-create value through **shared digital infrastructure**.

What Is Ecosystem Orchestration?

Ecosystem strategy centers on creating **platforms** that connect diverse participants—such as **suppliers**, **customers**, **startups**, and **third-party developers**—without owning every component. These platforms are built on **APIs**, **data-sharing protocols**, and **modular**

services, allowing for **rapid market expansion**, **scalable innovation**, and **capital-light growth**.
For example, **Ping An's OneConnect platform** extends its fintech capabilities to external banks and insurers, transforming Ping An from a vertically integrated financial services provider into an **ecosystem orchestrator**. Similarly, **Amazon Web Services (AWS)** enables thousands of businesses to build on Amazon's cloud infrastructure, while **Apple's App Store** empowers a global developer ecosystem to innovate on top of iOS.

Strategic Advantages of Ecosystem Orchestration

* **Scalability**: Ecosystems grow exponentially as more participants join and contribute.
* **Speed**: Innovation cycles accelerate as external partners bring new ideas and capabilities.
* **Capital Efficiency**: Firms can expand without owning physical assets or hiring large internal teams.
* **Resilience**: Distributed value creation reduces dependency on any single node in the system.

The Trade-Off: Control vs. Agility

While ecosystem orchestration offers **agility and scalability**, it comes with challenges:

* **Governance complexity**: Managing standards, incentives, and trust across diverse participants.
* **Reduced control**: Less direct oversight over customer experience, quality, and compliance.
* **Platform risk**: Ecosystem success depends on maintaining relevance and network effects.

In contrast, **vertical integration** provides **tight control**, **predictable quality**, and **strategic insulation**—but at the cost of **flexibility**, **speed**, and **capital intensity**.

The table below summarizes the key differences between the two approaches:

Table 6.5. Vertical Integration vs. Ecosystem Orchestration

Dimension	Vertical Integration	Ecosystem Orchestration
Ownership	Internal control of upstream/downstream activities	External partners connected via platforms
Strategic Focus	Control, efficiency, quality assurance	Scalability, innovation, speed
Capital Intensity	High (requires asset ownership and internal capabilities)	Low to moderate (leverages external contributors)
Flexibility	Low (rigid structures, slower to adapt)	High (modular, adaptive, partner-driven)
Innovation Model	Internal R&D and process optimization	Open innovation through partner ecosystems
STEM Tools Used	Digital twins, supply chain simulations	API platforms, network analytics, ecosystem simulation engines
Example	Tesla's vertically integrated EV production	Ping An's OneConnect, Apple App Store, Amazon Marketplace

Blended Models: The Future of Strategic Scope

Many firms are now adopting **hybrid models**, combining **vertical integration** in core areas (e.g., proprietary technology, customer experience) with **ecosystem orchestration** in adjacent or fast-moving domains. For instance, **Apple** tightly controls its hardware and OS but enables a vast ecosystem of third-party apps and accessories.

In a STEM-powered world, the decision between integration and orchestration is no longer binary. Firms use **AI-driven trade-off models**, **ecosystem simulations**, and **real-time performance dashboards** to dynamically adjust their strategic scope—balancing **control**, **agility**, and **value creation** across a fluid business landscape.

In the next section, we explore how **AI and advanced analytics** are transforming "make-or-buy" decisions—helping firms evaluate whether to build capabilities internally, partner with others, or acquire them through M&A.

6.4.3 AI in Make-or-Buy Decisions

The classic **make-or-buy decision**—whether to build a capability internally, partner with a third party, or acquire it through M&A—has long been a cornerstone of corporate strategy. Traditionally, these decisions were based on **cost comparisons**, **core competency assessments**, and **qualitative trade-offs**. However, in today's fast-moving, data-rich environment, such static evaluations are no longer sufficient.

AI and advanced analytics now augment make-or-buy decisions by introducing **real-time intelligence**, **predictive modeling**, and **systemic foresight** into the process. These tools help firms evaluate not only **cost and speed**, but also **strategic fit**, **innovation potential**, and **ecosystem compatibility**.

AI enhances make-or-buy decisions in several key ways:

- **Analyzing cost-benefit scenarios** using **real-time operational data** rather than static estimates. AI models can simulate the total cost of ownership, time-to-market, and scalability under different sourcing strategies.
- **Predicting hidden risks** such as **supplier dependency**, **intellectual property leakage**, or **quality degradation** by analyzing historical patterns, anomaly detection, and external risk signals (e.g., geopolitical instability, regulatory shifts).
- **Forecasting innovation potential**, helping firms determine whether **internal development**, **strategic partnership**, or **acquisition** will deliver faster and more sustainable competitive advantage. AI can assess R&D velocity, patent activity, and market readiness to inform these decisions.

For example, when evaluating **AI-powered healthcare services**, **Ping An** uses algorithmic assessments of **internal R&D speed**, **external vendor maturity**, and **ecosystem compatibility**. These insights guide whether the company should **build the solution in-house**, **partner with a healthtech startup**, or **acquire a proven player**.

Other firms, like **Microsoft**, use AI to evaluate whether to develop new cloud capabilities internally or acquire them through strategic M&A—as seen in its acquisition of Nuance Communications to accelerate its healthcare AI offerings.

These decisions are further supported by:

- **Digital twins** that simulate the operational and financial impact of different sourcing strategies.
- **AI-powered trade-off models** that weigh cost, control, speed, and strategic alignment.
- **Ecosystem orchestration frameworks** that assess how a new capability fits within the broader platform or partner network.

In the modern economy, **corporate boundary decisions are no longer static**. Firms must continuously re-evaluate what they own, what they outsource, and what they orchestrate—based on **real-time data**, **market dynamics**, and **strategic priorities**.
AI enables this **fluid reconfiguration** by turning the make-or-buy decision into a **living model**—one that evolves with the business, adapts to external shocks, and aligns with long-term value creation.

In the next section, we explore how STEM tools help maintain **strategic coherence** across diversified businesses—ensuring that the corporate portfolio remains aligned with long-term goals, even as it grows more complex and dynamic.

6.5 Strategic Coherence and Corporate Advantage

As organizations expand through **diversification**, **acquisitions**, and **ecosystem participation**, a critical challenge emerges: how to maintain **strategic coherence** across an increasingly complex portfolio of businesses. Strategic coherence refers to the **alignment of business units, resources, and activities** around a clear **corporate-level logic** that enhances overall performance and enables the firm to operate as more than the sum of its parts.

In classical corporate strategy, coherence was maintained through mechanisms such as **synergy targets**, **shared services**, and **strategic business unit alignment**. These approaches relied on **manual oversight**, **periodic reviews**, and **top-down coordination**. While effective in stable environments, they often fail in today's **fast-moving, data-saturated, and decentralized ecosystems**, where misalignment can emerge quickly and silently.

In modern organizations, coherence can deteriorate due to:

* **Strategic drift**, where business units pursue goals that diverge from corporate priorities.
* **Resource dilution**, as capital and talent are spread too thin across unrelated initiatives.

* **Misaligned incentives**, where performance metrics and cultural norms conflict across units.

To address these risks, firms now rely on **STEM-enabled tools** to **detect incoherence early** and **reinforce alignment systematically**. These tools transform coherence from a static structure into a **dynamic capability**, continuously monitored and adjusted in real time.

Key STEM tools for strategic coherence include:

* **Real-time strategic dashboards** that track the performance contribution of each business unit relative to corporate objectives.
* **AI pattern detection algorithms** that identify misalignment signals—such as deviations from target customer segments, strategic KPIs, or brand positioning.
* **Systemic coherence audits** using **network analysis**, which visualize inter-unit synergies and highlight areas of value leakage or duplication.

For example, **Ping An** uses live dashboards and AI-based monitoring tools to assess the alignment between its **insurance**, **healthcare**, and **fintech divisions**, ensuring they contribute to the firm's broader **ecosystem vision**. If a unit begins to drift—e.g., by targeting a different customer base or launching products outside the strategic scope—alerts are triggered, and corrective actions are simulated before implementation.

Maintaining strategic coherence also involves **modeling synergies** across units using **system dynamics**. These models help quantify how growth in one business unit affects others, enabling coordinated resource allocation and avoiding cannibalization or bottlenecks.

In short, strategic coherence is no longer a matter of structure—it is a matter of **systemic intelligence**. STEM tools empower firms to:

* **Monitor alignment continuously**, not just during annual reviews.
* **Simulate the impact of strategic decisions** on coherence before executing them.
* **Adapt governance and incentives** to reinforce shared goals across diverse units.

In the next section, we explore how these tools are applied to **model synergies**, **detect misalignment**, and **ensure that every business unit contributes to an integrated, high-performing corporate system**.

6.5.1 How STEM Tools Ensure Coherent Corporate Strategy

In a **STEM-powered corporation**, strategic coherence is no longer maintained through **manual oversight**, **periodic reviews**, or **executive intuition**. Instead, it is

supported by a suite of **data-enabled governance systems** that continuously monitor, diagnose, and reinforce alignment across business units. These tools transform coherence from a static organizational principle into a **dynamic, measurable capability**.
Key STEM tools include:

* **Real-time strategic dashboards**: These dashboards integrate performance data from across business units and map it against corporate-level objectives. They allow leaders to see, at a glance, which units are contributing to strategic goals, which are lagging, and where misalignment may be emerging. Dashboards can include metrics such as revenue contribution, innovation velocity, ecosystem engagement, and customer overlap.
* **AI pattern detection algorithms**: These algorithms analyze large volumes of internal and external data to identify early signs of strategic drift. For example, if a business unit begins targeting a different customer segment, deviates from agreed KPIs, or shifts its messaging, AI can flag these changes before they become systemic issues. This enables proactive intervention and course correction.
* **Systemic coherence audits using network analysis**: Network analysis tools visualize the relationships and interdependencies between business units—such as shared technologies, customer bases, or operational platforms. These audits help identify whether **synergies are materializing** or whether **value leakages**, **duplication**, or **conflicting initiatives** are undermining coherence.

For example, **Ping An** uses a combination of **live dashboards**, **AI-based monitoring**, and **network analytics** to continuously assess the alignment between its **insurance**, **healthcare**, and **fintech divisions**. These tools ensure that each unit contributes to the firm's broader **ecosystem strategy**, rather than operating in isolation or pursuing divergent goals.

This approach offers several strategic advantages:

* **Early detection of misalignment**: Instead of waiting for quarterly reviews or performance slumps, firms can identify and address coherence issues in real time.
* **Quantification of strategic fit**: STEM tools allow firms to measure how well each unit aligns with corporate logic—not just financially, but in terms of customer experience, innovation, and ecosystem contribution.
* **Continuous feedback and adjustment**: Coherence becomes a living metric, updated continuously and used to guide resource allocation, organizational design, and strategic planning.

In essence, STEM tools enable firms to **operationalize strategic coherence**—making it visible, measurable, and actionable. This is especially critical in diversified organizations, where complexity can easily lead to fragmentation. By embedding coherence into the digital infrastructure of the firm, leaders can ensure that **every business unit is rowing in the same direction**, even as the portfolio evolves.

In the next section, we explore how STEM tools help **model synergies and strategic fit** across business units—turning abstract synergy goals into quantifiable, system-level insights.

6.5.1 How STEM Tools Ensure Coherent Corporate Strategy

In a **STEM-powered corporation**, strategic coherence is no longer maintained through **manual oversight**, **organizational charts**, or **periodic executive reviews**. Instead, it is supported by a suite of **data-enabled governance systems** that continuously monitor, diagnose, and reinforce alignment across business units. These tools transform coherence from a static principle into a **dynamic, measurable, and adaptive capability**.

Modern coherence management relies on five key STEM tools, each serving a distinct purpose in aligning strategy, operations, and culture:

Table 6.6. STEM Tools for Enhancing Strategic Coherence

STEM Tool	Purpose	Application in Corporate Strategy
Real-Time Dashboards	Continuous performance tracking	Monitor business unit alignment to corporate goals
AI Pattern Detection	Identify early signs of misalignment	Detect strategic drift across business units
Network Analysis	Visualize inter-business synergies	Uncover cross-unit dependencies and synergy pathways
System Dynamics Modeling	Simulate feedback loops and long-term impacts	Forecast systemic effects of resource allocation changes
NLP Analytics	Analyze language patterns and cultural signals	Detect shifts in organizational focus or value alignment

Let's explore how these tools work together to ensure coherence:

* **Real-time strategic dashboards** integrate performance data from across business units and map it against corporate-level objectives. These dashboards provide visibility into how each unit contributes to strategic goals—such as innovation, customer growth, or ecosystem engagement—and flag deviations early.
* **AI pattern detection algorithms** analyze internal and external data to identify subtle signs of misalignment. For instance, if a business unit begins targeting a different customer segment, shifts its messaging, or underperforms on strategic KPIs, AI can detect these anomalies and trigger alerts for review.
* **Network analysis** tools visualize the relationships between business units—such as shared technologies, customer bases, or operational platforms. This helps identify whether **synergies are materializing** or whether **value leakages**, **duplication**, or **conflicting initiatives** are undermining coherence.
* **System dynamics modeling** allows strategists to simulate how changes in one part of the portfolio affect others. For example, reallocating resources to a high-growth unit may unintentionally weaken support functions or cannibalize adjacent offerings. Modeling these feedback loops helps avoid unintended consequences.
* **Natural Language Processing (NLP) analytics** are increasingly used to analyze internal communications, strategic reports, and employee feedback. These tools detect shifts in tone, focus, or cultural alignment—providing early warnings of strategic drift or fragmentation.

For example, **Ping An** uses a combination of these tools to monitor the alignment between its **insurance**, **healthcare**, and **fintech divisions**. Real-time dashboards track performance against ecosystem goals, AI detects deviations in customer targeting, and network analysis reveals whether cross-unit synergies are being realized. This integrated approach ensures that each unit contributes to the firm's broader **ecosystem vision**, rather than operating in silos.

In essence, STEM tools enable firms to **operationalize strategic coherence**—making it **visible**, **quantifiable**, and **actionable**. This is especially critical in diversified organizations, where complexity can easily lead to fragmentation. By embedding coherence into the digital infrastructure of the firm, leaders can ensure that **every business unit is aligned**, **every initiative is synergistic**, and **every decision reinforces the corporate strategy**.

In the next section, we explore how STEM tools help **model synergies and strategic fit** across business units turning abstract synergy goals into quantifiable, system-level insights that guide smarter portfolio decisions.

6.5.2 Modeling Synergies and Strategic Fit Across Business Units

Strategic coherence thrives on synergies—the ability of business units to create more value together than they could independently. In classical corporate strategy, synergy was often treated as a **qualitative aspiration**, especially during M&A deals, where projected synergies were used to justify acquisitions. However, these estimates were frequently **overstated**, **poorly tracked**, or **never realized**.

In the STEM-powered era, firms are moving beyond hypothetical synergy projections and adopting **quantitative, system-level approaches** to model, monitor, and optimize synergies across their portfolios. This shift transforms synergy from a **one-time assumption** into a **continuously validated performance metric**.

STEM tools help model and quantify synergies more precisely by:

* **Building synergy simulations** that predict **combined performance gains** across business units. These simulations incorporate shared customer bases, cross-selling potential, operational overlaps, and innovation spillovers.
* **Applying system dynamics models** to capture **feedback loops** between units—for example, how customer growth in a digital platform business might drive demand in a logistics or payments unit, or how R&D in one unit accelerates product development in another.
* **Using AI to forecast synergy erosion**, identifying when initial integration benefits begin to fade due to cultural misalignment, shifting market conditions, or resource cannibalization. These models can flag when synergies are weakening and recommend corrective actions.

For example, a firm like **Alphabet** can use system dynamics to model how investments in its AI research division (DeepMind) enhance capabilities in Google Cloud, Search, and YouTube. Similarly, **Ping An** uses synergy simulations to evaluate how its healthcare and insurance units reinforce each other through shared data, customer trust, and digital infrastructure.

These tools enable firms to:

* **Quantify synergy potential** before making diversification or M&A decisions.
* **Track realized synergies** over time, using real-time performance data.
* **Adjust resource allocation** dynamically to reinforce high-synergy pathways and phase out underperforming linkages.

This approach shifts synergy analysis from a **static, pre-deal justification** to an **ongoing, data-validated performance management system**. It also supports **strategic fit**

assessments, helping firms evaluate whether a new business unit aligns with the firm's core capabilities, customer base, and ecosystem logic.

In a world of increasing complexity, modeling synergies is not just about maximizing efficiency—it's about ensuring that the **corporate portfolio functions as an integrated, value-creating system**. STEM tools make this possible by turning synergy into a **measurable, actionable, and continuously optimized dimension of strategic coherence**.

In the next section, we explore how analytics and natural language processing (NLP) help detect **strategic misalignment**—ensuring that early warning signs of drift, duplication, or fragmentation are identified and addressed before they erode corporate advantage.

6.5.3 Detecting Strategic Misalignment Using Analytics

One of the most insidious risks in diversified corporations is the **creep of strategic misalignment**—when business units begin to **pursue divergent goals**, **duplicate efforts**, or **undermine the overall corporate mission**. This misalignment often emerges gradually and invisibly, especially in large, decentralized organizations where autonomy is encouraged but coordination is weak.

In the past, such issues were typically discovered **too late**—during performance reviews, post-merger audits, or after market share erosion. Today, **advanced analytics** and **AI-powered monitoring systems** allow firms to **detect misalignment early**, enabling proactive intervention and realignment.

Key STEM tools and techniques include:

* **Early detection of misalignment signals**: Advanced analytics platforms continuously scan for **diverging investment patterns**, **conflicting brand positions**, or **contradictory incentive structures** across business units. For example, if one unit is investing heavily in premium branding while another pushes cost leadership, the system flags a potential strategic contradiction.
* **Natural Language Processing (NLP)**: NLP algorithms analyze **internal communications**, **strategic reports**, and even **meeting transcripts** to detect shifts in tone, focus, or cultural alignment. These tools can identify when a business unit's language begins to drift from the corporate narrative—signaling potential fragmentation or cultural divergence.

* **Interactive strategic alignment simulators**: These virtual tools allow corporate leaders to **test resource allocation adjustments**, **organizational restructures**, or **strategic pivots** in simulated environments before implementing them. This reduces the risk of unintended consequences and supports evidence-based decision-making.

For example, a global conglomerate might use NLP to analyze quarterly reports from its regional divisions. If the language in one region increasingly emphasizes short-term profitability while the corporate strategy emphasizes long-term innovation, the system can flag this as a potential misalignment. Similarly, if incentive structures in one unit reward siloed performance while others are measured on cross-unit collaboration, analytics can highlight the inconsistency.

These tools ensure that **corporate strategy remains coherent**, even as **market conditions shift**, **new technologies emerge**, and **organizational structures evolve**. They allow firms to:

* **Monitor alignment continuously**, not just during annual reviews.
* **Diagnose the root causes** of misalignment—whether structural, cultural, or strategic.
* **Simulate corrective actions** before deploying them in the real world.

In a world of **complex portfolios** and **ecosystem participation**, strategic coherence is not a fixed state—it is a **dynamic capability**. With **STEM-powered monitoring**, **simulation**, and **predictive tools**, modern corporations can **actively shape coherence**, ensuring that **every business unit contributes to an integrated, high-performing corporate system**.

In the final section, we summarize the chapter's key insights and provide reflective exercises to help readers apply **STEM-based corporate strategy thinking** to real-world cases and decision-making scenarios.

6.6 Chapter 6 Wrap-Up: Corporate Strategy Reinvented for a STEM World

Corporate strategy in the 21st century is undergoing a profound transformation. While the core questions of **where to compete**, **how to expand**, and **how to allocate resources** remain essential, the tools, processes, and thinking behind those decisions have shifted dramatically. No longer dependent on **static frameworks** and **annual portfolio reviews**, corporate strategy has become a **living system**, powered by **real-time data**, **predictive analytics**, **simulation models**, and **AI-enhanced decision-making**.

This chapter explored how organizations like **Ping An** use **STEM-powered tools** to reinvent traditional corporate strategy decisions:

* **Growth strategies** are increasingly guided by **predictive models** that anticipate industry shifts and customer demand.
* **M&A decisions** are supported by **AI screening engines**, reducing bias and improving post-deal performance predictions.
* **Portfolio management** has evolved into a **dynamic, feedback-driven process**, with continuous rebalancing informed by live market signals.
* **Vertical integration and outsourcing** decisions now integrate **digital twin simulations** and **ecosystem orchestration models**.
* **Strategic coherence** is no longer managed manually but **monitored continuously**, with **analytics tools identifying drift and misalignment** in real-time.

The evolution of corporate strategy mirrors the broader shift of strategy as a whole—**from linear planning to adaptive orchestration**, **from intuition to intelligence**, and **from fragmented decisions to systemic coherence**.

In the STEM world, corporate strategists must become **modelers, orchestrators, and foresight architects**, able to **navigate complexity**, **balance trade-offs**, and **shape ecosystems** with precision and agility. This approach ensures that corporate decisions are not only **faster** but also **smarter and more resilient**, preparing organizations to thrive in **volatile, uncertain, and rapidly evolving markets**.

6.6.1. Discussion Questions

1. How does predictive analytics enhance the accuracy of diversification decisions compared to traditional market research?
2. In what ways can AI reduce risks and uncover hidden opportunities in M&A strategy? Provide an example.
3. Discuss how ecosystem orchestration might offer more flexibility than vertical integration in industries affected by rapid technological change.
4. Explain how system dynamics modeling can prevent resource misallocation in a diversified portfolio of businesses.
5. How can digital twins be used to simulate the impact of vertical integration decisions on operational efficiency and customer experience?
6. Compare the strengths and limitations of using the BCG Matrix versus AI-driven portfolio clustering tools in a multi-business strategy.

7. Describe how real-time dashboards and AI pattern detection can help corporate leaders maintain strategic coherence across rapidly expanding organizations.
8. What are some potential risks of relying too heavily on algorithmic decision-making at the corporate strategy level, and how can firms mitigate them?
9. How could STEM-enabled corporate strategy approaches help firms respond faster to external shocks such as geopolitical risks or industry disruptions?
10. Reflect on a company you are familiar with: what corporate strategy decision (diversification, divestiture, vertical integration, or ecosystem orchestration) could have been improved using STEM tools? Explain how.

6.6.2 Micro-Case: Zara's Data-Driven Corporate Expansion Strategy

As Zara scaled from a successful fashion retailer to a **multi-brand fashion and lifestyle conglomerate**, it faced critical corporate strategy decisions: **Should it expand vertically by controlling more of its supply chain? Should it diversify into non-fashion categories like home décor or beauty? Or should it orchestrate a broader digital ecosystem without heavy asset investment?**

Rather than relying solely on market reports or leadership instinct, Zara's **Chief Strategy Officer** adopted a **STEM-powered corporate strategy framework**:

* Zara built a **digital twin of its full business system**, simulating the financial and operational impacts of **forward integration into last-mile delivery services**.
* Using **AI-driven demand forecasting**, Zara evaluated entry opportunities into **adjacent lifestyle markets** like cosmetics and home furnishings, identifying segments with the **highest synergy potential**.
* A **real-time portfolio dashboard** monitored the performance of Zara's growing brand family, flagging emerging underperformance trends early and guiding **resource reallocation decisions**.
* Instead of pursuing heavy vertical integration in logistics, Zara adopted a **platform ecosystem model**, creating partnerships with **logistics tech startups** while focusing internal investment on **digital retail experiences** and **fast-cycle design teams**.

Through this **data-driven approach**, Zara expanded intelligently—avoiding **capital-heavy traps**, maximizing **synergies across product categories**, and maintaining **strategic coherence** across its portfolio.

Discussion Prompt:

1. If you were Zara's corporate strategist, how would you use **predictive analytics**, **AI-based market scans**, and **system dynamics modeling** to decide between **deepening vertical integration** or **expanding the product portfolio**?
2. How could Zara balance **ecosystem flexibility** with **strategic control** in its next growth phase?

6.6.3 Interactive Exercise: Designing a Data-Driven Corporate Portfolio

You have been appointed as the **Chief Strategy Officer (CSO)** of **NovaSphere Holdings**, a rapidly growing technology conglomerate. NovaSphere operates four distinct business units:

* **Unit A**: Cloud Infrastructure Solutions
* **Unit B**: Consumer Smart Devices
* **Unit C**: AI-Enabled Healthcare Services
* **Unit D**: Digital Content and Streaming Platform

Your corporate strategy team has access to a **real-time data dashboard** powered by **AI algorithms** and **predictive analytics**. This dashboard provides:

* **Financial Metrics**: Revenue growth rate, EBITDA margin, capital efficiency
* **Innovation Indicators**: R&D intensity, new product pipeline strength, patent filings
* **Market Disruption Signals**: Emerging competitor activity, technology disruption index, regulatory change forecasts
* **Synergy Scores**: Degree of operational and market synergy between units

Your Tasks:

1. **Portfolio Decision-Making**
    * **Analyze the dashboard data** (hypothetical or classroom-provided numbers):
        * Which business unit shows the strongest growth and innovation pipeline?
        * Which unit is at risk due to market disruption or margin erosion?
    * **Decide for each unit**:
        * **Grow**: Allocate additional capital, expand product lines, or invest in geographic expansion.
        * **Hold**: Maintain current investment levels, optimize operations, and monitor performance.
        * **Divest**: Plan to exit the business or prepare for a strategic sell-off to free up capital.
2. **New Sector Expansion**
    * Based on **AI-driven market forecasts**, select **one new sector** (e.g., **renewable energy**, **ed-tech platforms**, **autonomous transportation**) that complements NovaSphere's current capabilities.

* Justify your selection using **predictive analytics insights**, considering:
    * Market growth potential
    * Capability adjacencies
    * Long-term disruption resilience

3. **Strategic Coherence Monitoring**
    * Propose a **system dynamics-based feedback loop**:
        * What key **coherence indicators** (e.g., cross-unit customer overlap, shared platforms, unified brand perception) will you monitor?
        * How will you **detect early misalignment signals** as the portfolio evolves?
        * How often will you **simulate resource reallocation scenarios** to maintain synergy?

This exercise helps future strategists **experience the power of STEM tools in portfolio strategy**, encouraging **quantitative thinking**, **systems perspective**, and **adaptive leadership skills**.

6.6.4. Recommended HBR Case Studies for Chapter 6

❖ **Ping An: Pioneering Smart Ecosystems in Finance and Healthcare**

* **Overview**: This case explores how Ping An transformed from a traditional insurance company into a diversified, AI-powered ecosystem spanning finance, healthcare, and smart city services. It examines the firm's use of digital twins, predictive analytics, and platform orchestration to drive strategic coherence and cross-sector synergies. The case challenges students to evaluate Ping An's corporate strategy, particularly its use of STEM tools to manage complexity and growth.

* **Citation**: Zhu, Feng, Anthony K. Woo, and Nancy Hua Dai. "Ping An: Pioneering the New Model of 'Technology-driven Finance'." Harvard Business School Case 620-068, April 2020. (Revised November 2020.)

❖ **Danaher Corporation**

* **Overview**: Originally published in 2008 and revised in 2015, this 32-page case examines the corporate strategy of Danaher, a diversified industrial conglomerate. It focuses on the Danaher Business System (DBS), a structured approach to continuous improvement, portfolio management, and value creation. The case is ideal for exploring how a multi-business firm balances organic and acquisition-led growth, and how system-level thinking supports strategic coherence.

* **Citation**: Citation: Anand, B., Collis, D. J., & Hood, S. (2015). Danaher Corporation (Case No. 708-445). Harvard Business School.

6.6.5 Strategy-in-Film: Learn by Watching

Corporate strategy today is no longer about static expansion but about **dynamic portfolio management**, **ecosystem building**, and **data-driven resource allocation**. The films below highlight how organizations navigate **diversification choices**, **integration dilemmas**, and **complex multi-business environments**. As you explore corporate strategy in a STEM context, use these narratives to reflect on the **risks and rewards of strategic scope decisions**.

Movie / Documentary	Strategic Theme Highlighted
The Founder	Aggressive corporate expansion, vertical integration, ecosystem control, and power shifts in strategy execution
Moneyball	Data-driven resource allocation, challenging traditional portfolio decisions under budget constraints
Too Big to Fail	Strategic decision-making at the corporate level under extreme systemic risk
Inside Bill's Brain	Strategic philanthropy, portfolio diversification into social impact areas, and balancing profit and purpose
The Social Network	Platform strategy evolution, transitioning from product focus to ecosystem orchestration
The China Hustle	Risks in rapid corporate expansion, the importance of due diligence, and detecting strategic misalignment
Barbarians at the Gate	Corporate mergers, acquisition strategy, and value destruction risks in diversification decisions
Margin Call	Failures of corporate oversight, risk management, and short-term versus long-term corporate strategy tension
Ford v Ferrari	Strategic focus versus corporate interference: the consequences of misaligned corporate objectives
Zara: The Story of the World's Richest Man	Vertical integration and agile corporate systems enabling global dominance through supply chain mastery

References

Adamides, E. D. (2015). Linking operations strategy to the corporate strategy process: a practice perspective. *Business Process Management Journal*, *21*(2), 267-287.

Adner, R. (2017). Ecosystem as structure: An actionable construct for strategy. *Journal of Management*, *43*(1), 39-58.

Andrews, K. R., & David, D. K. (1987). *The concept of corporate strategy* (Vol. 3). Homewood, IL: Irwin.

Ansoff, H. I. (1958). A model for diversification. *Management Science*, *4*(4), 392-414.

Arasti, M., Khaleghi, M., & Noori, J. (2017). Corporate-level technology strategy and its linkage with corporate strategy in multi-business companies: IKCO case study. *Technological Forecasting and Social Change*, *122*, 243-252.

Autio, E. (2022). Orchestrating ecosystems: a multi-layered framework. *Innovation*, *24*(1), 96-109.

Bohn, J. R., & Stein, R. M. (2009). *Active credit portfolio management in practice*. John Wiley & Sons.

Bowman, E. H., & Helfat, C. E. (2001). Does corporate strategy matter?. *Strategic Management Journal*, *22*(1), 1-23.

Caldart, A. A., & Ricart, J. E. (2004). Corporate strategy revisited: a view from complexity theory. *European Management Review*, *1*(1), 96-104.

Chaudhary, A., & Gill, S. K. (2024). The AI-Enhanced Transformation: Unveiling the Synergy of Digital Twins and Artificial Intelligence. In *Harnessing AI and Digital Twin Technologies in Businesses* (pp. 400-410). IGI Global.

Choi, J., Menon, A., & Tabakovic, H. (2021). Using machine learning to revisit the diversification–performance relationship. *Strategic Management Journal*, *42*(9), 1632-1661.

Christensen, H. K., & Montgomery, C. A. (1981). Corporate economic performance: Diversification strategy versus market structure. *Strategic Management Journal*, *2*(4), 327-343.

Eisenhardt, K. M., & Piezunka, H. (2011). Complexity theory and corporate strategy. *The SAGE handbook of complexity and management*, 506-523.

Gouveia, S., de la Iglesia, D. H., Abrantes, J. L., & López Rivero, A. J. (2024). Transforming Strategy and Value Creation Through Digitalization?. *Administrative Sciences*, *14*(11), 307.

Halttunen, K., Slade, R., & Staffell, I. (2023). Diversify or die: Strategy options for oil majors in the sustainable energy transition. *Energy Research & Social Science*, *104*, 103253.

Hanelt, A., Bohnsack, R., Marz, D., & Antunes Marante, C. (2021). A systematic review of the literature on digital transformation: Insights and implications for strategy and organizational change. *Journal of Management Studies*, *58*(5), 1159-1197.

Hunjra, A. I., Hanif, M., Mehmood, R., & Nguyen, L. V. (2021). Diversification, corporate governance, regulation and bank risk-taking. *Journal of Financial Reporting and Accounting*, *19*(1), 92-108.

Markides, C. C. (1997). To diversify or not to diversify. *Harvard Business Review*, *75*(6), 93-186.

Mbanefo, C. C., & Grobbelaar, S. S. (2024). Unveiling the core elements of platform ecosystem development: a systemic lens for value co-creation in small and medium enterprises and orchestrators. *Management Review Quarterly*, 1-44.

Menz, M., Kunisch, S., Birkinshaw, J., Collis, D. J., Foss, N. J., Hoskisson, R. E., & Prescott, J. E. (2021). Corporate strategy and the theory of the firm in the digital age. *Journal of Management Studies*, *58*(7), 1695-1720.

Pidun, U. (2019). Corporate Strategy Process. In *Corporate Strategy: Theory and Practice* (pp. 253-278). Wiesbaden: Springer Fachmedien Wiesbaden.

Prahalad, C. K., & Bettis, R. A. (1986). The dominant logic: A new linkage between diversity and performance. *Strategic management journal*, *7*(6), 485-501.

Ramanujam, V., & Varadarajan, P. (1989). Research on corporate diversification: A synthesis. *Strategic Management Journal*, *10*(6), 523-551.

Sadeghi, V. J., Garcia Perez, A., Vrontis, D., & Bedford, D. (2024). Digital resilience, new business models and international entrepreneurship in the era of knowledge-economy. *Journal of Enterprise Information Management*, *37*(5), 1401-1417.

Scholes, K., Johnson, G., & Whittington, R. (2002). *Exploring corporate strategy*. Hoboken, NJ, USA: Financial Times Prentice Hall.

Tabim, V. M., Ayala, N. F., & Frank, A. G. (2024). Implementing vertical integration in the industry 4.0 journey: which factors influence the process of information systems adoption?. *Information Systems Frontiers*, *26*(5), 1615-1632.

Thomas, L. D., Ritala, P., Karhu, K., & Heiskala, M. (2024). Vertical and horizontal complementarities in platform ecosystems. *Innovation*, 1-25.

Wang, X., Yang, E., Zheng, C., & Yuan, S. (2024). Effects of vertical integration on the healthcare system in China: a systematic review and meta-analysis. *Health Policy and Planning*, *39*(1), 66-79.

CHAPTER 7

Financial Strategy Analytics

"The smartest strategy isn't always bold or cautious it's analytically balanced ethically grounded and dynamically adaptive"

by MAK

Learning Outcomes

By the end of this chapter, you will be able to:

Like the falcon, the strategist must soar high to see the full landscape, dive with precision, and adapt mid-flight—balancing risk, ethics, and opportunity with every move

- ❖ Explain the strategic importance of financial forecasting in dynamic environments and identify the limitations of traditional deterministic methods.
- ❖ Apply real-options thinking to evaluate strategic investments under uncertainty and recognize how this method enhances flexibility in decision-making.
- ❖ Interpret Monte Carlo simulation outputs and describe how this technique supports scenario-based financial strategy and risk modeling.
- ❖ Assess capital allocation decisions using data-driven frameworks, including ethical considerations and multi-objective trade-offs.
- ❖ Compare traditional financial tools with STEM-enabled approaches, highlighting how algorithms, simulations, and modeling improve capital strategy.
- ❖ Analyze real-world case examples of firms that leverage financial strategy analytics to optimize resource deployment and competitive agility.
- ❖ Design strategic recommendations for capital investment using STEM tools such as real-options analysis and Monte Carlo simulation.

In an era where uncertainty is the norm, not the exception, financial strategy can no longer rest on static spreadsheets or backward-looking ratios. Strategic leaders must move beyond deterministic tools to embrace probabilistic thinking, advanced modeling, and dynamic simulation. Financial decisions are no longer isolated from technological change; they are deeply intertwined with it. This chapter explores how STEM-enabled tools transform financial strategy from a control function into a predictive, adaptive, and value-creating engine.

Traditional financial planning often assumes a single, most likely future. But today's volatile business environment demands more than accuracy; it demands agility. Firms must assess not only *what* to invest in, but *how* to sequence, scale, and stage those investments under changing assumptions. That's where financial strategy analytics comes in. By integrating approaches like real-options analysis and Monte Carlo simulation, organizations can model uncertainty, stress-test assumptions, and make capital allocation decisions with confidence and clarity.

This chapter unpacks how finance becomes a strategic partner when paired with analytics. You'll learn how forward-looking firms are using simulations to optimize product launch timing, how real-options logic supports investments in R&D, and why scenario modeling beats gut instinct when markets shift rapidly. From budgeting and valuation to strategic trade-offs and resource deployment, financial strategy analytics empowers decision-makers to see risk not as a threat, but as a variable to model, manage, and strategically master.

In the pages ahead, we will blend financial theory with computational methods and real-world business cases. Whether you're advising a startup or steering a global enterprise, these tools will prepare you to make financial decisions that are not just fiscally sound but strategically transformative. Welcome to the frontier where financial strategy meets predictive intelligence.

Opening Case: Ping An's Financial Intelligence Engine

When most people think of a financial services company, they picture conservative balance sheets, quarterly reports, and legacy systems that move more slowly than the markets they serve. Ping An, one of China's largest and most disruptive financial conglomerates, shattered that mold. Rather than treat finance as a back-office function, Ping An turned it into a strategic engine powered by analytics, simulations, and artificial intelligence.

By the mid-2010s, Ping An had already transformed its core business model into a data-driven ecosystem spanning insurance, banking, healthcare, and cloud services. But its approach to **financial strategy** was where its transformation became truly remarkable. While competitors debated capital deployment using static models and deterministic forecasts, Ping An built an internal simulation platform that ran thousands of probabilistic scenarios daily, allowing the firm to model risk, returns, and resilience in real time.

One major leap came when the company adopted **real-options reasoning** for evaluating its R&D initiatives and digital platforms. Instead of committing to multi-billion-dollar tech projects with fixed assumptions, Ping An treated each innovation as a strategic option, with the flexibility to expand, defer, pivot, or abandon based on real-time data. This logic not only preserved capital but also gave the company a first-mover advantage in high-risk, high-reward sectors like AI-driven diagnostics and blockchain-based settlements.

Behind the scenes, Ping An's financial strategy team ran **Monte Carlo simulations** to stress-test investment portfolios under thousands of economic and regulatory scenarios. These models incorporated data from partner ecosystems, AI forecasts, and global market signals, allowing Ping An to optimize capital buffers, reallocate investments dynamically, and even identify arbitrage opportunities faster than most hedge funds.

Critically, this wasn't just about maximizing returns. Ping An embedded **ethical capital allocation principles** into its algorithms, ensuring that investments are aligned with broader societal outcomes such as healthcare access, sustainable infrastructure, and fintech inclusion. Its "Finance + Technology + Ecosystem" model wasn't just a slogan; it was an engineered advantage.

As you begin this chapter, consider what it means to lead financial strategy in a world where uncertainty can be simulated, and opportunity is modeled as a range, not a point. Ping An's case invites you to reimagine finance not as a constraint on strategy, but as a launchpad for it.

Play the Strategist

1. How did Ping An's use of real-options logic change the way it approached strategic investments, and what are the benefits of this approach over traditional valuation methods like NPV?
2. What role did Monte Carlo simulations play in Ping An's ability to forecast and adapt under uncertainty? How might this technique enhance capital decision-making in other industries?
3. Ping An embedded ethical considerations into its financial algorithms. In your view, how should firms balance financial optimization with broader societal impact when allocating capital?

7.1 Real-Options Thinking in Strategic Finance

In the dynamic, uncertain, and technology-driven environment that defines today's competitive landscape, traditional financial valuation methods often fall short. Techniques such as net present value (NPV) and internal rate of return (IRR), while widely adopted, rely heavily on static assumptions and deterministic projections. These approaches typically treat investment decisions as binary choices, either fully investing or not investing at all. However, strategic decisions in an uncertain world require more nuanced approaches, capable of capturing the value of flexibility, adaptation, and proactive risk management. This is where real-options thinking transforms financial strategy into a dynamic, analytically driven discipline.

7.1.1 Moving Beyond Static Valuation

Traditional valuation methods have inherent limitations when dealing with uncertain outcomes. The common practice of discounting expected cash flows at a predetermined rate assumes stable markets and predictable competitive dynamics. This assumption rarely holds true in highly volatile environments like emerging technologies, biotech innovations, or digital platforms. Real-options thinking moves beyond this simplistic view by recognizing that strategic investments often present opportunities, not obligations, to expand, defer, pivot, or exit based on evolving information and market conditions.

Real-options logic frames investments as strategic options, analogous to financial options in securities markets. A real option gives a company the right, but not the obligation, to pursue future strategic opportunities. For example, a firm investing in research and development (R&D) obtains an option to commercialize a technology in the future, depending on market readiness, regulatory conditions, and competitor actions. This flexible approach better aligns valuation practices with the reality of modern strategic decision-making.

7.1.2 Strategic Value Under Uncertainty

The strategic value of real-options thinking lies in its explicit recognition and incorporation of uncertainty. Real options help managers quantify flexibility and adaptability as tangible sources of strategic value. Consider a biotechnology company investing in early-stage research for a new vaccine. Traditional methods might undervalue this investment due to uncertainty in clinical trials or market adoption. However, real-options analysis would explicitly model the flexibility to scale up, modify, or discontinue the project based on incremental discoveries and data points. This transforms uncertainty from a threat into a manageable and potentially profitable factor in the decision calculus.

Real-options logic also encourages staged investment strategies, incremental commitments contingent on future milestones. Companies like Tesla or Amazon continuously apply real-options principles by initially investing small amounts in experimental technologies or business models, scaling their investments only after initial assumptions have been validated through market feedback. Thus, real-options thinking transforms strategic uncertainty into actionable insights, fostering dynamic capital allocation and adaptive resource management.

7.1.3 Categories of Real Options in Strategy

Real options can take multiple forms, each reflecting different strategic scenarios:

- **Option to Defer**: Delaying an investment until market conditions or technological maturity improve.
- **Option to Expand**: Scaling up production or market presence if initial efforts prove successful.

* **Option to Contract or Abandon**: Scaling down or discontinuing projects if results fall short of expectations.
* **Option to Switch**: Shifting resources between alternative products, technologies, or markets.
* **Compound Options**: Combinations of multiple strategic decisions over a sequence of stages.

For instance, technology companies frequently hold expansion options, making initial smaller investments in emerging markets and technologies, reserving the option to expand rapidly if demand grows faster than expected. Conversely, energy companies often manage abandonment or contraction options, strategically downsizing or divesting projects as environmental policies or energy markets shift.

7.1.4 STEM Applications of Real Options

Real-options valuation methods can be greatly enhanced through STEM-based analytics, integrating sophisticated modeling and simulation tools that quantify strategic options explicitly. Techniques like decision tree analysis, Monte Carlo simulations, and computational optimization algorithms help strategists visualize the potential value of flexibility, quantify risk exposure, and systematically evaluate scenarios before investing significant resources.

For example, firms investing in new technologies, such as renewable energy or autonomous vehicles, leverage simulation and predictive analytics to quantify the value of various strategic options. Python and R-based software packages now allow managers to integrate vast datasets, scenario probabilities, and dynamic cash-flow modeling into real-options valuation, delivering insights at the speed and complexity required by modern markets.

Strategic Option in Electric Aviation

Imagine an aviation firm evaluating entry into the emerging electric aircraft market. Given uncertainties around battery technology, regulatory approval, and market adoption, traditional valuation methods may suggest caution. However, by applying real-options thinking, the firm can quantify the strategic flexibility embedded in early-stage investments:

* **Initial investment:** Minimal R&D, reserving the option to scale up if battery breakthroughs occur.
* **Expansion option:** Ready-to-execute contracts and supplier agreements triggered upon regulatory milestones.
* **Pivot or abandonment option:** Predefined triggers that halt further investments if technical or market barriers become insurmountable.

This structured, options-based strategy enables the firm to capture potential upside while systematically managing downside risks, something traditional Net Present Value (NPV) approaches often fail to accommodate.

The comparison below illustrates how real-options logic differs from traditional NPV in evaluating strategic investments under uncertainty:

Table 7.1: Comparison of Traditional NPV vs. Real Options Logic

Feature	Traditional NPV	Real Options Logic
Decision Framework	Fixed, immediate decision-making	Flexible, staged decision-making
Risk Treatment	Single discount rate, static risk	Dynamic, explicit modeling of uncertainty
Investment Timing	Immediate investment assumed	Timing is treated as a strategic choice
Adaptability to Market Change	Limited flexibility	High flexibility, built-in response
Valuation Methodology	Deterministic cash flows	Probabilistic modeling and simulation
Use of Computational Tools	Basic spreadsheet models	Advanced simulations and analytics
Strategic Application	Mature, stable environments	Highly volatile, dynamic markets

Examples: Real Options in Action

* **Tesla:** Invests incrementally in battery technologies, holding options to scale rapidly as technologies mature.
* **Merck:** Manages pharmaceutical R&D investments as real options, continually re-evaluating based on clinical trial outcomes.
* **Netflix:** Holds options to expand or discontinue original content based on data-driven audience analytics.

With this comprehensive articulation, students grasp not only the theory but also the strategic applicability and STEM-enhanced practicality of real-options thinking.

In an uncertain world, strategic financial decisions are options, not obligations. Real-options thinking equips leaders to value flexibility and navigate uncertainty

7.2 Monte Carlo Simulations for Financial Forecasting

Strategic decisions are fraught with uncertainty, from fluctuating market conditions to shifting consumer behaviors and volatile regulatory environments. In such contexts, point estimates and static projections can be misleading or even dangerous. Monte Carlo simulation (MCS), a powerful computational technique rooted in probability theory, offers a way to model uncertainty with precision and clarity. Rather than predicting a single outcome, it generates thousands of possible futures, revealing the full distribution of risks, returns, and outcomes that strategic leaders must consider.

In financial strategy, Monte Carlo simulations have become indispensable. They enable organizations to assess a range of potential investment outcomes, quantify risk exposure, stress-test capital structures, and support resilient decision-making. Whether evaluating the future value of a new product line, modeling cash flows for a major acquisition, or analyzing market volatility in portfolio strategy, Monte Carlo methods bring probabilistic rigor to otherwise ambiguous scenarios.

7.2.1 Fundamentals of Monte Carlo Simulation

At its core, Monte Carlo simulation involves four key steps:

1. **Define the Model**: Establish a financial or strategic model (e.g., future revenues, investment returns) with input variables that are uncertain.
2. **Assign Probability Distributions**: Instead of using fixed input values, assign probability distributions (e.g., normal, triangular, lognormal) to variables like costs, demand, interest rates, or market growth.
3. **Generate Simulated Scenarios**: Use random sampling techniques to generate thousands, or even millions, of simulations across the input distributions.
4. **Analyze the Output**: Aggregate the simulation results to assess the full probability distribution of outcomes, mean, standard deviation, percentiles, and tail risks.

For example, a firm considering a $50 million investment in a renewable energy project might model future revenues based on uncertain electricity prices, carbon credit policies, and technology costs. Monte Carlo simulation allows the firm to evaluate not only the average expected return but also the likelihood of negative cash flows, breakeven timelines, or extreme downside risks.

7.2.2 Strategic Applications in Finance

Monte Carlo simulation supports a wide range of strategic financial decisions:

* **Capital Budgeting**: Evaluate the range of net present values (NPVs) for new projects, including upside potential and downside risk.

* **Portfolio Optimization**: Simulate different asset allocations under various market conditions to identify optimal risk-return profiles.
* **Cash Flow Forecasting**: Predict operating cash flows under various scenarios, accounting for market volatility and cost uncertainties.
* **Pricing Strategy**: Assess the profitability of pricing models under fluctuating demand or cost structures.
* **Supply Chain Resilience**: Model cost impacts of supply chain disruptions or delays under different geopolitical and environmental scenarios.

These applications shift financial decision-making from deterministic spreadsheets to dynamic models, where uncertainty is a central variable, not an afterthought.

7.2.3 Integration with STEM Capabilities

Monte Carlo simulations represent a prime example of STEM-powered strategy. They rely on:

* **Mathematics**: Statistical distributions, stochastic processes, and confidence intervals.
* **Engineering and Computer Science**: Algorithm development, simulation coding (often in Python or R), and computational optimization.
* **Data Science**: Use of historical datasets to inform distributions, machine learning to refine inputs, and dashboards to visualize outputs.

Many firms now integrate simulation models into live financial dashboards. Cloud platforms and enterprise AI systems allow continuous simulation runs, feeding real-time updates into scenario planning tools. This real-time adaptability enables firms to make decisions faster and more confidently, even as conditions evolve. The table below contrasts Monte Carlo simulation with traditional forecasting approaches, highlighting its strategic advantages:

Table 7.2: Monte Carlo Simulation vs. Traditional Forecasting

Feature	Traditional Forecasting	Monte Carlo Simulation
Input Values	Fixed point estimates	Probability distributions
Output	Single outcome or average forecast	Full distribution of potential outcomes
Uncertainty Handling	Ignored or simplified	Explicitly modeled and quantified
Decision Support	Static reports	Scenario-rich insights and sensitivity maps
Tools Used	Spreadsheets, basic regression	Simulation engines, R/Python, analytics
Strategic Usefulness	Low in volatile contexts	High—enables resilient and flexible planning

Strategic Planning for Retail Expansion

Consider a global retail chain evaluating whether to expand operations into three emerging markets. Instead of relying on average GDP growth forecasts or fixed demand assumptions, the firm applies Monte Carlo simulations. It assigns probability distributions to variables like local wage growth, competitor pricing actions, exchange rates, and regulatory compliance costs. After running 10,000 simulations per market, the firm learns:

* **Market A** has the highest average return but a 20% chance of loss due to regulatory risks.
* **Market B** shows stable mid-range returns with low variance.
* **Market C** offers low average returns but strong upside under certain demand scenarios.

Armed with this insight, executives do not simply choose the "highest return" market. Instead, they structure a **portfolio strategy**, combining markets to optimize total risk-return exposure, and they design flexible market entry plans with contingency clauses built on simulated worst-case outcomes.

Strategic Snapshot Examples: Simulation in Action

* **Maersk**: Uses MCS to simulate route profitability and fuel cost risks across 100+ shipping scenarios.
* **Pfizer**: Runs clinical trial forecasts with MCS to assess timelines, approval probabilities, and capital exposure.
* **Unilever**: Applies simulations to model climate-related impacts on agricultural supply chains and raw input costs.

Monte Carlo simulations elevate strategic finance from reactive control to proactive scenario design. As the business environment becomes more uncertain, leaders equipped with this tool can confidently navigate volatility, identify resilience strategies, and optimize financial decisions with a full understanding of what lies within and beyond their control.

7.3 Ethical and Strategic Allocation of Capital

In an era where sustainability, transparency, and stakeholder capitalism define the boundaries of corporate legitimacy, the allocation of capital is no longer a purely financial exercise. It is a strategic, ethical, and systemic decision with long-term consequences. Capital allocation shapes not only a firm's profitability but also its social footprint, innovation trajectory, and legitimacy in the eyes of investors, regulators, and the public. This section examines how modern firms must evolve from shareholder-only logics to broader frameworks for strategic, ethical capital deployment, supported by advanced data systems and STEM-based analysis.

The essence of capital allocation lies in deciding where to invest, how much to invest, and when to divest. These decisions influence a firm's growth engines, shape its risk profile, and impact everything from employee welfare to environmental sustainability. In a STEM-enabled strategy context, ethical capital allocation integrates predictive models, ESG data systems, AI-driven risk assessments, and scenario testing, not only to maximize returns but to optimize for long-term value creation, system resilience, and stakeholder trust.

> Monte Carlo simulations don't predict the future, they illuminate its possibilities. In doing so, they empower leaders to act with clarity in the face of complexity.

7.3.1 Traditional Capital Allocation: Power and Pitfalls

Traditionally, capital allocation was framed through the lens of economic efficiency and financial returns. Using tools like hurdle rates, discounted cash flow (DCF), and internal rate of return (IRR), managers prioritized projects that yielded the highest short-term return on investment. Capital was concentrated in "core" business lines or areas with immediate growth prospects, often favoring scale and efficiency over adaptability or resilience.

While efficient in stable markets, this approach often ignores systemic risk, underestimates long-term liabilities, and fails to account for environmental and social externalities. For example, a fossil fuel investment might outperform in the short term, but pose significant long-run costs due to stranded assets, litigation risk, or reputational damage. Moreover, rigid capital allocation frameworks discourage experimentation, deprioritize innovation, and foster organizational inertia.

7.3.2 The Strategic Pivot: Capital as a Moral and Adaptive Resource

Modern strategy demands a rethinking of capital, not as a passive input, but as an **intentional force** for adaptive change. Ethical capital allocation requires firms to consider the second- and third-order effects of financial decisions. These include impacts on climate, labor conditions, data privacy, community equity, and algorithmic fairness.

For instance, investing in AI systems without governance safeguards may lead to short-term productivity gains but long-term regulatory and ethical risks. Conversely, capital investments in decarbonization infrastructure, employee training, or community development may have delayed returns, but build long-term resilience, legitimacy, and competitive insulation.

The strategic pivot requires three key shifts:

* **From efficiency to resilience**: Capital is directed toward initiatives that absorb shocks and enable flexibility.
* **From shareholder to stakeholder lens**: Allocation considers diverse interests, from employees and customers to ecosystems and future generations.
* **From fixed allocation to adaptive portfolios**: Continuous rebalancing of capital based on dynamic modeling, real-time performance, and scenario signals.

7.3.3 STEM Tools for Ethical Allocation

STEM capabilities now offer organizations the ability to rigorously quantify ethical and strategic implications of capital decisions:

* **ESG Dashboards**: Platforms integrate environmental, social, and governance (ESG) metrics with financial KPIs to guide investment screening.
* **Multi-Criteria Decision Analysis (MCDA)**: These mathematical tools weigh competing criteria—like ROI, carbon impact, and social equity, when allocating funds.
* **Machine Learning for Bias Detection**: Algorithms identify patterns of unequal capital distribution across regions, demographics, or product lines.
* **System Dynamics Modeling**: Simulates long-term feedback loops between capital flows, system performance, and unintended externalities.

For example, a global logistics company using system modeling may discover that underinvestment in digital worker training leads to longer-term inefficiencies and churn, offsetting the savings from initial automation investments. Ethical capital modeling helps correct these myopic incentives by embedding long-term, systemic thinking into financial processes.

The table below contrasts traditional capital allocation with an ethical-strategic approach, highlighting the shift toward more holistic, data-driven frameworks:

Table 7.3: Traditional vs. Ethical-Strategic Capital Allocation

Dimension	Traditional Approach	Ethical-Strategic Approach
Objective Function	Maximize short-term financial return	Maximize multi-dimensional long-term value
Time Horizon	Quarterly to annual	Multi-year to generational
Stakeholder Focus	Shareholders	Multiple stakeholders (internal & external)
Risk Management	Market and financial risks only	Systemic, ethical, reputational, and sustainability risks
Tools Used	ROI, NPV, IRR	ESG dashboards, MCDA, simulations, scenario planning
Data Scope	Financial metrics only	Integrated ESG, ethics, system feedback
Feedback Mechanisms	Annual reviews or budget cycles	Continuous, real-time adjustment based on data loops

7.3.4 Case Example: Capital Allocation at Ping An

Ping An, one of the world's most technologically advanced insurers, provides a compelling case of ethical and strategic capital allocation. The company invests not only in insurtech and AI capabilities but also in health-tech, eldercare platforms, and smart city initiatives, balancing return expectations with social outcomes.

Using AI-driven capital management systems, Ping An monitors both the financial and social return on its investments. Its Smart Healthcare division receives capital not just for profitability but for its contribution to public health access. Ping An also employs ESG rating algorithms that dynamically adjust funding across business units based on shifting stakeholder expectations and real-time feedback from regulatory and civil society actors.

Such practices show how ethical capital deployment can become a living system, adaptive, data-driven, and anchored in long-term purpose.

Strategic Snapshot: Capital Allocation in Practice

* **Microsoft**: Introduced an internal carbon fee that reallocates capital from high-emission units to green innovations.
* **BlackRock**: Integrates ESG risk into capital decisions across its trillion-dollar portfolio, pressuring firms to disclose climate risks.
* **Danone**: Reclassified as an "Entreprise à Mission" to legally bind capital decisions to social and environmental objectives.

Strategic finance in the STEM era is not just about numbers. It's about responsibility, resilience, and redefining the logic of value. Ethical capital allocation is no longer a moral luxury; it is a strategic necessity. It aligns purpose with precision and embeds integrity into the architecture of growth.

7.4 Predictive Credit Scoring and Strategic Lending

> Capital is not just a resource, it's a signal of your priorities. Ethical allocation builds systems that endure, not just portfolios that perform.

Modern credit scoring has moved far beyond static ratios and backward-looking reports. In the age of artificial intelligence and big data, **predictive credit scoring** is transforming how financial institutions evaluate risk, assess borrowers, and execute lending strategies. But what makes this transformation truly strategic is how scoring algorithms not only assess default probability but **shape lending behavior,** directing capital flows, influencing business models, and even altering the socioeconomic landscape.

This section explores how predictive credit systems, fueled by data analytics, behavioral modeling, and machine learning, create both competitive advantage and systemic responsibility. It also examines the strategic implications of **algorithmic lending**, particularly when deployed at scale across underserved or evolving markets.

7.4.1 From Traditional Scoring to Predictive Intelligence

Traditional credit scoring models, such as FICO or internal bank scores, have historically relied on a limited set of variables: income, debt-to-income ratios, payment history, and outstanding balances. These models are mostly linear, rule-based, and rooted in historical data. Their key limitations are well-documented:

- Poor adaptability to non-traditional borrowers (e.g., freelancers, migrants, small entrepreneurs)
- Blind spots for emerging risks or behavioral shifts
- Inflexibility in dynamic economic contexts

Predictive scoring models, by contrast, draw from **diverse data sources,** ranging from real-time transaction logs and e-commerce patterns to utility payments, mobile behavior, and even social signals. These models use **machine learning algorithms** (e.g., decision trees, ensemble models, neural networks) that **learn patterns**, update over time, and identify non-linear relationships across borrower profiles.

What emerges is not just a score, but a **dynamic, context-sensitive probability forecast**, embedded into lending decisions, risk pricing, and portfolio strategies.

The table below highlights the key differences between traditional and predictive credit scoring approaches:

Table 7.4: Traditional vs. Predictive Credit Scoring

Feature	Traditional Scoring	Predictive Credit Scoring
Data Sources	Credit bureau, income, liabilities	Transactional, behavioral, mobile, utility, platform
Methodology	Rule-based, static models	Machine learning, real-time updates
Borrower Inclusion	Primarily a formal economy	Broader inclusion (e.g., gig workers, SMEs)
Risk Interpretation	Binary cutoffs	Probabilistic distributions, confidence intervals
Strategic Use Case	Risk assessment	Strategic customer acquisition, pricing, and cross-sell

7.4.2 Strategic Impacts of Credit Analytics

The strategic role of credit analytics extends far beyond underwriting. Firms increasingly **embed predictive credit scores into broader business models**, shaping which markets they serve, how they structure products, and how fast they can scale. Consider three strategic use cases:

- **Customer segmentation**: Predictive scores enable micro-segmentation, allowing firms to tailor loan terms, interest rates, or onboarding flows to specific borrower profiles.
- **Dynamic risk-based pricing**: Lenders can fine-tune interest rates and credit limits based on evolving risk indicators rather than fixed bands, improving margin precision.

- **Portfolio resilience**: Aggregated credit forecasts at the portfolio level allow firms to rebalance exposure in real-time as economic conditions or user behavior shift.

For fintechs and challenger banks, predictive credit capabilities are not just tools; they are **core differentiators**. The speed and accuracy of credit decisions impact customer experience, regulatory compliance, and growth velocity.

7.4.3 Ethical Risk and Bias in Algorithmic Lending

Yet the promise of predictive scoring is shadowed by a strategic dilemma: **bias in algorithmic decision-making**. Credit scoring systems may unintentionally reflect or amplify historical inequities, denying access to capital for certain racial, geographic, or income groups. The opacity of machine learning ("black box" models) further complicates accountability.

Strategic leaders must address these risks through:

- **Explainable AI (XAI)**: Using interpretable models (e.g., SHAP values) to justify why a credit decision was made.
- **Fairness-aware modeling**: Incorporating fairness metrics (e.g., demographic parity, equal opportunity) during model training.
- **Audit loops**: Regular internal and external audits of scoring algorithms for performance, bias, and drift.
- **Policy overlays**: Introducing human-in-the-loop review for borderline or high-impact decisions.

Firms that fail to govern predictive scoring systems risk not only reputational damage and regulatory backlash but also **systemic exclusion,** undermining market trust and long-term growth.

Ping An's AI-Driven Credit Strategy

Ping An's strategic use of predictive scoring showcases a sophisticated fusion of AI, customer insights, and systemic thinking. The firm uses an **AI-based credit engine** that draws from thousands of behavioral signals, health records, insurance payments, and app usage patterns, to predict not only creditworthiness but **financial health trajectories**.

This allows Ping An to dynamically adapt loan terms, design alternative credit products, and offer **graduated financing pathways** for emerging customers. Strategic lending is not just a gatekeeper function; it becomes a **lifecycle enabler**, deepening customer loyalty and unlocking long-term value.

At the same time, Ping An subjects its models to **regular fairness checks** and maintains human oversight protocols in high-risk domains like healthcare lending or rural microfinance. The result is an ecosystem where algorithmic speed meets ethical precision.

Predictive credit scoring redefines how capital reaches people and firms. When embedded into a lending strategy, it becomes a strategic lever, shaping growth trajectories, inclusion outcomes, and systemic trust. The challenge for tomorrow's strategist is not just to adopt these tools, but to wield them responsibly, modeling not only **who to lend to**, but also **how and why** lending shapes the world.

7.5 Strategic Budgeting and Adaptive Finance Systems

Budgeting, long considered a backbone of corporate control, is undergoing a fundamental transformation. In traditional settings, budgets serve as static forecasts, limiting flexibility in volatile markets. In a STEM-enabled strategy, however, budgeting becomes **adaptive,** a real-time system of financial sense-making, responsive to external signals, internal simulations, and strategic pivots.

This section explores how organizations are shifting from rigid annual budgeting toward **dynamic, analytics-enabled finance systems** that support iterative strategy execution. Strategic budgeting today is no longer just a financial process; it is a **strategic intelligence tool** for managing ambiguity, enabling foresight, and optimizing resource deployment at the speed of change.

> In predictive lending, the goal isn't just to predict default, it's to design financial systems that adapt, include, and evolve.

7.5.1 From Fixed Budgets to Strategic Adaptability

Traditional budgeting models operate on fixed annual cycles. Resources are allocated early, assumptions are locked in, and deviations are treated as failures rather than signals. This approach creates several tensions:

* Misalignment with fast-changing markets
* Inability to pivot resources during crises or opportunities
* Encouragement of "use-it-or-lose-it" behaviors rather than efficient allocation

In contrast, **adaptive finance systems** leverage data, simulation models, and feedback loops to **continuously update financial plans**. These systems embrace uncertainty, treating budget assumptions as hypotheses to be tested and refined. Tools such as **rolling forecasts**, **zero-based budgeting (ZBB)**, and **driver-based models** allow finance leaders to respond in weeks, not quarters.

The table below contrasts traditional budgeting with adaptive finance systems, highlighting the shift from rigidity to responsiveness:

Table 7.5: Traditional Budgeting vs. Adaptive Finance Systems

Dimension	Traditional Budgeting	Adaptive Finance Systems
Time Horizon	Annual, fixed cycle	Rolling, real-time
Core Assumptions	Static, top-down	Dynamic, data-driven
Scenario Flexibility	Limited	High—models multiple contingencies
Strategic Responsiveness	Slow	Fast, iterative
Tools Used	Spreadsheets, variance reports	AI forecasting, real-time dashboards, digital twins

7.5.2 STEM Tools Powering Adaptive Finance

Adaptive budgeting is made possible through a suite of STEM-based tools that support strategic agility:

> In adaptive strategy, a budget isn't a cage, it's a compass. The goal is not control, but continuous calibration.

* **Predictive Analytics**: Forecasting revenue, costs, and macroeconomic trends using machine learning models (e.g., time series ensembles, gradient boosting).
* **Simulation Engines**: Monte Carlo simulations to model budget outcomes across thousands of plausible futures.
* **Digital Twins**: Virtual replicas of business units or products that simulate financial performance under various operational scenarios.
* **Dynamic Dashboards**: Integrated visualization tools that show real-time budget vs. actual trends, risk flags, and forward projections.

These tools do more than optimize spreadsheets; they transform the budget into a **living model of organizational strategy**.

7.5.3 Strategy-as-Hypothesis: Budgeting for Learning

One of the most profound shifts in adaptive finance is the idea of treating strategy as a hypothesis and budgeting as a **structured learning loop**. Rather than assume strategic choices will succeed, firms allocate resources in stages, measuring early signals and adjusting quickly. This logic parallels startup methodologies such as **lean experimentation**, now applied to corporate budgeting.

Key features of this mindset include:

* **Minimum Viable Budgeting (MVB)**: Allocate a small initial budget to test new strategic initiatives, increasing investment only with proven traction.
* **Trigger-Based Adjustments**: Pre-define metrics or events that automatically reallocate budgets (e.g., market demand spike triggers more marketing funds).
* **Post-Mortem Analytics**: Systematically analyze budget deviations as insights, not failures, improving the quality of future assumptions.

Unilever's Dynamic Investment Engine

Unilever, the global consumer goods giant, exemplifies this shift. The company has deployed a **digitally integrated investment model** where marketing, supply chain, and sustainability budgets are **re-evaluated monthly** using predictive demand forecasts, climate data, and social sentiment. Product lines are backed by **financial digital twins** that allow decision-makers to simulate price elasticity, packaging costs, and carbon impacts before allocating funds.

This approach has increased capital efficiency and allowed Unilever to respond to global shocks (e.g., COVID-19, supply chain disruptions) more nimbly than competitors. Strategic budgeting, for Unilever, is now **a platform for agility, not a constraint on action.**

The shift to adaptive finance systems empowers organizations to **treat change as a signal, not noise**. By integrating budgeting into a broader framework of strategic intelligence, powered by data, simulations, and feedback, leaders create a financial foundation for real-time learning, risk-aware innovation, and operational resilience.

Strategic budgeting in a STEM world is not simply about balancing spreadsheets; it's about **orchestrating capital, capacity, and confidence in motion.**

7.6. Chapter 7 Wrap-Up: Financial Intelligence in a STEM-Driven Strategy

As we close this chapter, it becomes evident that financial strategy is no longer limited to cost control or balance sheet stewardship. In today's volatile, data-rich, and fast-moving environment, **financial intelligence** is a strategic muscle—one that allows organizations to allocate capital dynamically, model uncertainty, and align finance with innovation and impact.

This chapter redefined financial strategy as an adaptive, analytics-powered capability. We explored how **real-options logic** helps leaders treat uncertainty as opportunity; how **Monte Carlo simulations** offer clarity amidst volatility; how **predictive credit scoring** reorients lending toward strategic inclusion and agility; and how **adaptive budgeting** transforms finance into a responsive, hypothesis-driven system.

The firms we examined, Ping An, Unilever, and others, don't merely use data to track financial outcomes. They use it to **shape financial futures**. Finance becomes the language of experimentation, foresight, and competitive edge.

We also examined how financial decisions embed ethical and systemic consequences. Whether through algorithmic lending, ESG-aligned portfolios, or budget trade-offs in crisis settings, the future of finance lies not just in precision, but in **purposeful precision,** in making decisions that are fair, flexible, and future-ready.

As you progress in your strategic thinking journey, remember: STEM-enabled financial strategy is not about replacing human intuition, but **enhancing it with clarity, computation, and creative foresight.** Strategic leaders of tomorrow must speak the language of both capital and code, and use it to build systems that adapt, learn, and lead.

7.6.1. Discussion Questions

1. How does real-options logic change the way firms approach long-term strategic investments?
2. What are the key advantages of Monte Carlo simulations in financial forecasting?
3. How can predictive credit models drive inclusion and innovation in underserved markets?
4. What risks arise from relying on algorithmic credit scoring without ethical oversight?
5. In what ways does adaptive budgeting enable strategic agility?
6. How can digital twins be used beyond operations—within finance and capital allocation?
7. What are the limitations of traditional NPV-based investment decision-making?
8. How does financial intelligence support sustainability-driven strategy?
9. What STEM tools are essential for future finance leaders?
10. How can finance departments shift from control centers to learning engines?

7.6.2. Micro-Case: Zara's Financial Foresight

Zara, now leading strategic finance for a global healthcare technology firm, faces her most complex challenge yet: financing a next-gen diagnostics platform while navigating economic downturns and supply shocks.

Her task? **Orchestrate capital flows** across three continents, balancing R&D risk, ESG investment pressure, and shifting consumer demand. She starts by implementing a **real-options logic** to structure R&D investments as modular, kill-or-scale projects. Using Monte Carlo simulations, she models financial outcomes across multiple rollout timelines and economic conditions.

Zara integrates a **predictive scoring algorithm** to identify hospital systems likely to default on financing, adjusting contract terms accordingly. She also creates a dynamic budget tied to key triggers, like regulatory approval stages and patient adoption rates, allowing funds to reallocate monthly.

To ensure ethical foresight, Zara establishes a "bias audit loop" for her team's algorithms and invites external advisors to monitor their predictive credit scoring tool. In investor calls, she doesn't just share net present value metrics; she shares **financial resilience indicators** based on live simulations.

Zara's financial strategy illustrates how **quantitative logic, ethical awareness, and real-time decision architecture** now define modern finance. It's not just about numbers. It's about **navigating the future with structure, speed, and purpose.**

Micro-Case Discussion Questions: Zara's Financial Foresight

1. **How does Zara's use of real-options logic improve her ability to finance innovation amid uncertainty?**
   What traditional financial pitfalls does this approach help avoid?
2. **What are the advantages and potential ethical risks of integrating predictive credit scoring into hospital financing decisions?**
   How could Zara ensure fairness while maintaining financial viability?
3. **Zara implemented a trigger-based adaptive budget tied to regulatory and adoption milestones.**
   What are the strengths and limitations of such a budgeting method in high-stakes innovation?
4. **How does Zara balance financial transparency with technical complexity in investor communications?**
   What lessons can be drawn from her approach to stakeholder alignment?

7.6.3. Interactive Exercise: Budgeting in an Uncertain World

Scenario:
You are the CFO of a multinational logistics company operating across Southeast Asia and Latin America. Geopolitical tensions and supply chain disruptions have created financial volatility. Your board is demanding a strategic reallocation of capital for the next 6 months.

Your Resources:

* Monte Carlo simulation results showing delivery time variance and revenue volatility.
* Predictive credit scores of SME clients in local markets
* Real-options model for expanding a regional warehouse

Your Task:
Using tools from this chapter (e.g., adaptive budgeting, predictive credit analytics, simulation), develop a **3-point capital allocation plan** that balances:

* Strategic opportunity
* Financial resilience
* Ethical risk

Your plan should include trigger conditions for budget adjustments and at least one mechanism for stakeholder transparency.

7.6.4. Recommended HBR Case Studies for Chapter 6

❖ Upstart's Upshot: Is Fintech Lending Fair?

* **Overview:** The case "Upstart's Upshot: Is Fintech Lending Fair?" explores the intersection of financial innovation and ethical responsibility in the rapidly evolving world of AI-driven lending. It centers on Alliant Credit Union, which is considering a partnership with Upstart Holdings, Inc., a fintech company that uses machine learning algorithms to assess creditworthiness. Upstart's model incorporates both traditional financial data and alternative variables, such as education and employment history, to expand access to credit. While the company claims its approach improves loan approval rates and reduces defaults, it has also faced scrutiny over potential bias and fairness issues, particularly regarding the use of educational background in credit decisions. The case challenges students to evaluate whether Upstart's model aligns with fair lending laws, how to balance innovation with equity, and what role financial institutions should play in ensuring algorithmic accountability. It's ideal for discussions on AI ethics, regulatory compliance, and inclusive finance.
* Citation: Appel, I., & Sesia, A. (2023). *Upstart's Upshot: Is Fintech Lending Fair?* (Case No. UV8787-PDF-ENG). Darden School of Business.

❖ Fintech Solutions: Under Pressure

* **Overview:** Published on July 16, 2023, and authored by Julie Gosse and Kartik Rao, this 10-page case focuses on Priyanka Bhat, a project manager at Fintech Solutions, who is facing a critical team conflict during a high-stakes client engagement in October 2022. The central issue involves Jada Goodwin, a talented and client-favored business analyst who has been accused by several team members of creating a hostile work environment. As tensions rise, Bhat must decide how to restore team cohesion, ensure timely project delivery, and address the interpersonal dynamics—while also navigating the complexities of remote work, diversity and inclusion, and conflict management. This case is ideal for discussions on team leadership, managing difficult employees, and balancing performance with workplace culture.

* **Citation:** Gosse, J., & Rao, K. (2023). *Fintech Solutions: Under Pressure* (Case No. W33636-PDF-ENG). Ivey Publishing.

❖ CoVenture: Financing Innovations in Fintech with Asset-Backed Credit

* **Overview:** Published on May 14, 2024, and authored by David S. Scharfstein and Robert Ialenti, this 10-page case explores how CoVenture, an investment firm, innovates in the fintech space by providing asset-backed credit to early-stage companies. The case focuses on CoVenture's unique approach to structured finance, where it partners with fintech startups that need capital to fund loans to their own customers. Rather than taking equity stakes alone, CoVenture structures credit facilities backed by the underlying assets (such as consumer or small business loans), allowing fintechs to scale while managing risk. Students are asked to evaluate the risk-return profile of this model, the due diligence process, and how CoVenture balances innovation with financial discipline. The case is ideal for discussions on alternative lending, venture debt, and financial structuring in the fintech ecosystem.
* **Citation:** Scharfstein, D. S., & Ialenti, R. (2024). *CoVenture: Financing Innovations in Fintech with Asset-Backed Credit* (Case No. 224089-PDF-ENG). Harvard Business School.

❖ NorthCentral Bank: Navigating the Challenges of Fintechs, Neobanks, and Cryptocurrencies

* **Overview:** Published on November 8, 2023, and authored by Hubert Pun, Felipe Restrepo, and Dollar Luo, this 13-page case focuses on Jonathan Gray, the Head of Digital Assets at NorthCentral Bank. He is tasked with determining how the bank should respond to the growing demand for cryptocurrency investment products and the broader disruption caused by fintechs, neobanks, and blockchain technologies. As younger, digitally native consumers increasingly seek crypto-friendly financial services, Gray must evaluate three strategic options:
    * Maintain the status quo and avoid entering the crypto space.
    * Offer investment-only products related to digital assets.
    * Become a full ecosystem player, offering wallets and enabling native blockchain transactions.

  Each path presents distinct risks and opportunities, and the case challenges students to consider how a traditional financial institution can remain competitive in a rapidly evolving digital landscape.

* **Citation:** Pun, H., Restrepo, F., & Luo, D. (2023). *NorthCentral Bank: Navigating the Challenges of Fintechs, Neobanks, and Cryptocurrencies* (Case No. W33866-PDF-ENG). Ivey Publishing.

7.6.5. Strategy-in-Film: Learn by Watching

Financial strategy is more than numbers; it is a critical enabler of innovation, resilience, and long-term value creation. The films below highlight how organizations manage risk, allocate resources, and make strategic financial decisions under uncertainty. As you engage with financial strategy analytics in this chapter, use these narratives to reflect on how data-driven financial choices shape organizational success in complex environments.

Movie / Documentary	Strategic Theme Highlighted
The Big Short	Financial forecasting, risk modeling, and market mispricing
Margin Call	Decision-making under financial pressure
Moneyball	Value-based investing, financial tradeoffs, and analytics
Too Big to Fail	Capital strategy and systemic financial risk
Inside Job	Ethics in finance, systemic bias in modeling
Startup.com	Budgeting, venture finance, capital burn dynamics
The China Hustle	Predictive signals and fraud detection
Enron: The Smartest Guys in the Room	The danger of short-termism and opaque financial systems
Boiler Room	Risk and deception in aggressive financial strategy
Trust No One: The Hunt for the Crypto King	Decentralized finance and algorithmic risk

References

Ahmed, S., Alshater, M. M., El Ammari, A., & Hammami, H. (2022). Artificial intelligence and machine learning in finance: A bibliometric review. *Research in International Business and Finance*, *61*, 101646.

Ahsan, T., Al-Gamrh, B., & Mirza, S. S. (2022). Economic policy uncertainty and sustainable financial growth: does business strategy matter?. *Finance Research Letters*, *46*, 102381.

Al Shiam, S. A., Hasan, M. M., Pantho, M. J., Shochona, S. A., Nayeem, M. B., Choudhury, M., & Nguyen, T. N. (2024). Credit risk prediction using explainable AI. *Journal of Business and Management Studies*, *6*(2), 61.

Arslanian, H., & Fischer, F. (2019). *The future of finance: The impact of FinTech, AI, and crypto on financial services*. Springer.

Bahoo, S., Cucculelli, M., Goga, X., & Mondolo, J. (2024). Artificial intelligence in Finance: a comprehensive review through bibliometric and content analysis. *SN Business & Economics*, *4*(2), 23.

Banerjee, B. (2015). *Fundamentals of financial management*. PHI Learning Pvt. Ltd..

Benaroch, M., & Kauffman, R. J. (1999). A case for using real options pricing analysis to evaluate information technology project investments. *Information Systems Research*, *10*(1), 70-86.

Bertsimas, D., & Kallus, N. (2020). From predictive to prescriptive analytics. *Management Science*, *66*(3), 1025-1044.

Breuer, W., & Ruiz de Vargas, S. (2021). Some key developments in international financial management. *Journal of Business Economics*, *91*(5), 595-615.

Cao, S. S., Jiang, W., & Lei, L. G. (2024). Applied AI for finance and accounting: Alternative data and opportunities. *Pacific-Basin Finance Journal*, *84*, 102307.

Chen, Y., Calabrese, R., & Martin-Barragan, B. (2024). Interpretable machine learning for imbalanced credit scoring datasets. *European Journal of Operational Research*, *312*(1), 357-372.

Edo-Osagie, E. (2025). Advanced Budgeting and Dynamic Allocation Strategies for Maximizing Financial Resilience Amid Economic Volatility and Market Shifts. *International Journal of Research Publication and Reviews*, *6*(3), 1060-1077.

Gallagher, T. J. (2022). *Financial management: principles and practice*. SAGE Publications.

Gambacorta, L., Huang, Y., Qiu, H., & Wang, J. (2024). How do machine learning and non-traditional data affect credit scoring? New evidence from a Chinese fintech firm. *Journal of Financial Stability*, *73*, 101284.

Giudici, P., & Raffinetti, E. (2023). SAFE Artificial Intelligence in finance. *Finance Research Letters*, *56*, 104088.

Guo, J., Fang, H., Liu, X., Wang, C., & Wang, Y. (2023). FinTech and financing constraints of enterprises: Evidence from China. *Journal of International Financial Markets, Institutions and Money*, *82*, 101713.

Hammett, L. M., & Mixter, K. (2017). Adaptive finance to support post-disaster recovery. *Yale Center for Business and the Environment: New Haven, CT, USA*.

Hawkesworth, I., & Klepsvik, K. (2013). Budgeting levers, strategic agility, and the use of performance budgeting in 2011/12. *OECD Journal on Budgeting*, *2013*(1).

Heesen, B. (2015). *Effective strategy execution: improving performance with business intelligence*. Springer.

Hillier, D., Grinblatt, M., & Titman, S. (2023). *Financial Markets and Corporate Strategy: European Edition, 3e*. McGraw Hill.

Hilpisch, Y. (2020). *Artificial intelligence in finance*. O'Reilly Media.

Hitt, M. A., Jackson, S. E., Carmona, S., Bierman, L., Shalley, C. E., & Wright, M. (2017). *The Oxford handbook of strategy implementation*. Oxford University Press.

Hurlin, C., Pérignon, C., & Saurin, S. (2024). The fairness of credit scoring models. *Management Science*.

Khezri, B. (2022). Structure: Synergizing Governance and Operational Channels. In *Governing Continuous Transformation: Re-framing the Strategy-Governance Conversation* (pp. 51-62). Cham: Springer International Publishing.

Lee, C. F., Lee, J. C., & Lee, M. (2022). *Corporate finance and strategy: An active learning approach*. World Scientific.

Li, X., Sigov, A., Ratkin, L., Ivanov, L. A., & Li, L. (2023). Artificial intelligence applications in finance: a survey. *Journal of Management Analytics*, *10*(4), 676-692.

Maheshwari, S. N., Maheshwari, S. K., & Maheshwari, S. K. (2024). *Management Accounting and Financial Control*. Sultan Chand & Sons.

McLeish, D. L. (2011). *Monte Carlo simulation and finance* (Vol. 276). John Wiley & Sons.

Milana, C., & Ashta, A. (2021). Artificial intelligence techniques in finance and financial markets: a survey of the literature. *Strategic Change*, *30*(3), 189-209.

Mun, J. (2006). *Modeling risk: Applying Monte Carlo simulation, real options analysis, forecasting, and optimization techniques* (Vol. 347). John Wiley & Sons.

Mun, J. (2012). *Real options analysis: Tools and techniques for valuing strategic investments and decisions* (Vol. 320). John Wiley & Sons.

Nallakaruppan, M. K., Balusamy, B., Shri, M. L., Malathi, V., & Bhattacharyya, S. (2024). An Explainable AI framework for credit evaluation and analysis. *Applied Soft Computing*, *153*, 111307.

O'Grady, W., Akroyd, C., & Scott, I. (2017). Beyond budgeting: distinguishing modes of adaptive performance management. In *Advances in management accounting* (Vol. 29, pp. 33-53). Emerald Publishing Limited.

Puschmann, T. (2017). Fintech. *Business & Information Systems Engineering*, *59*, 69-76.

Ronyastra, I. M., Saw, L. H., & Low, F. S. (2024). Monte Carlo simulation-based financial risk identification for industrial estate as post-mining land usage in Indonesia. *Resources Policy*, *89*, 104639.

Ranjan, P., Dahiya, S., Khunger, A., & Pandiya, D. K. (2024). AI in Finance optimization: A review of current technologies and future potential. *International Journal of Global Innovations and Solutions (IJGIS)*.

Ross, S. (2015). *Ebook: Fundamentals of Corporate Finance*. McGraw Hill.

Ruppert, D. (2004). *Statistics and finance: An introduction* (Vol. 27). New York: Springer.

Sampat, B., Mogaji, E., & Nguyen, N. P. (2024). The dark side of FinTech in financial services: a qualitative enquiry into FinTech developers' perspective. *International Journal of Bank Marketing*, *42*(1), 38-65.

Sant'Anna, D. A., & Figueiredo, P. N. (2024). Fintech innovation: Is it beneficial or detrimental to financial inclusion and financial stability? A systematic literature review and research directions. *Emerging Markets Review*, *60*, 101140.

Schwartz, E. S., & Trigeorgis, L. (Eds.). (2004). *Real Options and Investment under uncertainty: classical readings and recent contributions*. MIT Press.

Shrestha, Y. R., Ben-Menahem, S. M., & Von Krogh, G. (2019). Organizational decision-making structures in the age of artificial intelligence. *California Management Review*, *61*(4), 66-83.

Talaat, F. M., Aljadani, A., Badawy, M., & Elhosseini, M. (2024). Toward interpretable credit scoring: integrating explainable artificial intelligence with deep learning for credit card default prediction. *Neural Computing and Applications*, *36*(9), 4847-4865.

Tigges, M., Mestwerdt, S., Tschirner, S., & Mauer, R. (2024). Who gets the money? A qualitative analysis of fintech lending and credit scoring through the adoption of AI and alternative data. *Technological Forecasting and Social Change*, *205*, 123491.

Trigeorgis, L., & Reuer, J. J. (2017). Real options theory in strategic management. *Strategic Management Journal*, *38*(1), 42-63.

Valle-Cruz, D., Fernandez-Cortez, V., & Gil-Garcia, J. R. (2022). From E-budgeting to smart budgeting: Exploring the potential of artificial intelligence in government decision-making for resource allocation. *Government Information Quarterly*, *39*(2), 101644.

Vernimmen, P., Quiry, P., & Le Fur, Y. (2022). *Corporate finance: theory and practice*. John Wiley & Sons.

Vijayagopal, P., Jain, B., & Ayinippully Viswanathan, S. (2024). Regulations and Fintech: A Comparative Study of the Developed and Developing Countries. *Journal of Risk & Financial Management*, *17*(8).

Xu, Y., Kou, G., Peng, Y., Ding, K., Ergu, D., & Alotaibi, F. S. (2024). Profit-and risk-driven credit scoring under parameter uncertainty: A multiobjective approach. *Omega*, *125*, 103004.

Zheng, X. L., Zhu, M. Y., Li, Q. B., Chen, C. C., & Tan, Y. C. (2019). FinBrain: when finance meets AI 2.0. *Frontiers of Information Technology & Electronic Engineering*, *20*(7), 914-924.

CHAPTER 8

International Strategy Through STEM

"Global strategy is no longer about choosing countries it's about sensing systems In a STEM powered world borders are not boundaries but data rich thresholds of opportunity"

by MAK

Learning Outcomes

By the end of this chapter, readers will be able to:

Like the sea turtle, the global strategist must navigate vast, shifting currents—guided not by borders, but by signals, systems, and the wisdom of long-term adaptation.

- ❖ Differentiate between traditional international strategy models (e.g., CAGE, Porter's Diamond) and data-driven STEM-based approaches to global expansion.
- ❖ Design a predictive, analytics-enabled international market entry strategy that incorporates digital infrastructure, ESG readiness, geopolitical risk, and climate exposure.
- ❖ Apply Hofstede and GLOBE cultural models using real-world datasets to assess cross-cultural leadership, communication, and organizational alignment challenges.
- ❖ Leverage machine learning tools (e.g., XGBoost, Gradient Boosting, Monte Carlo simulations) to prioritize countries and entry modes based on multidimensional strategic indicators.
- ❖ Develop and interpret real-time expansion dashboards using tools such as Power BI, Tableau, or R to monitor and adapt international strategies under changing conditions.
- ❖ Evaluate the integration of human judgment and algorithmic foresight in global decision-making, considering ethical, cultural, and strategic alignment across borders.

Globalization has entered a new era, one not simply defined by open borders or expanded markets, but by accelerated data, algorithmic insight, and technological asymmetry. In this world, international strategy can no longer rely on outdated checklists or generalizations about cultural distance. Instead, firms must harness STEM-based tools to navigate global complexity, anticipate shocks, and identify dynamic opportunities across diverse regions.

This chapter marks a turning point in our strategic journey. We move from understanding organizations and markets to understanding systems, networks of interdependent economies, infrastructures, climate impacts, regulations, and cultures. And rather than asking only *where* to expand, we now ask *how*, *when*, and *under what modeled uncertainties*. Market entry decisions are no longer a leap of faith; they can be informed by predictive algorithms, simulation tools, and real-time data platforms.

While classical frameworks such as the CAGE model or Porter's Diamond remain valuable, they are increasingly insufficient on their own. A STEM-powered strategist must go further: modeling digital maturity, integrating ESG readiness, and evaluating geopolitical volatility using satellite intelligence and climate risk indexes. Expansion strategies must also account for rapidly shifting consumer behaviors, automated logistics, and ethical dilemmas that vary by cultural context.

Just as importantly, international strategy requires understanding people, not in abstract, but through measurable, comparative models of culture, leadership, and communication. Tools like Hofstede's Dimensions and the GLOBE framework give us a way to model national business behaviors with clarity and purpose. When paired with machine learning, these cultural models become predictive, not just descriptive.

This chapter equips you to approach global expansion as both a science and an art. You'll build dashboards, model risk, compare countries with XGBoost or Monte Carlo simulation, and make decisions that balance mathematical rigor with cultural sensitivity. As the world grows more connected and complex, strategic leadership must become both globally aware and computationally fluent. Welcome to the frontier of international strategy, where STEM meets the world.

Opening Case: Ping An's Algorithmic Expansion into Southeast Asia

By 2025, Ping An had already established itself as a global pioneer in integrated finance, technology, and healthcare. But its ambitions didn't stop at dominating the Chinese market. The company set its sights on Southeast Asia, a region marked by rapid digital adoption, youthful populations, and a fragmented financial services landscape. Rather than relying on traditional market entry frameworks or anecdotal assessments, Ping An launched its expansion through a full-stack STEM strategy.

First, the company built a **Market Intelligence Dashboard** that fused real-time economic indicators, infrastructure readiness, mobile penetration, ESG performance, and regional climate risk data. Using predictive modeling, including **XGBoost** and **gradient boosting algorithms,** Ping An ranked ASEAN countries based on growth potential, regulatory predictability, and digital health gaps.

For cultural adaptation, the strategy team used **Hofstede's cultural dimensions** and **GLOBE leadership indices** to tailor fintech product launches by country. For example, in Malaysia, where uncertainty avoidance was high, Ping An emphasized platform security and regulatory compliance. In Vietnam, where future orientation and performance orientation scored high, the company marketed predictive financial planning tools powered by AI.

But data didn't just inform market selection; it shaped **entry mode strategy**. In Indonesia, Ping An opted for a **joint venture with a local super-app** after Monte Carlo simulations revealed high variance in user acquisition cost under solo expansion scenarios. In contrast, it launched a **greenfield digital bank** in Singapore, where infrastructure maturity and governance scores were optimal.

Internally, Ping An trained its international teams to **balance algorithmic foresight with human judgment**. Strategic decisions were not made solely by models; they were stress-tested by scenario planning, cultural advisors, and on-the-ground teams.

As a result, Ping An not only expanded profitably but became a benchmark for **STEM-powered globalization**. Its success demonstrated that modern international strategy is no longer about exporting core products; it's about modeling systems, optimizing entry paths, and harmonizing data with deep cultural insight.

Play the Strategist

1. How did Ping An combine predictive modeling with cultural frameworks to design tailored entry strategies across different countries?
2. What are the advantages of using machine learning models like XGBoost in international market selection? What are the risks?
3. Why might a Monte Carlo simulation change a firm's choice of entry mode? How does this differ from traditional financial analysis?
4. What lessons does Ping An's case offer for companies trying to expand into markets with volatile infrastructure or uncertain governance conditions?

8.1 STEM-Based Market Selection and Entry Strategy

In a global economy shaped by volatility, connectivity, and exponential technology growth, the process of entering foreign markets has become both more promising and more perilous. While the promise lies in access to new customers, talent pools, and innovation ecosystems, the peril arises from geopolitical risks, climate uncertainty, digital disparities, and cultural complexity. Traditional market entry strategies often lack the rigor to navigate this terrain. A modern, STEM-powered approach, by contrast, transforms market selection and entry strategy into a dynamic process of modeling, simulation, and data-driven decision-making.

This section introduces practitioners to the transformation of global expansion strategy through STEM disciplines, especially data science, systems modeling, and decision analytics. We examine how organizations can build intelligent market selection frameworks and choose entry modes not on intuition or tradition, but on evidence, foresight, and scenario testing.

8.1.1 Why Traditional Market Selection Falls Short

Historically, market selection relied on relatively static, retrospective frameworks. Analysts might consider GDP per capita, population size, foreign investment inflows, or cultural proximity to determine which countries offered the best growth opportunities. Tools like the **CAGE Distance Model** helped strategists map cultural, administrative, geographic, and economic distances between countries. While useful for comparative thinking, these tools rarely capture fast-moving disruptions or nonlinear system effects.

In today's interconnected and algorithmically reactive environment, relying solely on static indicators is a strategic liability. Consider this: A country might score high on GDP and consumer spending but be exposed to **critical weaknesses in data infrastructure, cybersecurity, or political continuity**. Another country might appear promising on paper, but face **invisible fragilities**, such as climate-induced migration or water stress, which only become apparent through dynamic modeling.

Moreover, the **speed and asymmetry of technological diffusion** further complicate the landscape. A country with low average income but high smartphone penetration and rapid fintech adoption may be more attractive to a digital-first company than a wealthier but analog-bound nation. These nuances are missed when companies don't integrate real-time, multidimensional data streams into their strategic analysis.

8.1.2 Building a Data-Rich Market Entry Dashboard

In the STEM era, effective market selection begins with designing a real-time, adaptive **Market Intelligence Dashboard**. This dashboard synthesizes variables from multiple domains to provide a holistic view of a country's readiness, risk, and reward profile. These variables are not arbitrary; they are chosen based on their predictive value in similar expansions across regions, and many are weighted and processed using **machine learning algorithms**.

For example, digital infrastructure metrics such as 5G coverage and cloud adoption rates help assess technological readiness, while ESG indicators and climate risk data provide insight into long-term sustainability and regulatory alignment. Consumer behavior, institutional stability, and talent availability further enrich the dashboard, enabling a multidimensional view of market viability.

The table below outlines the key domains, metrics, and data sources that power such a dashboard:

Table 8.1: Core Data Domains for Market Entry Intelligence Dashboards

Domain	Metrics Tracked	Data Sources
Digital Infrastructure	4G/5G coverage, fiber broadband access, and cloud adoption rates	GSMA, World Bank, OpenSignal
ESG Readiness	Corporate disclosure compliance, net-zero policies, and investor ESG sentiment	Refinitiv, CDP, WEF ESG dashboards
Climate Risk	Exposure to floods, droughts, sea-level rise, and energy resilience	ND-GAIN, UNEP, NASA EarthData
Regulatory Environment	Ease of doing business, data localization laws, and foreign ownership limits	World Bank, OECD, WTO
Consumer Readiness	Smartphone adoption, digital payments penetration, and e-commerce behavior	Statista, Google Mobility Reports
Institutional Stability	Corruption Perception Index, rule of law, government effectiveness	Transparency International, Worldwide Governance
Talent Pipeline	STEM graduates per capita, AI skill penetration, and R&D intensity	UNESCO, LinkedIn Economic Graph

Practitioners should learn how to use tools such as **Power BI, Tableau, or even R Shiny** to create dynamic visualizations and weight-adjusted country scores. More advanced analysts integrate supervised learning models (e.g., **XGBoost, LightGBM**) that have been trained on historical international expansions, flagging high-risk markets where others have failed, and surfacing hidden gems previously overlooked.

For example, suppose a health-tech startup wants to expand to Africa. A traditional analysis might prioritize South Africa due to GDP and regulatory maturity. However, a STEM-enabled dashboard might reveal that Kenya, with high mobile health penetration, strong innovation ecosystems (e.g., M-Pesa, Health-E-Net), and improved cyber regulation, offers a better fit for a digital-first model.

8.1.3 Predictive Analytics for Entry Mode Selection

Once high-potential markets are identified, the next critical question becomes: *How should we enter?* Entry modes range from **exporting, franchising, licensing, and strategic alliances** to **joint ventures, acquisitions, and greenfield investments**. Each mode varies in terms of control, investment cost, risk exposure, time to market, and adaptability.

Traditionally, companies made entry mode decisions using qualitative assessments: "acquire to gain speed," "partner to reduce risk," "go solo for full control." While these heuristics have some validity, they are often oversimplified and fail to reflect complex, path-dependent risks.

STEM-based Strategy Enhancement: Monte Carlo Simulations

A Monte Carlo simulation generates thousands of possible financial outcomes for each entry mode based on variable uncertainty in demand, regulatory approval time, currency fluctuations, and partner stability. Rather than giving one expected value, this simulation provides a **distribution** of potential outcomes. This allows firms to assess **value-at-risk (VaR), downside protection**, and optimal control strategies under uncertainty.

To support strategic decision-making, firms often integrate these simulations into **interactive dashboards** that compare entry modes, such as Greenfield investments, Joint Ventures (JVs), and Acquisitions, across multiple dimensions. These dashboards allow decision-makers to evaluate not just financial metrics, but also agility, compliance, and strategic alignment.

The table below illustrates how such a dashboard might compare entry options across key criteria:

Table 8.2: Entry Mode Comparison Dashboard—Greenfield vs. JV vs. Acquisition

Criteria	Greenfield	Joint Venture	Acquisition
Time to Market	Slow (12–24 months)	Moderate (6–12 months)	Fast (3–6 months)
Upfront Capital Investment	High	Medium	Very High
Strategic Control	Full	Shared	High
Cultural Fit Risk	Low	Medium	High
Data Sovereignty Compliance	Easier to align	Depends on the partner	Inherited
Scenario Volatility Impact	More resilient	Dependent on partner behavior	Exposed to legacy issues

By applying **multi-criteria decision analysis (MCDA)**, companies can visualize and rank entry options not just on cost, but on agility, stakeholder alignment, compliance, and scalability.

8.1.4 Integrating Geopolitical and Climate Risk Modeling

Global expansion strategies that ignore geopolitical and climate dimensions are not only outdated, they are dangerous. The 21st century has proven that national borders no longer contain risk; conflict in one region can send supply chains rippling, destabilize currencies, and trigger policy cascades in dozens of markets. For example, Russia's invasion of Ukraine in 2022 sent shockwaves across global energy and food systems, affecting markets far beyond Eastern Europe. Similarly, prolonged droughts in India or flash floods in Germany have demonstrated how climate volatility can shutter logistics hubs, disrupt customer demand, and erode the viability of expansion plans overnight. For strategic leaders, this means market entry decisions must now be made within a **system-of-systems view**, incorporating not just country-level attractiveness, but cross-border interdependencies and fragilities. STEM-enabled strategy elevates this practice by embedding **real-time risk signals and simulation tools** directly into the expansion planning process.

In practical terms, companies are beginning to implement **dual-risk dashboards** that combine geopolitical volatility indicators with long-term climate risk indices. For example, a retail giant considering Southeast Asia might juxtapose real-time civil unrest probabilities (from machine learning models scraping media sentiment and diplomatic briefings) with ten-year flood forecasts for key urban centers. This combination not only informs the "go/no-go" decision, but it also guides **which cities to prioritize**, **which supply chain configurations to adopt**, and **which insurance instruments to procure**. Moreover, integrating these insights early in the strategy process allows companies to build **resilience-based entry plans,** such as modular operations, flexible sourcing, and decentralized tech stacks—ensuring they are not blindsided by the next crisis. This is not crisis management; it is a foresight-driven expansion strategy.

Geopolitical Risk Modeling:

Companies increasingly rely on real-time geopolitical risk dashboards to track variables such as political stability, civil unrest, regional conflict, regulatory backlash, and social fragmentation. Tools like Verisk Maplecroft, Control Risks, and geopolitical AI from Stratfor and Dataminr use natural language processing (NLP) and machine learning to scan global news feeds, regulatory bulletins, and diplomatic cables. These systems can generate early-warning alerts for rising nationalist sentiment, populist trade policies, or mass protest movements—factors that may destabilize entry assumptions. Predictive modeling helps firms simulate the probability of policy reversals (e.g., tax hikes, data localization laws, expropriation risks) and assign market-specific risk weights to expansion viability scores.

Climate Risk Simulation:

The era of climate-resilient strategy demands dynamic modeling of physical risks such as sea-level rise, drought cycles, heat waves, and water scarcity. Using data from global sources like NASA EarthData, ND-GAIN, WRI Aqueduct, and World Bank climate indices, companies can simulate the vulnerability of operational assets, distribution channels, and urban consumption patterns. For instance, a fintech company evaluating Lagos, Jakarta, or Dhaka must analyze not only market size and mobile penetration, but also future flooding risk that could impair infrastructure and customer behavior. Predictive GIS models and scenario-based hazard mapping can inform decisions such as where to locate data centers, which last-mile partners to use, and how to structure customer onboarding timelines in volatile climates.

Integration with ESG Strategy:

Climate and geopolitical modeling also serve as enablers of ESG-aligned global strategy. Countries with strong climate adaptation frameworks, ESG-reporting mandates, and access to green finance tend to attract sustainable investors and long-term partners. Strategists now assess entry alignment with ESG goals as part of broader portfolio planning. For example, countries that rank higher on ESG transparency and have robust environmental regulations may also be early adopters of carbon disclosure standards—giving first-mover advantage to firms that prioritize sustainability. Conversely, entering a country with weak climate governance may pose reputational risks, especially for public companies or B-Corps committed to environmental impact standards.

Feedback Loops and Dynamic Monitoring:

Modern STEM-based risk models are not one-time assessments; they are continuous systems that integrate real-time signals and structured re-evaluation intervals. Market conditions are fed into decision dashboards with automated updates. If a regulatory change occurs or a major natural disaster hits a region, expansion plans can be paused, rerouted, or accelerated accordingly. This kind of **strategic agility** becomes a competitive advantage, especially in sectors like logistics, renewable energy, and finance, where geopolitical or climate risks can reshape operations in weeks.

Geopolitical and climate risk modeling must be elevated from a peripheral concern to a central pillar of market entry strategy. Companies that use predictive simulation, AI-based sentiment analysis, and dynamic ESG integration can proactively avoid systemic shocks and enter markets with strategic resilience. The firms that succeed globally in the next decade will not be those that simply seek opportunity, but those that can **see the full terrain** in motion.

8.2 Cultural Modeling with Hofstede and GLOBE Data

In an era of algorithmic expansion and digitally enabled global strategy, cultural misalignment remains one of the most persistent and underestimated causes of failure. Many firms, particularly STEM-driven organizations, make the mistake of treating international markets as mere technical systems, overlooking the deeply human, institutional, and value-laden contexts in which technology operates. The result? Misfired marketing campaigns, ineffective leadership transfers, low product adoption, or, worse, reputational backlash. To counter this, strategic managers must adopt **cultural modeling frameworks** that are as rigorous and predictive as their economic or technological models. Two of the most widely used—and highly complementary—approaches are **Hofstede's Cultural Dimensions** and the **GLOBE Study's Leadership and Cultural Profiles**.

These frameworks do not just describe cultural differences; they offer predictive signals for how people in different societies interpret authority, time, risk, communication, and innovation. When integrated with data science tools, they enable **localization-by-design,** allowing firms to align leadership styles, brand tone, incentive structures, and even UX interface design with culturally congruent expectations.

8.2.1 Hofstede's Cultural Dimensions: Strategic Applications

Geert Hofstede's six-dimensional framework remains one of the most influential lenses for cross-cultural management. It quantifies national cultures using indices that represent:

1. **Power Distance** – Acceptance of hierarchical order and inequality
2. **Individualism vs. Collectivism** – Degree to which personal or group goals prevail
3. **Masculinity vs. Femininity** – Preference for competition vs. consensus
4. **Uncertainty Avoidance** – Tolerance for ambiguity and change
5. **Long-Term Orientation** – Pragmatism and future-focused planning
6. **Indulgence vs. Restraint** – Societal attitude toward gratification and control

In a strategic context, these dimensions are not abstract—they're **actionable signals**. For example:

* In **high uncertainty-avoidance cultures** like Japan or France, market entry strategies should prioritize regulatory compliance, long-term service guarantees, and local certifications.
* In **collectivist societies** like South Korea or Colombia, platform strategies should focus on group features, family bundles, and word-of-mouth virality rather than solo user incentives.
* In **masculine cultures** like the United States or Germany, product positioning can emphasize achievement, innovation, and performance metrics; whereas **feminine cultures** like the Netherlands may respond better to community impact, work-life balance, and social value.

Strategists can now integrate Hofstede data into **machine learning models**, feeding national scores into decision trees or clustering algorithms to simulate cultural response patterns to different strategies, be it product rollout, marketing messaging, or team structuring.

8.2.2 GLOBE Study: Leadership Fit and Strategic Communication

While Hofstede focuses on societal values, the **GLOBE (Global Leadership and Organizational Behavior Effectiveness) Study** adds an invaluable layer: how those values translate into **leadership expectations and institutional behavior**. GLOBE identifies nine cultural dimensions and clusters countries based on shared leadership prototypes, such as **charismatic/value-based**, **team-oriented**, **participative**, and **autonomous** leadership.

Consider how GLOBE findings can inform global strategy:

* In **Confucian Asia** (China, South Korea, Singapore), leadership is expected to be **self-sacrificial, family-oriented, and status-conscious**. Strategic managers from flat-hierarchy cultures must recalibrate to show deference to seniority and engage in ceremonial communication.
* In **Nordic Europe** (Sweden, Finland, Denmark), **participative leadership** and egalitarianism are prized. A top-down expansion playbook would likely be met with resistance or apathy.
* In **Middle Eastern clusters**, high scores in **in-group collectivism** and **power distance** suggest that informal networks, loyalty, and hierarchy are central to market success, especially in B2B or government-facing strategies.

For strategic modeling, firms are now using GLOBE dimensions to simulate **organizational fit**, **team compatibility**, and **change management risk** in cross-border mergers or new market ventures. A leadership approach that resonates in Silicon Valley may fall flat in Jakarta unless recalibrated through this lens.

8.2.3 Practical Use: Building Culture-Responsive Strategy Models

To succeed in global markets, strategists must go beyond cultural awareness—they need to **embed cultural intelligence directly into their analytics pipeline**. This means translating cultural dimensions into measurable strategic variables that can inform everything from organizational design to product messaging.

For example, in high **power distance** cultures, centralized governance structures may be more effective, while in low power distance environments, decentralized decision-making may foster better engagement. Similarly, **uncertainty avoidance** influences how firms approach compliance and risk signaling, while **institutional collectivism** shapes how referral programs and brand loyalty mechanisms are designed.

By integrating cultural data from frameworks like **Hofstede** and **GLOBE** into strategy models, firms can tailor their market entry and operational approaches with greater precision. When combined with **A/B testing platforms**, **NLP tools for localized messaging**, and **user behavior analytics**, this approach enables real-time monitoring of cultural alignment and proactive strategy adjustment.

The table below outlines how specific cultural factors can be operationalized into strategic variables:

Table 8.3: Operationalizing Cultural Factors in Strategy Design

Cultural Factor	Strategic Variable	Data Source	Impact on Entry Strategy
Power Distance	Org design & decision rights	Hofstede	Centralized vs. decentralized governance
Uncertainty Avoidance	Risk mitigation & compliance	Hofstede	Regulatory navigation and assurance signals
Institutional Collectivism	Branding & referral programs	GLOBE	Group incentives and loyalty design
Performance Orientation	Product feature emphasis	GLOBE	Metrics, speed, innovation signaling
Future Orientation	Investment horizon and roadmap messaging	GLOBE/Hofstede	Short-term vs. long-term market engagement tactics
Humane Orientation	CSR framing, recruitment language	GLOBE	Emotional tone and community engagement

By fusing this structured data with A/B testing platforms, NLP tools for localized messaging, and user behavior analytics, firms can monitor cultural alignment in real-time and adjust strategy proactively.

8.2.4 STEM-Enhanced Cultural Intelligence: New Frontiers

Traditionally, cultural intelligence (CQ) was seen as the domain of HR or international relations experts. Today, however, we see a convergence between **cultural analytics** and **STEM-based strategic tooling**. For instance:

* **Natural Language Processing (NLP)** can detect cultural resonance in real-time—analyzing local customer reviews, social media chatter, or employee feedback to assess fit and emotional response.
* **Computer vision** tools can test visual content (e.g., colors, gestures, symbols) across cultures for unintentional offense or misalignment.
* **Digital ethnography** uses AI to observe user behavior in digital spaces, providing a window into cultural values expressed through UX navigation, content sharing, and device preferences.

More importantly, emerging **AI-powered strategy platforms** now offer plug-and-play modules that incorporate Hofstede and GLOBE data to dynamically generate localization recommendations. These tools give strategic leaders **on-demand cultural simulations**, allowing them to test strategy prototypes across virtual personas from different cultures, before real capital is deployed.

Culture is not a soft variable; it is a **strategic parameter**. For STEM-enabled strategists, integrating Hofstede and GLOBE data provides a quantifiable, repeatable way to operationalize culture across leadership, branding, product, and market entry design. The future belongs to firms that combine global vision with cultural fluency and know how to code both into their algorithms.

8.3 Forecasting Political Stability with Machine Learning Models

In today's volatile geopolitical landscape, political risk is no longer a peripheral concern; it is a central variable in global strategy. A sudden regulatory shift, a government collapse, or widespread civil unrest can disrupt supply chains, erode investor confidence, and derail expansion plans. Traditional methods of assessing political stability, such as expert judgment, country reports, or retrospective indexes, are increasingly inadequate. Strategic foresight now demands more than static analysis. It requires continuous, data-driven prediction. Machine learning (ML) provides this capability by transforming disparate political signals into actionable forecasts, offering a probabilistic understanding of where instability may emerge and when.

Unlike conventional risk assessments, machine learning models integrate dynamic data sources, ranging from media sentiment and protest frequency to economic performance and legislative trends, to produce forward-looking insights. These models do not eliminate uncertainty but instead offer strategic leaders a structured framework to quantify and model it. Forecasting political stability through ML helps organizations design flexible market entry plans, prepare contingency scenarios, and prioritize countries based on more than just economic potential.

Machine learning models also enable the automation of political risk tracking across multiple jurisdictions simultaneously. This is particularly valuable for firms managing complex global portfolios or operating across politically unstable regions. By setting algorithmic thresholds for risk indicators, organizations can implement early warning systems that automatically trigger internal reviews or strategic pauses. The ability to move from reactive to proactive political risk management can result in significant cost savings, reputational protection, and improved agility in strategic execution.

Moreover, integrating political risk forecasts into enterprise risk management (ERM) systems creates a unified view of strategic, operational, and financial threats. This alignment allows C-suite executives and boards to visualize how political risks interact with other variables, such as regulatory compliance, environmental risk, and currency fluctuations. As a result, political forecasting becomes a part of a holistic, enterprise-wide strategy rather than an isolated function.

8.3.1 Key Data Sources for Political Stability Modeling

Building reliable forecasting models for political stability begins with **curated, multi-source datasets**. Effective political risk modeling combines **long-term governance indicators** with **real-time signals**, enabling organizations to anticipate disruptions and proactively manage exposure.

For example, **World Bank Governance Indicators** provide a longitudinal baseline for assessing rule of law, regulatory quality, and corruption—critical for evaluating institutional reliability. Complementing this, **Freedom House** and **V-Dem** track civil liberties, press freedom, and electoral fairness, offering insight into the trajectory of democratic institutions.

To capture real-time dynamics, tools like the **GDELT Event Database** analyze global news events and sentiment, while **UN and IMF crisis trackers** monitor fiscal and legal responses to global shocks. Additionally, **social media and search data** offer early signals of unrest, protest planning, or shifts in public sentiment—often surfacing before traditional media coverage.

The table below summarizes these key sources and their strategic applications:

Table 8.4: Strategic Data Sources for Political Stability Forecasting

Data Source	Variable Type	Strategic Application
World Bank Governance Indicators	Rule of law, regulatory quality, corruption	Longitudinal baseline for political reliability
Freedom House / V-Dem	Civil liberties, press freedom, electoral fairness	Trends in democratic resilience or erosion
GDELT Event Database	Event-based news tracking and tone analysis	Sentiment and frequency mapping of political events
UN & IMF Crisis Policy Trackers	Fiscal and legal responses to global shocks	Monitoring volatility triggers in policy domains
Social Media & Search Data	Online discourse, protest planning, and dissatisfaction	Early detection of unrest and citizen sentiment shifts

Combining structured indicators (e.g., rule of law) with unstructured data (e.g., tweets or news articles) allows for more nuanced and responsive forecasting. For instance, an uptick in negative media sentiment ahead of an election in a politically fragile country can serve as a precursor to unrest, giving companies time to activate contingency plans.

Firms can also enrich their datasets by incorporating proprietary data sources, such as in-country interviews, satellite imagery, or internal incident reports. Integrating local intelligence helps validate global indicators and contextualize risk models with cultural and sector-specific insights. This hybrid approach enhances both the accuracy and relevance of political forecasting systems.

8.3.2 Machine Learning Techniques in Political Forecasting

The complexity of political systems lends itself well to a range of machine learning approaches, each tailored to a different aspect of forecasting. Key techniques include:

* **Random Forests**: Ideal for classifying countries into risk tiers and identifying feature importance. These models can reveal, for example, that press freedom decline is more predictive of future unrest than GDP per capita in some regions.
* **Gradient Boosting (e.g., XGBoost)**: Suitable for high-accuracy predictions in nonlinear data environments. Often used to fine-tune predictions of election volatility or regime change likelihood.
* **Natural Language Processing (NLP)**: Applied to media and legislative texts to detect changes in political rhetoric, polarizing narratives, or shifts in public discourse.
* **Time Series Forecasting (ARIMA, LSTM)**: Effective for modeling periodic risk cycles, such as elections or protest waves, and forecasting future probability distributions.

Ensemble methods, which combine several techniques, are especially useful for political forecasting. By leveraging the strengths of different models, ensemble methods offer improved robustness and accuracy. For example, a model might combine NLP sentiment analysis with structured governance indicators to enhance both depth and precision.

Interpretability is another key consideration. Political analysts and executives must be able to understand and act on model outputs. Techniques such as SHAP (Shapley Additive Explanations) can help illuminate which variables most influenced a prediction, adding transparency and confidence to strategic decisions.

8.3.3 Strategic Applications of Political Risk Forecasting

Predictive political modeling has broad utility across strategic domains. Key applications include:

* **Entry Market Ranking**: ML-generated political risk scores feed into composite attractiveness indices, ensuring that high-growth countries with rising instability are deprioritized or approached with caution.
* **Supply Chain Contingency Planning**: Forecasts of labor strikes, transport disruption, or social unrest help pre-emptively adjust distribution strategies or supplier networks.
* **Capital Investment Decisions**: Political volatility forecasts feed into risk-adjusted ROI calculations, affecting discount rates, project timelines, and hedging strategies.
* **Scenario Testing**: By simulating moderate and extreme political events, teams can stress-test international expansion plans and ensure resilience under plausible worst-case scenarios.

In sectors like energy, healthcare, and telecommunications, where long-term infrastructure investments are common, the ability to forecast political cycles and regulatory reforms becomes critical. Companies can proactively engage with stakeholders, prepare lobbying strategies, or structure deals in ways that insulate them from political transitions.

Political forecasting can also influence corporate diplomacy and reputation management. Anticipating a governance crisis allows firms to position themselves as socially responsible actors or to decouple from controversial regimes before reputational damage occurs.

8.3.4 From Prediction to Action: Embedding Insights into Strategy

The strategic value of machine learning lies in its ability to support not just awareness, but action. Organizations are increasingly deploying ML-based outputs through:

* **Visual Dashboards**: Displaying current risk levels across target regions alongside business exposure data.
* **Monte Carlo Simulations**: Modeling long-term volatility effects on cost structures, customer demand, and regulatory burdens.
* **Strategic Control Rooms**: Centralized teams that monitor real-time data feeds from news, social platforms, and intelligence sources, issuing alerts when key thresholds are crossed.

Advanced firms are experimenting with **digital twin models of target countries**, using predictive insights to simulate strategic moves in a controlled digital environment. These simulations help refine decisions related to timing, resource allocation, and risk transfer mechanisms, before actual market entry occurs.

Embedding these insights into enterprise planning cycles ensures that strategy remains responsive. Political stability forecasts are now often reviewed in quarterly board meetings, used to update investment theses, and integrated into dynamic budgeting models. Firms that treat political risk forecasting as a living system, rather than a static checklist, are better equipped to seize opportunities and mitigate shocks.

Cross-functional alignment is essential. Insights from political forecasting models should inform marketing, legal, operations, and talent strategies. For example, a forecasted rise in labor unrest might trigger accelerated automation plans, stakeholder outreach, or legal contingency design. This integration ensures that insights don't remain siloed but instead cascade into strategic readiness across the enterprise.

Machine learning transforms political forecasting into a continuous, adaptive capability. It offers strategic leaders a structured lens to quantify the unquantifiable and make expansion decisions with deeper awareness of volatility, legitimacy, and systemic fragility.

8.4 Strategic Intelligence in Global Expansion: AI-Enhanced Market Entry Planning

As businesses continue to expand across borders, the stakes for choosing the right entry strategy have never been higher. Traditional heuristics such as GDP growth or market size are no longer sufficient for global decision-making. Instead, successful global expansion increasingly relies on strategic intelligence, powered by AI, predictive analytics, and system modeling. By combining these tools, firms can simulate market entry decisions, forecast competitive dynamics, and optimize timing and resource allocation in a structured, data-driven manner.

Entry into foreign markets involves a multifaceted risk-return calculus that encompasses political, economic, cultural, regulatory, and technological dimensions. In the past, market selection often depended on historical financials or executive intuition. Today, however, a STEM-based approach allows decision-makers to build sophisticated market-entry models that learn, adapt, and simulate the probable outcomes of different strategies. These include direct investment, franchising, joint ventures, and strategic alliances, each with different implications for risk sharing, control, and speed.

Machine learning algorithms can assess the historical success rates of various entry modes across industries and geographies, accounting for dozens of contextual variables. For instance, an algorithm might reveal that joint ventures perform better in heavily regulated markets, whereas franchising thrives in culturally homogenous, consumer-driven economies. This analytical layer empowers strategists to move beyond generalizations and tailor their approach to each country's specific ecosystem.

8.4.1 AI-Driven Market Entry Strategy Mapping

AI-enhanced strategic planning enables firms to automate and enhance traditional frameworks like CAGE (Cultural, Administrative, Geographic, Economic) and PESTLE by embedding them into data-rich simulations. For example:

* **Cultural Distance Models**: Natural Language Processing (NLP) can mine local media, social platforms, and consumer reviews to assess cultural compatibility and adjust localization strategies accordingly.
* **Administrative Forecasting**: AI models using World Bank, IMF, and local legislative data can predict changes in trade policies, labor laws, or foreign investment regulations.
* **Economic Stress Testing**: Economic indicators like inflation, interest rate fluctuations, and debt levels are run through time-series models to simulate future macroeconomic stress scenarios.

These simulations are often visualized in interactive dashboards where strategists can adjust variables and instantly observe how market attractiveness scores change. Such tools not only reduce blind spots but also enable decision-makers to test entry strategies under varied assumptions, ranging from currency devaluation to sudden regulatory shifts.

8.4.2 Optimizing Entry Timing with Predictive Analytics

Timing is often as crucial as the mode of entry. AI-enhanced forecasting tools help identify optimal entry windows by analyzing:

- **Consumer sentiment trends** via real-time social listening platforms
- **Competitor activity patterns**, including expansion signals or product announcements
- **Seasonal economic fluctuations** that affect demand, labor availability, or regulatory review cycles

For instance, a firm might use clustering algorithms to detect that Q2 of a given year historically shows the most favorable convergence of consumer demand, exchange rate stability, and labor market openness in a target market. Armed with this insight, they can schedule launch activities accordingly, reducing ramp-up costs and boosting adoption.

Moreover, predictive models can simulate competitor countermoves and saturation risks. If three major players are forecasted to enter a market within a short window, firms can delay or pre-empt expansion with pricing, branding, or channel strategies. This form of "algorithmic timing" turns market entry into a dynamic, game-theoretic exercise.

8.4.3 Entry Mode Selection and Strategic Fit

Choosing between entry modes, such as wholly owned subsidiaries, joint ventures, franchising, or strategic alliances, has historically relied on a firm's internal resources, control preferences, and risk appetite. With STEM-enhanced modeling, these choices become quantifiable and scenario-tested. AI tools can estimate the long-term profitability, risk exposure, and adaptability of each mode under multiple regulatory, cultural, and economic contingencies.

For example, a decision-tree model might weigh:

* **Joint venture feasibility** based on political openness and legal infrastructure
* **Franchise potential** given brand strength, supply chain scalability, and training infrastructure
* **Greenfield investment viability** considering capital intensity, labor cost, and real estate dynamics

These models can also be integrated with ESG metrics to prioritize sustainable market entry strategies. A firm might opt for alliances with local sustainable enterprises or adapt entry modes to minimize its environmental footprint.

8.4.4 System Modeling for Global Strategy Alignment

Global strategy does not operate in silos. An entry decision in one country can have ripple effects across sourcing, finance, and brand positioning. System modeling enables strategic leaders to simulate these interdependencies. Using digital twins or causal loop diagrams, firms can test how decisions in one market affect:

* Global supply chain resilience
* Cross-border tax optimization
* Brand consistency and cultural resonance
* Capital allocation and opportunity cost

For instance, entering two adjacent markets with competing cultural values might require differentiated product strategies, increasing marketing complexity. Alternatively, expanding into a geopolitically tense region might increase regulatory scrutiny elsewhere.

By capturing these dynamics through system modeling, firms can identify leverage points, reduce unintended consequences, and align global expansion with overarching strategic themes—such as digital transformation, sustainability, or diversification.

AI, predictive analytics, and system modeling empower organizations to make global expansion decisions with higher precision, strategic alignment, and resilience. Entry strategies are no longer static templates but intelligent, adaptable frameworks shaped by dynamic data and interdisciplinary foresight.

8.5. Chapter 8 Wrap-Up: Strategic Expansion in a Borderless World

In this chapter, we reframed global expansion as a dynamic, data-informed strategic journey—no longer limited to static market entry models or intuition-based decisions. The modern landscape of international business demands precision, adaptability, and systemic awareness. Strategic expansion has become less about crossing borders and more about navigating interconnected systems in real time.

We explored how predictive analytics, AI simulations, and system modeling now form the backbone of intelligent global strategies. These tools empower decision-makers to simulate market entry scenarios, evaluate climate and geopolitical risk, and forecast long-term sustainability metrics across diverse geographies. Rather than rely solely on legacy frameworks, firms now engage in living strategy design—iterative, evidence-driven, and stakeholder-aware.

Crucially, this chapter emphasized that international growth strategies must embed foresight, ESG alignment, and resilience planning from the outset. We introduced tools that allow for proactive stress-testing of entry modes, adaptability to climate volatility, and transparent communication with global partners. Strategic expansion in today's volatile world is not just about ambition—it's about coherence, simulation, and continuous recalibration.

Throughout, we examined real-world and hypothetical applications, highlighting how leading firms integrate these insights to build strategies that are not only globally ambitious but also contextually grounded. The shift from fixed blueprints to adaptive strategic ecosystems defines the new norm in borderless business.

8.5.1. Discussion Questions

1. How does predictive modeling improve the accuracy of international market selection?
2. What role does climate foresight play in modern global expansion strategy?
3. How can entry mode decisions be influenced by AI and scenario simulations?
4. What risks emerge if a global expansion strategy ignores system-level interdependencies?
5. How can ESG metrics be integrated into global expansion strategies for stakeholder alignment?
6. In what ways does strategic timing, powered by data, shape success in new markets?
7. How does cultural analytics contribute to international brand adaptation?
8. What new KPIs should firms track when launching in multiple countries?
9. How do geopolitical dashboards reshape expansion plans in unstable regions?
10. Which part of the global strategy process benefits most from system modeling?

8.5.2. Micro-Case: Zara's Algorithmic Leap into Latin America

Zara, a global leader in fast fashion, faces growth saturation in European and Asian markets. To sustain momentum, it turns to Latin America—a region brimming with opportunity but marked by economic volatility and diverse regulatory climates.

Rather than rely on traditional expansion tactics, Zara deploys a proprietary AI-enhanced global strategy engine. This tool integrates climate risk projections, real-time social sentiment analysis, and supply chain stress indicators. Predictive analytics identify 30 high-potential urban zones, factoring in political events, natural disaster patterns, and regional GDP growth rates.

Zara models multiple entry modes for each country, franchising, joint ventures, or wholly owned subsidiaries, and tests each mode's performance under different simulated conditions. In Chile, a stable economic climate and robust retail infrastructure suggest a wholly owned model. In Colombia, joint ventures offer flexibility and local insight. In Venezuela, franchise partnerships minimize exposure.

To reinforce stakeholder trust, Zara integrates a real-time ESG dashboard accessible to investors, tracking labor standards, energy sourcing, and waste metrics in each market. The transparency initiative strengthens brand reputation and preempts regulatory scrutiny.

Weekly sprints across cross-functional teams refine decisions as new data flows in. By integrating cultural sentiment analysis, Zara tailors its product offerings and in-store experiences to each locale, avoiding one-size-fits-all expansion.

Zara's expansion is no longer a linear rollout. It's an orchestrated simulation—adaptable, ethical, and deeply informed by systems thinking.

Discussion Questions

1. How does Zara's algorithmic strategy engine enhance decision-making in unpredictable markets?
2. What are the implications of using ESG dashboards for stakeholder engagement during expansion?
3. How does cultural sentiment analysis improve Zara's localization strategy?
4. What challenges might arise from coordinating cross-functional teams across multiple geographies?
5. How does this case exemplify the shift from static expansion plans to living, adaptive strategy models?

8.5.3. Interactive Exercise: Global Risk Simulation

Scenario: You're leading strategic expansion into Brazil, Egypt, and Indonesia. You have access to:

* Climate vulnerability models
* Political stability indexes
* Consumer sentiment via social analytics

Task: Using at least two tools introduced in this chapter (e.g., system modeling, scenario simulation), create a strategic blueprint for expansion that includes:

* Market prioritization

* Entry mode selection
* Risk mitigation and adaptation plan
* ESG integration metrics

This simulation mirrors real-world decision-making under uncertainty, powered by predictive strategy and interdisciplinary foresight.

8.5.4. Recommended HBR Case

❖ How Should We Diversify Our Supply Chain? (HBR Case Study)

- **Overview**: Authored by Krishna G. Palepu, this fictionalized case centers on Kshore, a Chinese appliance manufacturer with $150 million in annual revenue. In the wake of COVID-19 disruptions, Kshore is under pressure from major clients like Walmart to reduce shipping time, costs, and environmental impact. The company's leadership is considering whether to diversify its supply chain by expanding production to Mexico, closer to its North American markets. During a visit to Monterrey, Mexico, Kshore's CEO and COO are impressed by the professionalism of local factories but remain uncertain about logistics, staffing, and long-term viability. The case presents a dilemma: should Kshore invest in nearshoring to Mexico, explore other international options, or double down on its existing operations in China? Two expert commentaries follow the case, offering contrasting perspectives on how Kshore should proceed—making it ideal for classroom discussion on global strategy, supply chain resilience, and geopolitical risk.
- **Citation**: Palepu, K. G. (2023). *Case study: How should we diversify our supply chain?* Harvard Business Review, November–December 2023.

❖ From PingAn's inspiration to HDFC ERGO's journey: Building an insurance-based ecosystem

- **Overview**: Published on April 1, 2024, and authored by Michael Jacobides, Anubhab Goel, Dalbert Ma, and Aryan Mittal, this case explores how HDFC ERGO, a leading Indian insurance provider, drew inspiration from Ping An, China's pioneering insurance-based ecosystem, to craft its own digital transformation strategy. The case begins by examining Ping An's evolution into a multi-product, multi-actor ecosystem, emphasizing its integration of technology, healthcare, and financial services. It then shifts focus to HDFC ERGO's efforts to build a similar ecosystem tailored to the Indian market. Rather than directly replicating Ping An's model, HDFC ERGO developed a mindshare-centric strategy, aiming to engage customers more frequently and meaningfully beyond traditional insurance touchpoints.
    - Key themes include:

* Strategic ecosystem design
* Digital innovation in insurance
* Customer engagement through value-added services
* Leveraging India Stack and local market insights

The case also highlights the development of the "Here" app, which integrates healthcare services and reflects HDFC ERGO's broader ambition to become a daily-use platform rather than a once-a-year insurance provider. This case can be taught independently or alongside its companion case, "HDFC ERGO: A Product Ecosystem Built on Mindshare", which focuses more on the strategic rationale and early steps of the transformation.

- **Citation**: Jacobides, M., Goel, A., Ma, D., & Mittal, A. (2024). *From PingAn's inspiration to HDFC ERGO's journey: Building an insurance-based ecosystem* (Case No. LBS362). London Business School.

❖ Haier Europe: Bringing RenDanHeyi for All

- **Overview**: Published on June 13, 2023, and authored by Felipe L. Monteiro and Anne-Marie Carrick, this 29-page case explores how Haier Europe, a subsidiary of Haier Smart Home (part of the Chinese Haier Group), is working to implement the RenDanHeYi management model across its European operations. RenDanHeYi, developed by Haier's CEO Zhang Ruimin, is a radically decentralized organizational model that empowers micro-enterprises within the company to act autonomously, directly connected to user needs. The model has been credited with driving Haier's rapid growth and innovation in China. The case focuses on the post-acquisition integration of Candy, an Italian appliance brand, and the challenges of applying RenDanHeYi in a European context, amid rising inflation, the war in Ukraine, and increasing pressure for environmental sustainability. It raises questions about cultural adaptation, organizational transformation, and strategic alignment in a multinational setting.
- **Citation**: Monteiro, F. L., & Carrick, A.-M. (2023). *Haier Europe: Bringing RenDanHeYi for All* (Case No. IN1938-PDF-ENG). INSEAD.

❖ Starbucks: Creating the ambience and rekindling people's thirst by bringing the brew to the world

- **Overview**: Published on October 10, 2022, and authored by Kumaran Rajaram, this 13-page case study examines how Starbucks, the iconic American coffeehouse chain, has strategically positioned itself as a premier roaster, marketer, and retailer of specialty coffee on a global scale. The case is designed to help students and professionals understand the macro and competitive environments that influence Starbucks' strategic decisions. It uses frameworks such as PESTLE (Political,

Economic, Social, Technological, Legal, and Environmental) and Porter's Five Forces to analyze the company's external and internal challenges.

- Key themes include:
    - Global expansion and localization
    - Brand positioning and customer experience
    - Sustainability and ethical sourcing
    - Competitive dynamics in the food and beverage industry

The case encourages critical thinking about how Starbucks can continue to innovate and grow while maintaining its brand identity and navigating complex international markets.

- **Citation**: Rajaram, K. (2022). *Starbucks: Creating the ambience and rekindling people's thirst by bringing the brew to the world* (Case No. NTU337-PDF-ENG). ABCC at Nanyang Technological University.

8.5.5. Strategy-in-Film: Learn by Watching

Competing on a global stage requires navigating diverse markets, regulatory environments, and cultural complexities. The films below capture how organizations expand internationally, adapt strategies across borders, and manage cross-cultural dynamics. As you explore international strategy in this chapter, use these narratives to reflect on how data, analytics, and systemic thinking inform smarter global decisions.

Movie / Documentary	Strategic Theme Highlighted
The Founder	Franchising, scalability, and global branding
The Constant Gardener	Global ethics, systemic failures, and risk exposure
Lion	Logistics, data trails, and cultural systems
Syriana	Geopolitical complexity and multinational strategy
The Year Earth Changed	Climate disruption and environmental interdependencies

References

Adamovic, M. (2023). Unlocking the cultural mosaic: a comparison of Hofstede and modern cultural value frameworks. *Cross-Cultural Business and Management: Perspectives and Practices, NA, 181.*

Amankwah-Amoah, J., Adomako, S., Danquah, J. K., Opoku, R. A., & Zahoor, N. (2022). Foreign market knowledge, entry mode choice, and SME international performance in an emerging market. *Journal of International Management*, *28*(4), 100955.

Benito, G. R., Cuervo-Cazurra, A., Mudambi, R., Pedersen, T., & Tallman, S. (2022). The future of global strategy. *Global Strategy Journal*, *12*(3), 421-450.

Bergamaschi, M., Bettinelli, C., Lissana, E., & Picone, P. M. (2021). Past, ongoing, and future debate on the interplay between internationalization and digitalization. *Journal of Management and Governance*, 1-50.

Beugelsdijk, S., Kostova, T., & Roth, K. (2017). An overview of Hofstede-inspired country-level culture research in international business since 2006. *Journal of International Business Studies*, *48*, 30-47.

Blut, M., Beatty, S. E., & Northington, W. M. (2022). Cultural personal values and switching costs perceptions: Beyond Hofstede. *Journal of Business Research*, *150*, 339-353.

Buckley, P. J., & Strange, R. (2015). The governance of the global factory: Location and control of world economic activity. *Academy of Management Perspectives*, *29*(2), 237-249.

Cavusgil, S. T., Knight, G., Riesenberger, J. R., Rammal, H. G., & Rose, E. L. (2014). *International business*. Pearson Australia.

Chintalapati, S. (2020). BankBuddy. Ai—Business Expansion and marketing dilemma: A case study to discuss the Ansoff growth matrix concepts combined with business expansion strategies for expanding into emerging markets. *Emerging Economies Cases Journal*, *2*(1), 44-53.

Collinson, S., Narula, R., & Rugman, A. M. (2020). *International business*. Pearson UK.

Eden, L., & Nielsen, B. B. (2020). Research methods in international business: The challenge of complexity. *Journal of International Business Studies*, *51*(9), 1609.

Ghemawat, P. (2003). Semiglobalization and international business strategy. *Journal of International Business Studies*, *34*, 138-152.

Hennart, J. F. (2022). How much is new in Brouthers et al.'s new foreign entry modes, and do they challenge the transaction cost theory of entry mode choice?. *Journal of International Business Studies*, *53*(9), 2116-2132.

Jiang, L., Chang, Z., Yao, W., Huang, L., & Zhou, S. (2024). Transcend local for global: ESG as a legitimacy signal in the global expansion of emerging multinational enterprises. *Finance Research Letters*, *69*, 106174.

Meyer, K. E., Li, J., & Brouthers, K. D. (2023). International business in the digital age: Global strategies in a world of national institutions. *Journal of International Business Studies*, *54*(4), 577.

Minkov, M., & Kaasa, A. (2022). Do dimensions of culture exist objectively? A validation of the revised Minkov-Hofstede model of culture with World Values Survey items and scores for 102 countries. *Journal of International Management*, *28*(4), 100971.

Moharrak, M., Nguyen, N. P., & Mogaji, E. (2024). Business environment and adoption of AI: Navigation for internationalization by new ventures in emerging markets. *Thunderbird International Business Review*, *66*(4), 355-372.

Narula, R., & Verbeke, A. (2015). Making internalization theory good for practice: The essence of Alan Rugman's contributions to international business. *Journal of World Business*, *50*(4), 612-622.

Nyssen Guillén, V. I., & Deckert, C. (2021). Cultural influence on innovativeness-links between “The Culture Map” and the “Global Innovation Index”. *International Journal of Corporate Social Responsibility*, *6*(1), 7.

Pucik, V., Björkman, I., Evans, P., & Stahl, G. K. (2023). *The global challenge: Managing people across borders*. Edward Elgar Publishing.

Shenkar, O., Tallman, S. B., Wang, H., & Wu, J. (2022). National culture and international business: A path forward. *Journal of International Business Studies*, *53*(3), 516-533.

Shenkar, O., Luo, Y., & Chi, T. (2021). *International business*. Routledge.

Stallkamp, M., & Schotter, A. P. (2021). Platforms without borders? The international strategies of digital platform firms. *Global Strategy Journal*, *11*(1), 58-80.

Teece, D. J. (2025). The multinational enterprise, capabilities, and digitalization: governance and growth with world disorder. *Journal of International Business Studies*, 1-16.

Venaik, S., Brewer, P., & Midgley, D. (2023). *Management research, international business, and national culture: Evaluating Hofstede and GLOBE*. Routledge.

Witt, M. A., Lewin, A. Y., Li, P. P., & Gaur, A. (2023). Decoupling in international business: Evidence, drivers, impact, and implications for IB research. *Journal of World Business*, *58*(1), 101399.

Chapter 9
Competitive Advantage In a Data-Driven Economy

"In the algorithmic age competitive advantage isn't inherited it's engineered through intelligent systems"

by MAK

Learning Outcomes

In the algorithmic age, advantage belongs not to the biggest or boldest, but to the most adaptive—those who, like the ant colony, learn, sense, and evolve as one intelligent system

After engaging with this chapter, readers will be able to:

- ❖ Explain how the foundations of competitive advantage have evolved in an economy increasingly shaped by data velocity, algorithmic decision-making, and real-time market adaptation.
- ❖ Differentiate between resource-based and capability-based strategies, recognizing how data infrastructure, feedback systems, and intelligent platforms are becoming central to long-term value creation.
- ❖ Apply the concept of dynamic capabilities, including sensing, seizing, and transforming, in the context of organizations leveraging AI and predictive systems for competitive differentiation.
- ❖ Utilize advanced modeling tools such as agent-based simulation, digital twins, and system dynamics to map, predict, and adapt to shifting competitive landscapes.
- ❖ Evaluate how firms strategically renew their competitive edge through continuous learning cycles and by anticipating the erosion of existing advantages.
- ❖ Analyze real-world case examples of organizations that compete using dynamic data assets and intelligent infrastructure, extracting strategic insights for other industries and sectors.
- ❖ Design strategic renewal loops that integrate foresight, performance monitoring, and data-enabled decision frameworks to adapt in rapidly evolving markets.

- Critically assess the ethical, operational, and systemic implications of using predictive analytics to secure or defend competitive advantage in a global, digital-first economy.

For decades, competitive advantage was seen as a stable outcome of superior resources, unique products, or operational efficiency. Strategy was built on moats, those defensible positions that protected firms from rivals. But in the algorithmic age, these moats are drying up faster than ever. The forces reshaping the business landscape, data acceleration, real-time analytics, platform competition, and intelligent systems, demand a new understanding of what it means to win, and more importantly, to keep winning.

In this new reality, competitive advantage is not a fixed state; it is a moving target. It is earned, lost, and re-earned in faster cycles. The firms that thrive are not necessarily those with the deepest war chests or the most patents, but those that can sense changes earlier, learn faster, and adapt continuously. The core currency of modern competition is not scale, but speed, not certainty, but agility, not efficiency, but intelligence.

Strategic advantage now lies in how well a firm builds and orchestrates **dynamic capabilities,** the capacity to sense shifts, seize emerging opportunities, and transform its value propositions before the competition does. These capabilities are no longer human-powered alone. Increasingly, they are **augmented by STEM tools**: AI that detects patterns beyond human foresight, digital twins that simulate outcomes in complex environments, and feedback loops that learn and self-adjust in real time.

In this chapter, we explore how competitive advantage is evolving in a data-driven economy. We examine how traditional strategy concepts like the resource-based view must be updated with frameworks that account for **data assets**, **digital infrastructures**, and **predictive intelligence**. We will look at how leading firms, from healthtech startups to global platforms, are shifting their strategic mindset from defending static advantages to designing **learning systems** that evolve alongside the environment.

More than a theoretical discussion, this chapter offers concrete models, tools, and cases that show how strategic renewal, system modeling, and continuous feedback can be hardwired into your competitive strategy. Whether you're scaling a startup, steering a legacy firm through digital transformation, or crafting policy in a fast-moving sector, the questions of "what advantage do we have?" and "how quickly can we renew it?" will define your strategic future.

Opening Case: Ping An's Intelligent Advantage—Competing Through Continuous Strategic Renewal

In the competitive world of financial services, Ping An has achieved what many legacy firms struggle to do: evolve continuously. Founded as a traditional insurance company, Ping An is now recognized globally not just for its financial scale, but for its mastery of **AI-powered strategic renewal**. Unlike firms that seek temporary dominance through product differentiation or pricing tactics, Ping An has embedded **real-time learning systems** into its core strategy, enabling it to generate and regenerate competitive advantage with speed and precision.

At the heart of Ping An's strategy is a data-driven ecosystem, linking insurance, healthcare, smart city solutions, and banking services. It's not just that Ping An has more data than its competitors; it's that they've **converted data into feedback loops**. These loops drive product design, customer engagement, operational optimization, and predictive decision-making across its business lines.

For instance, Ping An's "One Account" platform integrates over 30 services into a single user interface. Behind the scenes, machine learning models analyze behavior patterns to offer hyper-personalized services in real time. Their **AI Doctor** tool, trained on over a billion data points, improves patient triage and disease diagnosis. But what gives Ping An its real edge isn't just its AI, it's how the company continuously **tests, learns, and adapts** faster than the competition.

Strategic foresight teams simulate market shifts using digital twin environments. These simulations model everything from demographic transitions to climate risks to fintech regulation, enabling Ping An to reposition its offerings proactively. Meanwhile, performance metrics are fed back into strategic dashboards monitored by both human leaders and AI agents.

Rather than defending a fixed competitive position, Ping An treats advantage as **an active system**, one that must evolve as its environment changes. By institutionalizing strategic renewal, through dynamic capabilities, advanced modeling, and intelligent infrastructure, Ping An has shifted the basis of competition from strength to **adaptability**.

As markets become more volatile, customer expectations become more dynamic, and innovation cycles shorter, Ping An's example offers a blueprint for competing not once, but continuously.

Play the Strategist

1. How does Ping An shift the traditional concept of competitive advantage from static to dynamic?
2. What role do AI and data feedback systems play in Ping An's strategy renewal process?
3. How could digital twins or system modeling tools enhance strategic foresight in your organization?
4. What are the risks of relying heavily on algorithmic decision-making to maintain a competitive edge?

9.1 Competing on Data, Not Just Products

For decades, firms have built strategies around products, pricing, and positioning. Yet in today's algorithmically charged economy, **data has emerged as the ultimate strategic asset**. It doesn't just inform decisions, it **reshapes value chains**, **predicts behavior**, and **drives adaptation** in real time. In a world where speed and insight outperform legacy scale, firms no longer gain a lasting advantage by owning more resources; they gain it by extracting more **value and foresight** from data.

This section explores the new battleground of competitive advantage: the intelligent infrastructure of data flows, feedback loops, and adaptive learning systems. Leading firms, from industrial icons like John Deere to digital disruptors like Spotify, are no longer competing on products alone. They are building **data ecosystems** that generate insight faster than rivals, continuously improve customer experience, and anticipate strategic shifts before they emerge on a competitor's radar.

Importantly, these firms don't just collect data; they operationalize it. They move beyond dashboards and KPI tracking to embed data into **every strategic and operational decision**, treating analytics as a living system rather than a reporting tool. Moreover, the **velocity of data**, not just its volume, creates a competitive flywheel: real-time feedback powers experimentation, iteration, and market reconfiguration.

In this section, we examine how the **strategic architecture of data** changes the foundations of competition. We unpack how firms can move from static assets to dynamic, data-fueled ecosystems that evolve continuously. By transitioning from product-centric to **information-centric strategy**, firms gain a new form of competitive advantage, **one that is faster, smarter, and more sustainable** in a STEM-shaped world.

9.1.1 The Rise of Data as a Strategic Asset

In the 20th century, firms competed by optimizing tangible resources: production facilities, distribution networks, and customer relationships. In the 21st century, a new asset class has emerged, **data,** and it is reshaping the foundations of strategy. Data is no longer a byproduct of operations; it is the raw material of insight, foresight, and competitive agility. Organizations that recognize this shift are not just collecting data but architecting it into **strategic infrastructure** capable of generating continuous advantage.

What makes data strategic is not its volume, but its **ability to fuel learning**. When curated, structured, and modeled properly, data enables firms to detect patterns, anticipate customer needs, optimize internal processes, and simulate future scenarios. These capabilities are not abstract. Consider how **John Deere** transformed from a traditional equipment manufacturer into a data powerhouse. Through its precision agriculture platform, the company collects real-time field data from sensors embedded in tractors and combines it with satellite and weather inputs. This integrated intelligence allows farmers to optimize planting, watering, and harvesting with pinpoint accuracy, turning each decision into a competitive edge.

Moreover, data allows firms to shift from static to **dynamic resource orchestration**. Rather than relying on annual strategy cycles or fixed customer segmentation models, data-driven firms operate in **real-time loops**, constantly sensing, adjusting, and improving. In this new landscape, data is the asset that learns, scales, and self-improves, making it more strategic than any single product or brand.

9.1.2 Beyond Dashboards: Data as Competitive Infrastructure

Many firms remain stuck in a dashboard mindset, treating data as a reporting layer rather than a **strategic foundation**. But the most advanced competitors are embedding data capabilities across every function: operations, finance, product development, and strategy. This shift, from analytics as a service to data as an operating system, changes how firms plan, compete, and grow.

Take **Spotify**, for example. Its competitive advantage is not just its music library, but its **ability to personalize listening experiences** through behavioral data and AI algorithms. Behind every playlist is a continuous loop of user feedback, preference modeling, and real-time optimization. The firm's "Discover Weekly" feature has become a signature innovation, not because of content, but because of data orchestration. This invisible infrastructure is difficult to replicate because it reflects years of investment in architecture, algorithms, and organizational learning.

Similarly, **Siemens** uses industrial IoT platforms to convert machine performance data into actionable intelligence for predictive maintenance and energy efficiency. This embedded data infrastructure doesn't just support operations; it **enables new value propositions** for clients and new revenue streams for Siemens. It exemplifies how firms can move beyond data tools and build competitive systems.

For firms aspiring to lead in their industries, the lesson is clear: dashboards may inform, but **embedded data ecosystems differentiate**.

9.1.3 Data Velocity and the Feedback Advantage

In traditional strategy, competitive advantage often came from static positioning, being in the right place with the right product at the right time. But in a data-rich environment, **speed of feedback and action** becomes a central source of advantage. It's not just what data a firm has, but how **quickly and intelligently** it can learn from it and act.

This is the principle of **data velocity**: high-speed data flows enable real-time decision-making, continuous experimentation, and adaptive business models. In high-velocity firms, feedback loops are built into digital interfaces, operational systems, and customer interactions. These loops become flywheels, constantly refining strategy through action, measurement, and revision.

Consider **Shopify**, which tracks seller behavior, transaction patterns, and user flows across its platform in real time. This intelligence powers algorithmic recommendations, automated fraud detection, and dynamic feature deployment. The faster the feedback, the faster the platform evolves, and the more value it delivers.

In competitive terms, feedback velocity enables **hyper-adaptiveness**. When a firm can detect shifts in demand, sentiment, or risk before its rivals, it can pivot faster, allocate better, and serve smarter. Over time, this results in **performance divergence,** where firms with real-time systems learn and grow faster than those operating on lagging indicators.

9.1.4 Strategic Shifts: From Products to Data-Driven Ecosystems

The most profound shift in the data economy is strategic: from selling products to building **data-driven ecosystems**. Rather than competing on features or cost, firms compete by orchestrating interactions, learning from users, and delivering value that scales with data accumulation. In such ecosystems, value is co-created by users, devices, and algorithms.

A compelling example is how **Ping An** transformed from a traditional insurer into a digital ecosystem spanning health, finance, and smart city services. By integrating data from wearables, medical records, credit histories, and behavioral analytics, Ping An creates personalized offerings that evolve with each customer. This isn't just digital transformation, it's ecosystem strategy powered by data intelligence.

This approach makes advantage **self-reinforcing**: as more users engage, more data is generated; as more data is processed, better services emerge; as services improve, more users are attracted. This feedback-rich loop is difficult for competitors to match without equivalent data scope, integration, and learning infrastructure.

In effect, the firm ceases to be a seller of products and becomes a **curator of adaptive experiences**, constantly refined through data signals. In the platform age, it is the quality of data orchestration, not just the product, that drives growth, loyalty, and resilience. The table below contrasts traditional industrial advantage with the emerging logic of data-driven ecosystems:

> In the algorithmic era, data does not merely support advantage, it **generates, amplifies, and protects**

Table 9.1. Industrial Advantage vs. Data-Driven Advantage

Dimension	Traditional Advantage	Data-Driven Advantage
Strategic Asset	Physical products, IP	Real-time data, learning algorithms
Value Creation	Economies of scale	Personalization, adaptability
Competitive Focus	Efficiency, cost	Speed, insight, prediction
Strategy Cycle	Annual, top-down	Continuous, feedback-driven
Learning	Post-action review	Real-time experimentation

Foundations of Competitive Advantage in the Algorithmic Age

While the classic VRIO framework, Value, Rarity, Inimitability, and Organization, has long explained why some firms outperform others, it must be reinterpreted for a data-centric world. In the algorithmic age, competitive advantage is no longer rooted solely in physical assets or brand equity, but in the ability to collect, process, and learn from data at scale.

For instance, data becomes valuable when it enables firms to personalize experiences, predict behaviors, and optimize operations in real time. It is rare when it consists of proprietary datasets or uniquely trained machine learning models that competitors cannot access. It becomes inimitable when embedded in complex, path-dependent systems, such as custom data pipelines or feedback loops that evolve over time. And it is organizationally effective when the firm is structured to learn continuously, adapt rapidly, and deploy insights across functions.

The table below illustrates how each VRIO dimension is redefined in the context of intelligent data ecosystems:

Table 9.2. Reframing VRIO for Intelligent Data Ecosystems

VRIO Criterion	Data-Driven Application
Valuable	Data enables firms to personalize experiences, predict behaviors, and optimize systems, delivering immediate and strategic value.
Rare	Proprietary datasets and unique machine learning models are rare assets that drive competitive differentiation.
Inimitable	Embedded data pipelines, path-dependent algorithms, and integrated feedback systems are difficult for competitors to replicate.
Organization	No traditional asset, capital, brand, or scale can substitute for a firm's ability to learn and adapt in real time using data.

9.2 Dynamic Capabilities and AI-Based Differentiation

In an environment marked by volatility, digital acceleration, and evolving consumer expectations, traditional sources of competitive advantage, cost leadership, operational efficiency, and even brand equity are increasingly insufficient on their own. What sets enduring firms apart is their ability to **sense opportunities, seize them quickly, and transform their operations to stay relevant**. This triad, **sensing, seizing, and transforming,** is at the heart of what strategy scholar David Teece calls *dynamic capabilities.*

Unlike operational capabilities that focus on doing existing things well, dynamic capabilities empower organizations to **adapt, evolve, and innovate continuously**. They are not tied to a specific product or market but to the firm's strategic agility, its ability to orchestrate resources, respond to disruption, and reinvent itself when the environment demands it.

Enter artificial intelligence, machine learning, and predictive analytics. These technologies have rapidly evolved from narrow tools into strategic enablers that **enhance, accelerate, and automate** the development of dynamic capabilities. Firms are now using AI not just to make decisions, but to **refine their ability to make better decisions over time,** a capability that compounds with use and learning.

When AI is embedded into core strategic processes, product design, market sensing, customer segmentation, and risk management, it becomes more than an operational asset. It becomes a **differentiation engine**, continuously enhancing the firm's ability to outpace competition not by doing the same things faster, but by **learning faster, adapting faster, and innovating more precisely**.

This section explores how dynamic capabilities are being reimagined in the age of AI. We'll begin by defining the concept and then examine how AI tools, when deployed strategically, augment each stage of capability development. We will also introduce a framework that links Teece's dynamic capabilities model to real-world examples and AI use cases, providing a clear lens for how firms build sustainable advantage through intelligent transformation.

In this new landscape, **strategy is no longer about defending a fixed position;** it's about building an organization that can evolve its position, reshape its boundaries, and renew its logic continuously. Dynamic capabilities are the architecture for that evolution, and AI is the engine that powers it.

9.2.1 What Are Dynamic Capabilities?

Dynamic capabilities refer to an organization's ability to **purposefully adapt, reconfigure, and renew its internal competencies** in response to rapidly changing environments. First conceptualized by David Teece and his colleagues, this theory emerged as a refinement of the resource-based view (RBV), addressing a critical limitation of RBV: the assumption of relatively static markets. In contrast, dynamic capabilities are explicitly designed for **environments characterized by volatility, complexity, and discontinuous change**.

At their core, dynamic capabilities are not about *what* a firm owns but *how* it mobilizes and evolves what it owns. They represent the routines, processes, and learned behaviors that enable companies to sense new opportunities (or threats), seize them through strategic initiatives, and transform the organization to embed the changes sustainably. For example, Apple's ability to pivot from computers to mobile devices, and later, to services, reflects a mastery of dynamic capability. Likewise, Amazon's seamless transition from bookseller to global cloud service provider illustrates not just operational prowess but **capability transformation at scale**.

Importantly, dynamic capabilities are not confined to technology firms. They are equally essential in **manufacturing, healthcare, finance, logistics, and even non-profit sectors**. What distinguishes high-performing organizations is not their initial resource base, but their **reflexive ability to update, upgrade, and redeploy resources strategically over time**. In a STEM-powered strategy landscape, these capabilities are becoming increasingly codified and measurable through digital platforms, dashboards, and simulation tools.

In practical terms, dynamic capabilities often manifest through actions like early technology scouting, cross-functional innovation labs, agile development cycles, continuous reskilling programs, and strategic alliances that enable rapid experimentation. These actions may appear disconnected, but when orchestrated under a unifying strategic logic, they form the **adaptive backbone** of a resilient enterprise.

As we explore in the following sections, **AI technologies are not just tools for automation;** they are becoming embedded in the very processes that enable sensing, seizing, and transforming. By integrating dynamic capabilities with intelligent systems, firms can generate a sustainable edge that evolves as fast as the world around them.

9.2.2 AI as a Differentiation Engine

Artificial Intelligence is no longer just a support function; it is a core driver of strategic differentiation. At a fundamental level, AI enables firms to move beyond static segmentation, reactive decision-making, and lagging indicators. Instead, it empowers organizations to identify micro-opportunities, anticipate shifts, and generate value faster and more precisely than traditional strategic tools allow.

To understand how AI drives differentiation, consider its role across three interconnected areas: data ingestion, pattern recognition, and intelligent action. First, AI systems ingest massive volumes of structured and unstructured data, ranging from real-time customer behaviors to social sentiment and operational anomalies. Second, through machine learning algorithms, these systems detect complex patterns that would remain invisible to human analysts. Finally, AI recommends or executes intelligent actions, such as price adjustments, product recommendations, fraud detection, or operational shifts, at speeds that far surpass human capacity.

The strategic edge lies not just in automation but in **learning loops that improve with each interaction**. A firm using AI to personalize customer experiences becomes better at it over time. A manufacturer using AI for predictive maintenance reduces downtime continuously. This compounding advantage, **AI learning while acting,** is the foundation for scalable, defensible differentiation.

For example, Adobe has used its AI engine, Adobe Sensei, to embed intelligence into its suite of creative and marketing tools. What differentiates Adobe is not only the quality of its products but the **experience layer,** enhanced by AI, that adapts to user behavior in real time. Similarly, Flex, a global electronics manufacturer, uses AI-powered platforms to orchestrate its supply chain dynamically, transforming logistics from a cost center into a strategic asset. Unilever's AI-led marketing analytics deliver hyper-localized insights, allowing the firm to respond with culturally tailored campaigns at scale.

In these cases, AI becomes the differentiator not by replacing human decision-makers, but by **enhancing their precision, scope, and foresight**. When embedded into dynamic capabilities, AI transforms them from episodic responses into continuous strategic advantages. In this way, AI does not merely support differentiation; it becomes the **engine** that makes it intelligent, scalable, and future-ready.

9.2.3 From Resource Possession to Capability Orchestration

Traditional strategic thinking often emphasized *what* a firm owns, its factories, patents, brand equity, or workforce. This logic, rooted in the resource-based view, assumed that unique or rare resources would be the source of sustained advantage. But in today's turbulent environment, **ownership without orchestration is obsolete**. The real source of competitive advantage lies in how resources are *assembled, reconfigured, and mobilized,* often in real time, to meet emerging demands. This is the essence of capability orchestration.

Capability orchestration refers to the **strategic alignment and deployment of resources, both internal and external, to build adaptive advantage**. It's about synchronizing human talent, digital infrastructure, ecosystem partnerships, and organizational learning in a way that's context-specific and future-oriented. What distinguishes market leaders is not their resource depth, but their ability to **move, combine, and evolve resources into new configurations** faster than rivals can react.

Consider how Unilever builds global campaigns: it doesn't rely solely on internal marketing departments. Instead, it draws on a global partner network, AI-driven consumer insights, and agile experimentation loops to roll out market-specific strategies at scale. Or take Adobe, which orchestrates its vast suite of tools, data streams, and creative professionals through cloud-based platforms, allowing seamless updates, personalization, and cross-functional collaboration.

AI elevates orchestration to a new level. It enables firms to **coordinate at scale and speed** that human managers cannot achieve alone. For example, intelligent supply chain systems can rebalance sourcing, logistics, and delivery based on shifting weather patterns, consumer behavior, or geopolitical disruptions. Predictive HR systems can reallocate talent in real time to high-priority innovation projects. These are not abstract possibilities; they are **operational realities** in leading firms.

Moreover, orchestration includes external platforms and partners. In today's ecosystems, competitive advantage often depends on co-innovation, working with fintechs, startups, universities, or even competitors. AI-enabled orchestration tools can map these networks, recommend optimal collaborators, and monitor knowledge flows. The result is a strategic system that is both **modular and dynamic**, capable of responding to complexity without collapsing under it.

Ultimately, resource possession offers diminishing returns. Strategic advantage comes from **how resources are continually reconfigured and amplified** by AI, analytics, and ecosystem intelligence. Firms that master orchestration will not just survive disruption; they will shape it.

To further clarify this strategic shift, we can map Teece's dynamic capabilities framework, **Sensing, Seizing, and Transforming,** to real-world, AI-enabled actions. This framework helps us understand how firms move beyond static resources and begin orchestrating adaptive advantage:

Mapping Teece's Dynamic Capabilities to AI-Driven Strategic Actions

Teece's framework of dynamic capabilities, sensing, seizing, and transforming, has long served as a foundation for understanding how firms adapt and thrive in volatile environments. In the AI era, these capabilities are no longer abstract or theoretical; they are increasingly measurable, operational, and automatable.

AI enables organizations to scale orchestration, turning once-manual strategic processes into continuous, data-driven loops. For example, AI can detect weak signals in consumer behavior, simulate future scenarios, and reconfigure operations in real time. This makes dynamic capabilities not only more actionable but also more responsive to complexity and speed.

The table below illustrates how each of Teece's capabilities maps to specific AI-enabled actions and real-world applications:

Table 9.3: Operationalizing Teece's Dynamic Capabilities with AI

Teece's Capability	Definition	AI-Enabled Action	Illustrative Example
Sensing	Identifying and assessing new opportunities or threats in fast-moving environments	Real-time trend analytics, market listening tools, weak signal detection via ML	**Unilever** uses AI to detect early shifts in consumer demand at the regional level
Seizing	Mobilizing resources to capture emerging opportunities	AI-led scenario planning, demand-based resource reallocation, cloud-driven deployment	**Adobe** uses predictive analytics to tailor product features and updates to user behavior
Transforming	Reconfiguring the organizational structure, assets, or operations for sustained relevance	AI-driven workforce agility, modular digital architectures, ecosystem co-innovation	**Flex** continuously reconfigures its operations based on real-time input across global manufacturing networks

9.2.4 Embedding Adaptiveness into Strategy

In an environment characterized by volatility, ambiguity, and nonlinear change, strategy can no longer be a fixed, five-year roadmap. It must become a **living system—responsive,**

iterative, and data-informed. Embedding adaptiveness into strategy means building the capability to sense shifts, reconfigure quickly, and evolve continuously, rather than merely reacting to external events.

This shift requires a fundamental rethink of both structure and mindset. At the structural level, adaptive firms design strategy processes that incorporate **real-time data feedback, scenario modeling, and dynamic KPIs**. Decision rights are distributed, not centralized; cross-functional teams are empowered to act; and learning loops are hard-coded into strategic execution. At the mindset level, leadership fosters **a culture of experimentation, resilience, and psychological safety**, where uncertainty is embraced as a design constraint rather than a threat.

AI plays a central role in embedding adaptiveness. It functions as a **strategic radar system,** processing signals at scale, identifying weak patterns, and recommending options before humans even detect the shift. For instance, platforms like Palantir or C3.ai support an adaptive strategy by continuously integrating external and internal data streams to simulate multiple futures. They enable leaders to test hypotheses, pressure-test plans, and rehearse responses without incurring real-world risk.

This capability is most visible in firms that **build "strategy as code" systems,** digital platforms where the entire strategy cycle, from sensing to seizing to transforming (per Teece's framework), is supported by data and algorithms. These systems allow for ongoing adaptation, not episodic planning. For example, Unilever embeds feedback loops into its sustainability goals, allowing dynamic resource reallocation based on real-time environmental data. Adobe adjusts its innovation portfolio based on user behavior data collected across millions of creative professionals globally.

Adaptiveness also enables **strategic resilience**. When advantage erodes, as it inevitably does in fast-moving industries, adaptive firms don't panic; they pivot. They rely on digital twins, early-warning dashboards, and predictive signals to determine when to renew, retreat, or reinvent. This mindset moves strategy from **a document to a dialogue**, and from planning to *perceiving.*

Ultimately, embedding adaptiveness means designing a strategy as a **perpetual motion system**, not a blueprint. It's about building dynamic capabilities that operate across all levels of the organization, supported by technology, driven by data, and grounded in a strategic culture that welcomes change as a competitive edge.

9.3 Modeling Evolving Competitive Landscape

The strategic environment firms face today is not just volatile; it is **nonlinear, data-rich, and perpetually evolving**. In such a climate, traditional methods of competitor analysis, based on fixed market shares, static positioning maps, or annual strategic reviews, are no longer sufficient. Competitive landscapes now shift in real time, shaped by algorithms, digital ecosystems, geopolitical turbulence, and consumer sentiment flowing across platforms and borders. In this context, strategy must evolve from a reactive discipline into a **predictive and adaptive modeling practice**.

Modeling evolving competition requires a transformation in how firms perceive and simulate strategic threats and opportunities. The old assumption that industries are clearly bounded and slowly moving has been replaced by a recognition of **blurring sectors, unexpected rivals, and emergent behaviors**. Think of how Amazon became a healthcare player, or how TikTok's algorithmic model disrupted digital advertising and attention economics. These shifts weren't random; they were the outcome of complex systems dynamics, signaling that competitive intelligence must now move beyond tracking peers to **understanding and forecasting competitive systems.**

This section explores how firms can deploy **advanced modeling tools** to simulate competitive behavior, anticipate moves by rivals, and stress-test strategic options. Techniques such as agent-based modeling, system dynamics, digital twins, and AI-driven scenario engines provide powerful lenses through which firms can visualize how competitive forces evolve and interact. These models are not just academic exercises; they are now embedded in the strategy functions of firms like Amazon, Palantir, and Uber.

The key shift is this: instead of using data only to describe what has already happened, forward-looking firms use models to simulate what *might* happen, and why. The goal is not perfect prediction but **strategic foresight**: the capacity to interpret weak signals, explore alternative futures, and shape competitive advantage in real time. Modeling is the new mapmaking in a world of shifting terrain.

9.3.1 The Strategy Landscape Is No Longer Static

The days of relatively stable industries and clearly defined competitive boundaries are over. We now live in a world of **perpetual flux**, where emerging technologies, data abundance, and platform convergence redefine industry structures almost overnight. This makes the competitive landscape **fluid, porous, and unpredictable,** an environment where rivals may appear from adjacent industries or digital-native firms may pivot to dominate traditional spaces.

Classic models like Porter's Five Forces were revolutionary in their time, but they were designed for static analysis. Today's strategists must contend with **accelerating change cycles**, industry blurring, regulatory swings, consumer activism, and real-time technological disruption. Netflix doesn't just compete with Disney; it competes with TikTok, Fortnite, and even sleep. Google isn't just in search; it's shaping quantum computing, healthcare diagnostics, and autonomous driving. The nature of competition itself has become **dynamic, layered, and asymmetric**.

In this new environment, the strategist's job is not just to analyze what is, but to **simulate what could emerge**. Traditional benchmarks are losing relevance, and pattern recognition must now be powered by **data modeling, simulations, and machine learning** to capture hidden shifts and latent variables. The notion of "industry peers" has transformed into that of **ecosystem actors**, and many of these actors don't play by the same strategic rulebook.

Let's also consider the **time dimension** of competition. In a static world, an advantage lasted years. In today's reality, the average lifespan of a competitive edge is shrinking rapidly. The World Economic Forum reports that the average tenure of firms on the S&P 500 has dropped from 33 years in 1964 to less than 20 years today, and it's forecasted to shrink further. Strategic moves must now anticipate **continuous reshaping**, not occasional restructuring.

To help visualize this shift, we can introduce the following table:

Table 9.4. Static vs. Evolving Competitive Landscapes

Dimension	Traditional (Static) Strategy	Modern (Evolving) Strategy
Industry Boundaries	Clear and stable	Fluid and redefined by platforms and data flows
Competitor Set	Known and fixed	Emergent and unpredictable
Time Horizon	Long-term planning cycles	Continuous scanning and adaptive planning
Strategic Tools	SWOT, Five Forces	Agent-based modeling, system dynamics, digital twins
Sources of Advantage	Resources, market position	Capabilities, speed, ecosystem leverage
Role of Data	Descriptive and lagging	Predictive, real-time, and dynamic
Organizational Mindset	Efficiency and control	Experimentation and adaptation

This table reinforces the transformation in strategic thinking. Organizations today must develop **agile foresight,** the ability to iterate on models, monitor real-time shifts, and reposition strategies on the fly. This requires investing in **modeling capabilities** and cultivating a culture that treats change not as an exception, but as the default condition of modern competition.

9.3.2 Competitive Intelligence as Simulation, Not Surveillance

For decades, competitive intelligence (CI) was largely treated as a **surveillance function,** a process of gathering information about rivals, benchmarking best practices, and reporting trends to leadership teams. CI teams focused on monitoring market share, tracking product launches, attending trade shows, and collecting insights from industry reports. While valuable, this model is now insufficient for a world defined by algorithmic speed, platform dynamics, and nonlinear competition.

Today's leaders must evolve CI into a **strategic simulation function,** a dynamic system for forecasting possible futures, testing assumptions, and modeling competitor behavior in evolving scenarios. The shift from surveillance to simulation reflects a deeper truth: in a rapidly changing environment, **reactive intelligence** is too slow. By the time a traditional CI report lands on an executive desk, the insights may already be outdated. Instead, firms must embrace **predictive, iterative modeling environments** that allow for continuous adaptation and scenario testing.

Simulation-based CI reframes how firms view competition. It treats rival firms not as static entities, but as **adaptive agents,** each with unique rules, learning patterns, and interdependencies. Agent-based models (ABMs), for example, allow strategists to simulate thousands of potential decisions made by competitors under varying conditions. These models consider how changes in pricing, regulation, technology, or customer sentiment could ripple across an entire industry landscape, revealing **emergent threats and latent opportunities**.

Another leap forward is the use of **digital twin competitor models,**data-rich, AI-powered replicas of rival firms that integrate publicly available financials, patents, hiring trends, product evolution, and media sentiment. These twins don't just monitor, they anticipate. A digital twin of a competitor might reveal how a firm's recent R&D investments align with its talent acquisition and M&A activity, offering a window into future strategic moves before they materialize.

Simulation tools also shift CI from **rear-view mirror** analysis to **strategic foresight**. Rather than asking "What happened?" or "What are they doing now?", simulation-based CI explores: "What might they do next?", "What could disrupt them?", and "Where can we preemptively differentiate?"

To make this actionable, CI must become **embedded in strategic planning loops**. This means building cross-functional teams where data scientists, industry experts, and strategists collaborate in real time. It also means integrating simulation tools into war-gaming sessions, quarterly reviews, and even day-to-day product decisions.

Ultimately, simulation-based CI transforms competitive intelligence from a static report to a **living, learning system,** one that evolves as fast as the environment it seeks to understand. In this model, firms don't just monitor the game; they simulate the entire playing field.

9.3.3 Tools for Modeling Competition in a Data-Driven World

As competitive landscapes become increasingly complex and fluid, the ability to model dynamic interactions among firms, consumers, technologies, and regulations has become central to strategy. Modeling competition is no longer just about plotting market share on a two-by-two matrix or analyzing historical trends. It requires **sophisticated computational tools** that simulate interdependencies, non-linearities, feedback loops, and uncertainty.

Let's examine four of the most powerful modeling tools now reshaping how firms simulate competition:

1. Agent-Based Models (ABMs)

Agent-based models simulate interactions between autonomous agents—each with its own rules, goals, and behavioral logic. These agents could represent customers, firms, regulators, or even influencers. ABMs allow strategists to simulate how micro-level interactions give rise to macro-level patterns such as product diffusion, price wars, or alliance formation.

For instance, a telecommunications company can use an ABM to model customer churn based on social influence, pricing sensitivity, and regional factors. By tweaking parameters, strategists can test different retention tactics and identify tipping points before rolling out a campaign.

2. System Dynamics (SD)

System dynamics models visualize how variables such as supply, demand, investment, innovation, and policy interact over time. They are especially useful for understanding **feedback loops** and **delay effects,** like how a drop in R&D today might affect market position five years down the line.

Take the automotive sector. A system dynamics model could simulate how regulation on electric vehicle emissions, consumer adoption, battery supply chains, and charging infrastructure co-evolve over a decade. These models help identify where policy levers, strategic investments, or consumer shifts could reshape competitive dynamics.

3. Digital Twin Competitor Models

Unlike traditional models, digital twins create **live, data-driven replicas** of competitor firms using a blend of structured and unstructured data. These models ingest real-time inputs such as patent filings, hiring patterns, ESG disclosures, product updates, and analyst sentiment. Through machine learning, they forecast competitor decisions and simulate their impact on the market.

For example, a pharmaceutical firm might build a digital twin of a rival's R&D division to anticipate the next wave of drug approvals. This twin might integrate regulatory filings, job postings for specific therapeutic areas, and NLP analysis of investor calls.

4. Scenario Simulation Engines

These engines allow companies to simulate plausible futures based on varying assumptions—such as geopolitical disruptions, consumer value shifts, or emergent technologies. Scenario simulation is particularly valuable when making **irreversible decisions** like entering new markets or acquiring firms.

A retail conglomerate, for instance, might use simulation engines to explore scenarios such as inflation spikes, currency devaluation, or AI-driven personalization, and test how each affects its omnichannel strategy across geographies.

Scenario simulation engines empower organizations to explore **plausible futures** by modeling how different variables—such as geopolitical disruptions, consumer value shifts, or emerging technologies- might interact over time. These tools are especially valuable when making **irreversible strategic decisions**, such as entering new markets, launching major product lines, or acquiring firms. The table below compares four leading modeling approaches and their strategic applications:

Table 9.5: Comparative Summary of Competitive Modeling Tools

Modeling Tool	Key Inputs	Key Outputs	Strategic Benefits
Agent-Based Models	Behavioral rules, agent types, interaction rules	Emergent behaviors, tipping points	Simulate customer behavior, market evolution
System Dynamics	Stock/flow diagrams, causal loops, time delays	Feedback insights, trend forecasts	Identify leverage points, policy impacts
Digital Twin Competitor Models	External firm data, ML pipelines, knowledge graphs	Real-time forecasts, strategy prediction	Anticipate rival moves, spot innovation early
Scenario Simulation Engines	Macroeconomic data, trend indicators, risk factors	Multiple futures, stress tests	Plan under uncertainty, test robustness

These tools are not mutually exclusive; they are often most powerful when integrated. A strategy team might use system dynamics to model long-term trends, ABMs to simulate customer shifts, and digital twins to anticipate competitive reactions. The future of strategy will not be defined by guessing the next move, but by **modeling what's possible** and preparing for it with rigor and foresight.

9.4 Strategic Renewal and the Erosion Curve

Competitive advantage, once considered an enduring asset, is now increasingly ephemeral. In the STEM-driven strategic environment, advantages do not last; they erode, decay, and often disappear faster than they can be consolidated. This shift demands a new strategic mindset: one that treats **renewal as a continuous, algorithmically informed process**, not a reactive fix when performance falters.

Welcome to the age of the **Advantage Erosion Curve,** a visual and analytical framework that captures the temporal fragility of strategic positioning. As data cycles accelerate and innovation diffuses more quickly, the time window during which a firm can maintain a lead is shrinking. Whether it's a pricing edge, a feature innovation, or even a proprietary algorithm, today's differentiators are tomorrow's baseline expectations.

Strategic renewal, then, is not just about launching new products or restructuring divisions. It's about building systems that can **sense when an advantage is beginning to fade**, simulate different renewal pathways, and adapt faster than rivals. This involves not only foresight and modeling, but also integrating AI, simulation tools, and feedback mechanisms into the very architecture of the firm.

This section explores how STEM-enabled firms are using data, models, and simulations to monitor competitive decay in real time, and how they design **self-renewing ecosystems** that adapt before disruption hits. From Netflix's evolving content algorithms to Microsoft's product reinvention cycles, strategic renewal is becoming the heartbeat of long-term performance.

Rather than ask "how do we sustain our edge?", modern strategy asks:

"How do we renew ourselves before the market forces us to?"

9.4.1 Competitive Advantage Is Now Perishable

The assumption that competitive advantage can be sustained indefinitely is no longer viable. In an environment shaped by fast-moving technologies, shifting customer expectations, and global volatility, even the most formidable firms find their advantages fleeting. The idea of a "sustainable" edge, once the holy grail of strategic thinking, is being replaced by the concept of **transient advantage**, where success is measured by a firm's ability to innovate and pivot before its current model becomes obsolete.

Several forces are accelerating this perishability. First, **digital transparency** has increased dramatically. Today, competitors can reverse-engineer innovations faster, customers can compare offerings in real time, and feedback loops are instant. Second, **platform economics and open ecosystems** allow smaller players and startups to scale at unprecedented speed. What used to take a decade can now be achieved in a matter of months. Third, the **shortened product life cycles** across nearly every industry mean that firms must be in a constant state of reinvention or risk being leapfrogged by agile disruptors.

Firms that once enjoyed long periods of dominance, Kodak, Nokia, and Blackberry, serve as cautionary tales. Their failures were not due to a lack of resources or market share but an inability to respond to inflection points early enough. In each case, the competitive advantage they had built became a liability when they clung too long to a decaying model. These examples illustrate that even strong brands and robust capabilities can erode if they are not renewed continuously.

Importantly, **the perishability of advantage also implies a new strategic tempo**. Firms must operate on dual time horizons: one that delivers current value and another that anticipates future threats and opportunities. This demands investments not just in R&D or innovation labs, but in **strategic sensing systems**—the ability to detect early signals of change, whether they emerge from technology trends, regulatory shifts, or customer sentiment patterns.

In this new paradigm, the **most valuable capability is adaptability**. Firms that build systems to recognize when their current advantage is eroding and can initiate renewal before decline sets in are the ones that will dominate. It's no longer about whether a company has an advantage. The question is: **Can it see the edge of its advantage before others do, and move beyond it in time?**

9.4.2 Forecasting the Advantage Erosion Curve

In the past, strategic planning often assumed that competitive positions, once secured, could be held with modest adaptations. Today, the reverse is true. Every advantage comes with an expiration date. The most forward-looking firms actively monitor **when and how** their current advantage may start to erode. This mindset, forecasting the **Advantage Erosion Curve,** marks a fundamental shift from reactive to proactive strategy.

The Advantage Erosion Curve is a conceptual model that maps how a firm's edge in a given market decays over time, often due to external disruption (e.g., new entrants, changing customer preferences, regulatory shifts) or internal rigidity (e.g., inertia, overreliance on legacy systems). By recognizing patterns in this decline, firms can time strategic renewal **before** reaching crisis mode.

Several signals typically precede advantage erosion:

- **Margin compression** despite stable sales
- **Customer churn** to competitors offering enhanced experiences
- **Declining employee engagement** or talent exodus
- **Innovation lag**, visible in product obsolescence or missed opportunities
- **Increased reliance on price cuts or promotions** to maintain volume

By tracking such indicators—and integrating them with AI-driven early-warning systems—companies can estimate **where they are on the erosion curve** and how rapidly decline might accelerate.

To make strategic foresight operational, many firms are building **intelligent dashboards** that go beyond static KPIs. These dashboards integrate **rate-of-change metrics** and fuse internal performance data with external signals, such as social listening, competitor monitoring, regulatory intelligence, and trend forecasting tools.
The goal is not merely to detect decline after it happens, but to **anticipate inflection points**—giving leaders enough lead time to reallocate capital, redesign offerings, or pivot business models before competitive advantage erodes.

For example, a sudden spike in **support tickets** or a dip in **Net Promoter Score (NPS)** may signal early-stage customer dissatisfaction. Similarly, **market share volatility** or **price sensitivity** could indicate competitor-led erosion. When combined with signals from the broader ecosystem, like **platform migration trends** or **regulatory watchlists,** these indicators form a powerful early-warning system.

The table below outlines key categories of strategic indicators and the insights they provide:

Table 9.6: Strategic Indicators for Forecasting Advantage Erosion

Indicator Type	Example Metrics	Strategic Insight
Market Signals	Market share volatility, price sensitivity	Competitor or customer-led erosion
Internal Health Metrics	R&D velocity, cycle time, employee turnover	Organizational readiness to adapt
Ecosystem Shifts	Partnership fatigue, platform migration trends	Erosion due to inter-firm dynamics
Customer Signals	NPS decline, support tickets, social sentiment	Early signs of experience degradation
Tech/Regulatory Signals	Emerging tech adoption, regulatory watchlist	Foresight on non-market disruptions

These inputs can be fed into **simulation models** (see 8.3) or strategic renewal playbooks (see 8.4.3), helping organizations visualize multiple futures and identify the earliest point at which advantage erosion begins. Leaders can then create "trigger thresholds" where specific actions, such as increased R&D spending, spin-outs, or acquisitions, are pre-authorized and executed at speed.

In sum, forecasting the Advantage Erosion Curve isn't about predicting the future perfectly; it's about knowing when to **start letting go of the present**. It's a discipline that transforms strategic renewal from a one-time event into a continuous, data-informed capability.

9.4.3 Data-Driven Strategic Renewal Loops

In an environment where competitive advantages are increasingly perishable, static strategies are not only insufficient, they're dangerous. Firms must evolve from one-time transformations to continuous, iterative adaptation. This is the essence of **strategic renewal loops**: self-adjusting systems where data doesn't just monitor change, it drives it.

At its core, a **strategic renewal loop** is a cyclical process that detects signals of advantage decay, tests strategic alternatives, and implements changes with feedback-informed precision. Unlike traditional change management, which is often reactive and slow, these loops are **proactive**, **data-rich**, and **designed to operate at speed**.

Modern renewal loops typically unfold in four interdependent stages:

1. **Sensing Change** – using AI and analytics to detect weak signals and leading indicators of erosion
2. **Modeling Options** – generating, simulating, and scoring strategic alternatives based on predicted impact
3. **Executing Micro-Pivots** – deploying adaptive tactics through pilot initiatives, agile teams, or modular product design
4. **Learning and Recalibrating** – integrating outcome data to refine strategy and inform the next cycle

The goal is to **embed strategic flexibility** not as a trait of leadership, but as a structural property of the organization itself. These loops can be localized (e.g., in product development teams) or enterprise-wide (e.g., involving M&A, business model shifts, or supply chain overhauls).

Firms like **Netflix** exemplify this approach. With constant experimentation in content, pricing, and platform features, it operates in a near-perpetual renewal cycle. Data from user behavior drives both algorithmic content decisions and executive-level strategy shifts. Similarly, **Microsoft** uses telemetry from its cloud services to identify usage patterns, triggering adjustments in feature rollouts, marketing campaigns, or even acquisition strategies.

At the heart of this capability is a **Strategic Renewal Loop**—a closed feedback system that transforms signals into action and learning:

[Data Signals] → [Strategic Modeling] → [Pilot Execution] → [Feedback Integration] → ⟳

This loop ensures that strategy is not a static plan but a **living system**—constantly sensing, testing, and evolving. By institutionalizing this loop, organizations can respond to change not with disruption, but with **deliberate, data-informed renewal**.

To support this architecture, firms build **digital twin models of strategy,** virtual simulations of how different strategic choices might play out under varying conditions. These models, enriched with real-time data, allow firms to **fail safely** in simulated environments before scaling new strategies in the real world.

Equally important is governance. Strategic renewal loops require **decision protocols** that enable fast action without bypassing risk controls. Leading firms establish "renewal thresholds", pre-agreed KPIs or scenario triggers that activate renewal playbooks automatically. These playbooks include a range of tested strategic responses: sunset protocols for legacy offerings, alliance formation strategies, innovation sprints, or digital platform pivots.

Incorporating **machine learning**, these systems improve over time. Renewal loops become smarter with every iteration, learning which early signals correlate with performance decline, and which responses yield the best returns. In time, these loops begin to function as **autonomous strategy engines**—reducing cognitive load for executives while increasing organizational agility.

The outcome is not just adaptation, it's **strategic resilience**. Firms capable of running high-frequency, data-informed renewal loops are better positioned to absorb shocks, outpace rivals, and even **shape markets** rather than merely reacting to them.

9.5. Chapter 9 Wrap-Up: Competing on Intelligence, Not Just Products

As we close this pivotal chapter, it becomes clear that competitive advantage in the 21st century is no longer anchored in static assets or market positioning; it is dynamically constructed through **data, learning, and continuous orchestration**. The firms that thrive are those that move beyond competing on products or price and instead **compete on intelligence,** their ability to sense, simulate, and adapt in real time.

Throughout this chapter, we have redefined the foundations of competitive strategy by introducing a STEM lens to classical ideas. Data has emerged not just as a reporting tool but as a **strategic infrastructure,** a core input in building advantage. Whether it's the velocity of data informing micro-adjustments or digital platforms enabling ecosystems of co-creation, intelligence-driven competition is now the norm, not the exception.

We explored how **dynamic capabilities,** sensing, seizing, and transforming, are reengineered through AI, making strategic decision-making faster, more precise, and more adaptive. Strategic leadership today is less about guarding static resources and more about orchestrating capabilities in motion, across teams, geographies, technologies, and platforms.

We also delved into the new landscape of **competitive modeling**, moving beyond outdated competitor matrices to real-time simulations, digital twins, and agent-based strategy labs. This modeling capacity allows modern firms to explore not only what is happening, but what could happen, and prepare accordingly.

Finally, we addressed the **fragility of advantage** in a hypercompetitive, data-saturated world. The advantage erosion curve is steepening. But with embedded renewal loops, AI-powered sensing, and learning systems that iterate strategy at speed, firms can convert fragility into **resilience**. Those who master this renewal mindset will not only adapt to disruption, they will define the contours of the next competitive cycle.

As you continue through the rest of this textbook, remember: competitive advantage is no longer a destination; it's a dynamic process. It is constructed, deconstructed, and reconstructed continuously, driven by the flow of data, the rhythm of learning, and the vision of strategic intelligence.

8.5.1. Discussion Questions

1. How does shifting from products to data ecosystems reshape a firm's competitive logic?
2. What are the key differences between traditional capabilities and dynamic capabilities powered by AI?
3. Which modeling tools from this chapter could most help your organization anticipate future competition?
4. How can firms ensure strategic renewal becomes an embedded process, not a one-time event?
5. What are the ethical risks in using AI to drive advantage, and how should they be governed?
6. How can organizations balance speed and control when running renewal loops?
7. In what ways can ecosystem orchestration replace traditional vertical integration as a source of advantage?
8. How does data velocity influence a firm's feedback advantage? What are the risks of acting too slowly—or too quickly?
9. What are the implications of the advantage erosion curve for firms in regulated or slower-moving industries?
10. How would you integrate dynamic capabilities with scenario modeling to design a future-ready strategy?

9.5.2. Micro-Case: Zara's Data-Driven Renewal Loop

Zara, known for its fast-fashion agility, faces new pressures from eco-conscious consumers and rising competition from AI-native retail startups. Leadership tasks Zara with evolving beyond speed, toward **sustainable, data-driven competitiveness**.

The company launches a multi-pronged renewal loop:

- Real-time sustainability dashboards track product lifecycle metrics
- Predictive demand systems are recalibrated to reduce overproduction
- Customer feedback is captured through AI-curated design communities
- Strategic alliances are formed with circular fashion innovators and tech recyclers

But challenges emerge. The data ecosystem reveals misalignment between regional buyer behavior and global sustainability messaging. AI design tools receive backlash for replicating cultural aesthetics without context. Meanwhile, legacy infrastructure slows pilot execution in emerging markets.

Zara's strategy team revisits its **advantage erosion curve**, realizing speed is no longer sufficient; it must embed **purpose, intelligence, and modularity** into the brand. Using renewal modeling tools, they simulate a transition to "Just-in-Planet" manufacturing: a fully traceable, localized, and customer-co-designed supply chain.

Within 12 months, Zara doesn't just return to growth—it sets a new standard for ethical, data-driven fashion.

Discussion Questions

1. How did Zara's strategic renewal approach evolve beyond speed as a source of competitive advantage?
2. What role did real-time data and customer co-creation play in Zara's new value proposition?
3. What tensions did Zara face between global strategy and local market intelligence, and how could they be resolved?
4. How might the advantage erosion curve have influenced Zara's urgency and decision sequencing?
5. What feedback mechanisms would you add to ensure the success of Zara's "Just-in-Planet" model?
6. How can Zara balance AI-driven creativity with ethical cultural representation in global markets?
7. In what ways can Zara's renewal loop be made resilient to future disruption, environmental, technological, or societal?

9.5.3. Interactive Exercise: The Advantage Erosion Lab

You're the Chief Strategy Officer of a global consumer electronics firm whose wearable tech line is losing market share to a new startup. You suspect your advantage is eroding, but you're unsure where, how fast, and what your renewal options are.

Using tools from this chapter (e.g., advantage erosion modeling, digital twin simulation, renewal loops), develop a strategic renewal plan. Your response should:

- Estimate the firm's current position on the erosion curve
- Simulate at least two renewal strategies (e.g., feature redesign, ecosystem pivot, co-branding)
- Establish a renewal feedback loop using real-time metrics (e.g., customer sentiment, platform usage, churn risk)

9.5.4. Recommended HBR Case Studies

❖ Adobe: Building a Company the World Wants to Work For

* Overview: Published on March 20, 2024, and authored by Frances X. Frei, Anne Morriss, and Laura Tilghman, this 15-page case explores how Adobe has cultivated a workplace culture that consistently earns it recognition as one of the world's most desirable employers. The case centers on Adobe's efforts to align its people strategy with its business strategy, particularly as it navigates rapid growth, digital

transformation, and the evolving expectations of a global workforce. It highlights Adobe's approach to leadership development, diversity and inclusion, employee empowerment, and purpose-driven work. Students are invited to consider how Adobe's culture contributes to its innovation and performance, and how it can continue to evolve in a competitive talent landscape. The case is ideal for discussions on organizational behavior, human capital strategy, and values-based leadership.

* Citation: Frei, F. X., Morriss, A., & Tilghman, L. (2024). *Adobe: Building a Company the World Wants to Work For* (Case No. 324090-PDF-ENG). Harvard Business School.

❖ How Adobe Innovated Its Business Model and Blazed the Trail for the Digital Experience Platform

* **Overview:** Published on July 2, 2024, and authored by Sayan Chatterjee, this 18-page case from the Darden School of Business explores Adobe's bold and industry-defining transformation from a traditional software licensing model to a cloud-based subscription model. The case began in 2012, when Adobe announced that all future product releases would be available only through the Creative Cloud. At the time, this move was met with skepticism across the software industry. However, over the next decade, Adobe's strategy proved visionary—enabling faster customer onboarding, more flexible pricing, and continuous product updates. By 2024, Adobe had not only succeeded in its transition but had also become a leader in the Digital Experience Platform (DXP) space. The case examines the strategic decisions Adobe made during this period, including its investments in AI, data-driven personalization, and enterprise solutions. This case is ideal for discussions on business model innovation, digital transformation, and strategic agility in the face of market disruption.
* **Citation:** Chatterjee, S. (2024). *How Adobe Innovated Its Business Model and Blazed the Trail for the Digital Experience Platform* (Case No. UV8995-PDF-ENG). Darden School of Business.

❖ Perplexity: Redefining Search

* **Overview:** Published on March 20, 2025, and authored by Suraj Srinivasan, Michelle Hu, Sriraghav Srinivasan, and Radhika Kak, this 16-page case explores the meteoric rise of Perplexity AI, a startup that positioned itself as a disruptive "answer engine" in the global search market. By early 2025, Perplexity had grown into a company valued at $9 billion, challenging search giants like Google with its innovative approach to AI-powered information retrieval. The

company's platform combined custom-built indexes, real-time retrieval systems, and advanced language models to deliver fast, accurate, and ad-free answers to complex queries. Key innovations such as Deep Research, Pro Search, and Comet (an AI-native web browser) reflected Perplexity's ambition to redefine how users, both consumers and enterprises, interact with information. However, as the company scaled, it faced critical strategic decisions:

- Should it maintain its ad-free, user-centric model, or explore monetization through advertising?
- How could it preserve user trust while expanding into enterprise markets?
- What investments were needed to stay ahead of competitors like OpenAI and Google?

The case is ideal for discussions on AI strategy, business model innovation, platform competition, and scaling in high-tech markets.

- Citation: Srinivasan, S., Hu, M., Srinivasan, S., & Kak, R. (2025). *Perplexity: Redefining Search* (Case No. 125093-PDF-ENG). Harvard Business School.

❖ Zensar Technologies: From Living Digital to Living AI

- **Overview:** Published on November 1, 2024, and authored by Piyush Kumar, Geetika Shah, Vinodini Saihjpal, and Sushreeta Mohapatra, this 18-page case from the Indian School of Business explores the digital and AI-driven transformation of Zensar Technologies, a mid-sized Indian IT services firm with a global footprint. The case begins in 2016, when Sandeep Kishore took over as CEO and launched the "Living Digital" initiative—a strategic shift aimed at repositioning Zensar from a legacy IT services provider to a digital transformation partner. This involved upskilling employees, building proprietary platforms under the Return on Digital® initiative, and embedding digital-first thinking across the organization. As the market evolved, Zensar recognized the growing importance of artificial intelligence and launched the "Living AI" initiative to position itself as a leader in AI-as-a-Service (AIaaS). The case challenges students to evaluate whether Zensar can successfully pivot its business model to embrace AI at scale, while maintaining its core values and competitive edge.
- Citation: Kumar, P., Shah, G., Saihjpal, V., & Mohapatra, S. (2024). *Zensar Technologies: From Living Digital to Living AI* (Case No. ISB444-PDF-ENG). Indian School of Business.

9.5.5. Strategy-in-Film: Learn by Watching

In rapidly evolving markets, competitive advantage is no longer static; it must be continuously built, tested, and renewed. The films below illustrate how organizations gain, defend, or lose their edge through strategy, innovation, and responsiveness. As you engage with the dynamics of competitive advantage in this chapter, use these narratives to reflect on how data, speed, and adaptability separate winners from those left behind.

Movie / Documentary	Strategic Theme Highlighted
The Matrix	Simulation and reality—modeling alternate strategic worlds
Moneyball	Analytics-driven advantage in low-resource environments
The Great Hack	Data ownership, feedback loops, and algorithmic influence
Ready Player One	Digital platforms, ecosystem advantage, and adaptive behavior
The Social Dilemma	Feedback cycles and ethical tensions in algorithmic strategy
Black Mirror: USS Callister	Virtual environments and power dynamics in closed systems
Her	AI, empathy, and emergent user-system coevolution
Margin Call	Real-time strategic decision-making in collapsing environments
Ex Machina	Intelligence thresholds and the strategic leap of AI consciousness

References

Abumalloh, R. A., Nilashi, M., Ooi, K. B., Wei-Han, G., Cham, T. H., Dwivedi, Y. K., & Hughes, L. (2024). The adoption of metaverse in the retail industry and its impact on sustainable competitive advantage: moderating impact of sustainability commitment. *Annals of Operations Research*, *342*(1), 5-46.

Adewusi, A. O., Okoli, U. I., Adaga, E., Olorunsogo, T., Asuzu, O. F., & Daraojimba, D. O. (2024). Business intelligence in the era of big data: A review of analytical tools and competitive advantage. *Computer Science & IT Research Journal*, *5*(2), 415-431.

Agarwal, R., & Helfat, C. E. (2009). Strategic renewal of organizations. *Organization Science*, *20*(2), 281-293.

Ameen, N., Tarba, S., Cheah, J. H., Xia, S., & Sharma, G. D. (2024). Coupling artificial intelligence capability and strategic agility for enhanced product and service creativity. *British Journal of Management*, *35*(4), 1916-1934.

Bughin, J., Kretschmer, T., & Van Zeebroeck, N. (2021). Digital technology adoption drives strategic renewal for successful digital transformation. *IEEE Engineering Management Review*, *49*(3), 103-108.

Cavusgil, S. T., & Deligonul, S. Z. (2025). Dynamic capabilities framework and its transformative contributions. *Journal of International Business Studies*, *56*(1), 33-42.

Chatterjee, S., Rana, N. P., Tamilmani, K., & Sharma, A. (2021). The effect of AI-based CRM on organization performance and competitive advantage: An empirical analysis in the B2B context. *Industrial Marketing Management*, *97*, 205-219.

Climent, R. C., Haftor, D. M., & Staniewski, M. W. (2024). AI-enabled business models for competitive advantage. *Journal of Innovation & Knowledge*, *9*(3), 100532.

Cristofaro, M., Helfat, C. E., & Teece, D. J. (2025). Adapting, Shaping, Evolving: Refocusing on the Dynamic Capabilities–Environment Nexus. *Academy of Management Collections*, *4*(1), 20-46.

Cui, Y. G., van Esch, P., & Phelan, S. (2024). How to build a competitive advantage for your brand using generative AI. *Business Horizons*, *67*(5), 583-594.

Davenport, T. H. (2018). *The AI advantage: How to put the artificial intelligence revolution to work*. MIT Press.

Dubey, R., Bryde, D. J., Dwivedi, Y. K., Graham, G., Foropon, C., & Papadopoulos, T. (2023). Dynamic digital capabilities and supply chain resilience: The role of government effectiveness. *International Journal of Production Economics*, *258*, 108790.

Ellström, D., Holtström, J., Berg, E., & Josefsson, C. (2021). Dynamic capabilities for digital transformation. *Journal of Strategy and Management*, *15*(2), 272-286.

Fernandes, C. I., Hughes, M., Ferreira, J., & Veiga, P. M. (2023). Exploring the microfoundations of innovation: what they are, where they come from, and where they are going?. *European Business Review*, *35*(3), 356-396.

Heaton, S., Teece, D., & Agronin, E. (2023). Dynamic capabilities and governance: An empirical investigation of financial performance of the higher education sector. *Strategic Management Journal*, *44*(2), 520-548.

Helfat, C. E., Kaul, A., Ketchen Jr, D. J., Barney, J. B., Chatain, O., & Singh, H. (2023). Renewing the resource-based view: New contexts, new concepts, and new methods. *Strategic Management Journal*, *44*(6), 1357-1390.

Herold, S., Heller, J., Rozemeijer, F., & Mahr, D. (2023). Dynamic capabilities for digital procurement transformation: a systematic literature review. *International Journal of Physical Distribution & Logistics Management*, *53*(4), 424-447.

Hortovanyi, L., Szabo, R. Z., & Fuzes, P. (2021). Extension of the strategic renewal journey framework: The changing role of middle management. *Technology in Society*, *65*, 101540.

Hutzschenreuter, T., & Lämmermann, T. (2024). What Is Your AI Strategy? Systematically Integrating Self-Learning Technologies into Your Business Strategy. *Academy of Management Perspectives*, DOI: 10.5465/amp.2023.0243.

Jackson, I., Ivanov, D., Dolgui, A., & Namdar, J. (2024). Generative artificial intelligence in supply chain and operations management: a capability-based framework for analysis and implementation. *International Journal of Production Research*, *62*(17), 6120-6145.

Joussen, T. P., Kanbach, D. K., & Kraus, S. (2025). Enabling strategic change toward resilience: a systematic review from a dynamic capabilities perspective. *Strategic Change*, *34*(3), 373-405.

Kaipainen, J., & Aarikka-Stenroos, L. (2022). How to renew business strategy to achieve sustainability and circularity? A process model of strategic development in incumbent technology companies. *Business Strategy and the Environment*, *31*(5), 1947-1963.

Kemp, A. (2024). Competitive advantage through artificial intelligence: Toward a theory of situated AI. *Academy of Management Review*, *49*(3), 618-635.

Khan, S. H., Majid, A., & Yasir, M. (2021). Strategic renewal of SMEs: the impact of social capital, strategic agility and absorptive capacity. *Management Decision*, *59*(8), 1877-1894.

Krakowski, S. (2025). Human-AI agency in the age of generative AI. *Information and Organization*, *35*(1), 100560.

Krakowski, S., Luger, J., & Raisch, S. (2023). Artificial intelligence and the changing sources of competitive advantage. *Strategic Management Journal*, *44*(6), 1425-1452.

Leemann, N., & Kanbach, D. K. (2022). Toward a taxonomy of dynamic capabilities–a systematic literature review. *Management Research Review*, *45*(4), 486-501.

Maharani, I. A. K., Sukoco, B. M., Usman, I., & Ahlstrom, D. (2024). Learning-driven strategic renewal: systematic literature review. *Management Research Review*, *47*(5), 708-743.

Mehta, P., Chakraborty, D., Rana, N. P., Mishra, A., Khorana, S., & Kooli, K. (2025). AI-driven competitive advantage: the role of personality traits and organizational culture in key account management. *Journal of Business & Industrial Marketing*, *40*(2), 543-569.

Mikalef, P., Conboy, K., & Krogstie, J. (2021). Artificial intelligence as an enabler of B2B marketing: A dynamic capabilities micro-foundations approach. *Industrial Marketing Management*, *98*, 80-92.

Neiroukh, S., Emeagwali, O. L., & Aljuhmani, H. Y. (2024). Artificial intelligence capability and organizational performance: unraveling the mediating mechanisms of decision-making processes. *Management Decision*, DOI: 10.1108/MD-10-2023-1946.

Pitelis, C. N., Teece, D. J., & Yang, H. (2024). Dynamic capabilities and MNE global strategy: A systematic literature review-based novel conceptual framework. *Journal of Management Studies*, *61*(7), 3295-3326.

Prakash, D. (2024). Data-driven management: The impact of big data analytics on organizational performance. *International Journal for Global Academic & Scientific Research*, *3*(2), 12-23.

Raisch, S., & Fomina, K. (2025). Combining human and artificial intelligence: Hybrid problem-solving in organizations. *Academy of Management Review*, *50*(2), 441-464.
Robertson, J., Ferreira, C., Botha, E., & Oosthuizen, K. (2024). Game changers: A generative AI prompt protocol to enhance human-AI knowledge co-construction. *Business Horizons*, *67*(5), 499-510.
Rosa, A., Bento, T., Pereira, L., Costa, R. L. D., Dias, Á., & Gonçalves, R. (2022). Gaining competitive advantage through artificial intelligence adoption. *International Journal of Electronic Business*, *17*(4), 386-406.
Sá, T., Ferreira, J. J., & Jayantilal, S. (2025). Open innovation strategy: a systematic literature review. *European Journal of Innovation Management*, *28*(2), 454-510.
Schmitt, A., Raisch, S., & Volberda, H. W. (2018). Strategic renewal: Past research, theoretical tensions and future challenges. *International Journal of Management Reviews*, *20*(1), 81-98.
Van Zeebroeck, N., Kretschmer, T., & Bughin, J. (2021). Digital "is" strategy: The role of digital technology adoption in strategy renewal. *IEEE Transactions on Engineering Management*, *70*(9), 3183-3197.
Wingate, D., Burns, B. L., & Barney, J. B. (2025). Why AI Will Not Provide Sustainable Competitive Advantage. *MIT Sloan Management Review*, *66*(4), 9-11.
Yuan, D., Kim, J. K., & Gao, C. (2025). Adoption of Artificial Intelligence and Its Impact on Competitive Advantage: Mediated by Knowledge Management. *Journal of Information & Knowledge Management*, *24*(02), 2550003.

CHAPTER 10

Entrepreneurial Ecosystems Through The Stem Lens

"The most powerful strategies are no longer built within firms they are co created across ecosystems In a STEM driven world sustainable advantage belongs to those who can model complexity orchestrate collaboration and evolve systems not just organizations"

by MAK

Learning Outcomes

By the end of this chapter, readers will be able to:

The most resilient strategies are not built in silos, but in systems—like coral reefs, where diversity, interdependence, and co-creation shape the future of innovation.

- Understand the Strategic Role of Entrepreneurship Ecosystems: Define entrepreneurship ecosystems (EEs) and explain their impact on innovation, competitiveness, and national development through a STEM-based strategic lens.
- Analyze Country-Level Ecosystem Data: Evaluate multi-level ecosystem indicators and interpret how national-level entrepreneurship, innovation capacity, and institutional factors shape strategic outcomes.
- Use Ecosystem Strategy Tools: Apply advanced tools such as ecosystem heatmaps, interdependency matrices, and system modeling techniques (e.g., agent-based modeling, system dynamics) to assess and simulate the performance of entrepreneurial ecosystems.

- ❖ Model Ecosystems as Dynamic Systems: Distinguish between static and dynamic strategy models and demonstrate how to use simulations and digital twins to visualize policy interventions, innovation diffusion, and resilience in entrepreneurial systems.
- ❖ Design Ecosystem-Based Strategy for Sustainable Growth: Translate ecosystem insights into strategic design by identifying roles (e.g., orchestrator, contributor), aligning ecosystem goals, and balancing co-creation, governance, and adaptability.
- ❖ Differentiate Systemic Advantage from Competitive Advantage: Shift strategic focus from firm-level competition to systemic innovation and advantage creation, especially in the context of global sustainability, technological disruption, and inclusive entrepreneurship.

In today's hyperconnected and rapidly evolving global landscape, firms no longer operate in isolation. Strategic advantage is increasingly shaped by the networks they inhabit, the institutions they align with, and the collaborative systems they co-create. This chapter marks a pivotal turn in our journey, moving beyond firms and industries to explore strategy at the ecosystem level.

Entrepreneurship ecosystems (EEs) are complex systems of interrelated actors, institutions, resources, and conditions that enable or constrain entrepreneurial activity. But these ecosystems are not static; they pulse with feedback loops, policy shifts, digital disruptions, and shifting talent flows. In a STEM-enabled strategic environment, the ability to map, model, and mobilize these ecosystems is a core strategic capability.

Whether designing a startup city, investing in inclusive innovation, or engineering a digital venture hub, ecosystem strategy requires new thinking. Traditional strategic models, built for firms and industries, cannot capture the multi-actor, multi-level, and dynamic nature of ecosystems. Instead, we turn to simulation engines, agent-based models, and systemic visualizations that allow us to test scenarios, evaluate resilience, and guide co-creation across borders and sectors.

This chapter introduces a new frontier of strategic thinking: ecosystems as living, evolving systems of innovation and impact. We begin by unpacking the foundational components of entrepreneurial ecosystems, then move into frameworks for ecosystem intelligence, tools for dynamic modeling, and principles for designing systemic advantage. Along the way, we integrate insights from data science, complexity theory, public policy, and digital governance, reflecting the interdisciplinary nature of ecosystem strategy.

Strategic ecosystems are not only about economic gain. They are about inclusion, sustainability, and transformation. In a STEM world, strategy must move at the speed of systems, and those who can architect and adapt ecosystems will shape the future.

Opening Case: Ping An and the Ecosystem Strategy Revolution

When most people think of Ping An, they picture a large Chinese insurance company. But over the past decade, Ping An has quietly reinvented itself into one of the world's most sophisticated ecosystem architects, operating across health, finance, real estate, smart cities, and AI.

What makes Ping An exceptional is not its scale, but its strategy: it doesn't just participate in ecosystems—it designs them. Through platforms like **Good Doctor** (healthcare services), **Lufax** (financial advisory), and **OneConnect** (fintech solutions), Ping An has built self-reinforcing networks of users, data, services, and partners. These platforms are not add-ons to its core business; they are the foundation of a new strategic logic: **systemic advantage**.

Ping An's strategy evolved from serving customers to connecting actors, patients, doctors, banks, startups, regulators, and developers in ways that generate mutual value. For example, through Good Doctor, it connects over 300 million users with AI-powered diagnostics, a telehealth network, and pharmaceutical logistics. Every interaction feeds back into the company's data infrastructure, enhancing algorithms across its ecosystem. Similarly, OneConnect provides modular fintech tools to more than 600 banks and 80 insurance firms, not just to compete, but to expand Ping An's ecosystem reach.

What's revolutionary is how Ping An blends **agent-based thinking**, **feedback modeling**, and **scenario design** into its strategic process. It uses system dynamics to forecast how policy shifts, technology changes, or new entrants might reshape the competitive terrain, not just for itself, but for the entire ecosystem.

Ping An's transformation illustrates the future of strategy: not choosing between market segments, but designing the markets themselves. In a STEM-enabled world, firms that think like systems architects, integrating AI, data science, and complexity models, will lead not by power, but by connection.

Play the Strategist

1. How does Ping An's ecosystem approach differ from traditional corporate diversification strategies?
2. In what ways does Ping An use data and AI to sustain and scale its ecosystem advantage?
3. Why is Ping An considered an "ecosystem architect" rather than just a participant?
4. How can other firms in non-tech sectors apply similar ecosystem design principles?
5. What role do feedback and scenario modeling play in Ping An's strategic decisions?

10.1 Mapping the Architecture of Entrepreneurial Ecosystems

In the age of data, AI, and constant disruption, strategy is no longer a battle between firms; it is a contest between ecosystems. Entrepreneurial ecosystems (EEs) are not simply clusters of startups or innovation hubs; they are dynamic, interdependent systems of actors, institutions, technologies, and flows that collectively enable, or hinder, entrepreneurial success. From

Silicon Valley to Singapore, from Nairobi to Berlin, ecosystems have emerged as the strategic substrate upon which innovation, resilience, and competitiveness are built.

Understanding and mapping these ecosystems is now a critical task for strategists. Unlike traditional value chains, entrepreneurial ecosystems are non-linear, adaptive, and often unpredictable. They evolve over time, shaped by feedback loops, institutional quality, cultural norms, policy interventions, and technological breakthroughs. To navigate them, leaders need more than static models; they need an architectural mindset and a systems approach.

This section introduces the foundational elements of entrepreneurial ecosystems and explores how firms, governments, and stakeholders can visualize and analyze their architecture. It blends conceptual clarity with data-driven strategy, helping you decode how ecosystems function and where intervention or innovation can generate systemic impact. Whether you're a startup founder seeking product-market fit, a policymaker shaping local innovation policy, or a strategist orchestrating global partnerships, this section lays the groundwork for seeing and shaping entrepreneurial ecosystems as strategic systems.

Let's begin by unpacking the essential components that define entrepreneurial ecosystems, and why those components matter far beyond the startup pitch.

10.1.1 Core Components of the Entrepreneurial Ecosystem

At the heart of any entrepreneurial ecosystem lies a network of interrelated actors, institutions, and enablers that collectively shape the birth, growth, and sustainability of new ventures. Unlike traditional business environments, EEs are complex, adaptive systems in which no single actor dominates, and where success emerges from interactions across multiple layers of activity and influence. Understanding the core components of these systems is the first step in building or navigating an ecosystem strategically.

The Global Entrepreneurship Monitor (GEM) and other frameworks have helped categorize and systematize these components. While regional variations exist, the following elements commonly form the backbone of a healthy entrepreneurial ecosystem:

Entrepreneurial Actors and Talent

These are the people who create and grow businesses: entrepreneurs, innovators, and skilled workers. Human capital, particularly talent with technical, managerial, and creative expertise, is essential. Ecosystems that fail to attract, retain, or nurture this talent often stagnate, regardless of infrastructure or funding.

Markets and Demand

No ecosystem can thrive without access to markets. Whether local, national, or global, demand for innovative goods and services incentivizes entrepreneurial behavior. In many cases, the presence of large anchor firms or institutions (e.g., hospitals, universities, governments) can create demand-side pull and opportunity spaces.

Support Institutions

These include incubators, accelerators, co-working spaces, technical service providers, and business associations. While they don't generate innovation directly, they amplify entrepreneurial capacity by lowering barriers, offering mentorship, and facilitating connections.

Access to Finance

Venture capital, angel networks, crowdfunding platforms, and public funding mechanisms provide the financial lifeblood of new ventures. But more important than sheer capital is capital quality, smart money that comes with strategic guidance, long-term vision, and risk tolerance.

Culture and Norms

A culture that values risk-taking, tolerates failure, and celebrates entrepreneurial success is vital. Narratives, role models, and societal attitudes deeply influence whether individuals see entrepreneurship as viable or desirable. Cultural acceptance of innovation can accelerate ecosystem maturation.

Policy and Governance

Government policies, from tax codes and bankruptcy laws to education and immigration, play a powerful shaping role. Effective ecosystems typically feature responsive governance, regulatory clarity, and a pro-entrepreneurial posture across multiple levels of administration.

Infrastructure and Digital Connectivity

This includes transportation, broadband access, utilities, and workspace availability. In today's hyper-connected world, digital infrastructure is not optional; it is foundational to knowledge exchange, remote collaboration, and cross-border scaling.

Research and Knowledge Institutions

Universities, research labs, and think tanks generate not only knowledge but also talent and technology transfer. When tightly integrated into the ecosystem, they serve as innovation engines and intellectual anchors.

These components do not operate in isolation. They co-evolve through feedback loops, shared narratives, and policy interventions. A strength in one domain can catalyze momentum in others; for example, a vibrant culture of entrepreneurship may attract both investors and high-skilled immigrants, creating positive system-wide spillovers. The table below summarizes these components and their strategic roles:

Table 10.1. Entrepreneurial Ecosystem Components and Strategic Roles

Ecosystem Component	Strategic Role	Example Insight
Entrepreneurial Talent	Drives innovation, absorbs risk	Israel's STEM-focused education pipeline
Access to Finance	Fuels venture creation and scaling	Singapore's state-supported venture funds
Policy and Governance	Shapes incentives, reduces uncertainty	Estonia's e-residency and digital business laws
Culture and Norms	Encourages risk-taking and aspiration	Silicon Valley's "fail fast" mentality
Markets and Demand	Provides revenue and validation	Local procurement initiatives in Kenya
Support Institutions	Provides scaffolding and acceleration	MassChallenge's global acceleration model
Infrastructure	Enables collaboration and scale	Broadband penetration in South Korea
Knowledge Institutions	Seeds innovation, fuels talent supply	MIT's ecosystem role in Boston's tech cluster

Understanding the architectural logic of these components is crucial. But ecosystems are more than their parts; they are systems in motion. In the next section, we explore how these components interact and co-evolve within a systemic framework.

10.1.2 Mapping the Interdependent Ecosystem

In traditional strategic analysis, firms have often relied on simple models to map their environment, isolating factors like competition, customer behavior, and regulatory frameworks. But entrepreneurial ecosystems defy this reductionist logic. They are complex, dynamic, and deeply interconnected systems in which a shift in one node, such as policy or education, can ripple across the entire landscape. Mapping these ecosystems requires more than identifying components; it demands understanding their interdependencies, feedback loops, and emergent patterns.

A strategic ecosystem map moves beyond listing players. It illustrates how different elements—government policy, cultural norms, talent flows, investment availability, educational institutions, and infrastructure—mutually reinforce or constrain each other. For example, a well-designed policy that encourages startup investment can increase capital flows, which in turn stimulates more entrepreneurial activity, which may then prompt universities to create more tailored innovation programs. These feedback loops are crucial. Ecosystems evolve through interactions, not isolated interventions.

To make this complexity tractable for strategic modeling, we translate these interdependencies into systemic interaction frameworks. Below is a synthesized table that captures how key components of an entrepreneurial ecosystem influence and are influenced by each other. The final column highlights the type of systemic relationship, such as bidirectional loops, emergent behavior, or enabling structures. The table below captures how key components of an entrepreneurial ecosystem influence and are influenced by one another, along with the nature of their systemic relationships:

Table 10.2. Systemic Interactions Among Key Ecosystem Components

Ecosystem Component	Influences	Is Influenced By	Type of Interaction
Access to Finance	Startup growth, investor confidence, innovation pipeline	Support institutions; entrepreneurial readiness; regulatory policy	Bidirectional
Support Institutions	Investment readiness; business survival; ecosystem learning	Public funding; university–industry collaboration; entrepreneurial culture	Reinforcing loop
Culture & Societal Norms	Entrepreneurial intent, risk appetite, and inclusivity	Media narratives, visible success/failure stories, and education systems	Emergent pattern
Education & Research	Talent pipeline; innovation outputs; startup formation	Government policy, private R&D investment, and labor market needs	Recursive influence
Entrepreneurial Talent	Venture creation; ecosystem renewal; innovation cycles	Education system; mentorship access; social capital	Central engine
Policy & Governance	Ease of doing business; investor security; IP protection	Advocacy coalitions; global benchmarks; system performance data	Feedback and correction loop
Infrastructure & Platforms	Collaboration efficiency, speed of innovation, scaling potential	Urban development; digitalization levels; public–private initiatives	Enabling structure
Market Access	Revenue potential; internationalization; competition	Trade policy; startup exports; digital platforms	Amplifier of scaling
Networks & Social Capital	Knowledge spillovers; trust formation; cross-sector coordination	Networking events, shared spaces, and alumni communities	Connective tissue

This table is not meant to be exhaustive but rather illustrative. It shows that no ecosystem element functions in isolation. Strategy development in such systems must consider both direct and second-order effects. For instance, improving access to finance may not yield desired results if entrepreneurial talent is scarce or if culture discourages risk-taking. Similarly, a policy reform may fail to spark innovation if educational institutions are not aligned with startup needs.

By visualizing these interdependencies through a structured interaction map, or even computationally simulating them using agent-based or system dynamics tools, ecosystem leaders can identify leverage points, anticipate unintended consequences, and design smarter interventions. For strategy professionals, this shift from linear models to relational intelligence is not optional; it is essential.

10.1.3 Institutional Quality and Ecosystem Leadership

Entrepreneurial ecosystems thrive not merely on the presence of startups, accelerators, or venture capital, but on the **quality of institutions** that enable trust, reduce friction, and ensure the rules of the game are fair, predictable, and supportive of innovation. Institutional quality is the invisible infrastructure of an ecosystem, shaping everything from investor confidence to the ability of a startup to scale, fail, pivot, or exit. Without strong institutions, even the most well-funded ecosystems can unravel into fragmentation, corruption, or inertia.

High-quality institutions do more than regulate; they catalyze. They create a fertile environment where new ventures can take root and grow, where experimentation is encouraged but bounded by norms, and where knowledge is protected yet shared. These institutions include legal systems, intellectual property regimes, transparent bureaucracies, educational and research institutions, financial oversight bodies, and digital governance frameworks. When these elements work in harmony, they reduce uncertainty and transaction costs, making entrepreneurship less risky and more rewarding.

But institutional quality does not arise spontaneously. It requires leadership. **Ecosystem leadership** refers to the ability to shape not just firm-level strategies but the system-level dynamics that define entrepreneurial success. Effective leaders in ecosystems are not only founders or investors—they can be university presidents, policymakers, civic activists, corporate executives, or platform operators. What they share is a system-oriented mindset: they build bridges between actors, fill institutional voids, and design governance mechanisms that create alignment across diverse goals.

Consider the example of Estonia. Its transformation into one of the world's leading digital societies didn't stem from startup density alone. It was a product of deliberate institutional design—embedding digital identity, e-governance, and open-data policies into the core of how the country functions. This, in turn, enabled an ecosystem where trust, transparency, and agility fuel entrepreneurial activity across sectors.

In another example, Singapore's Economic Development Board has long acted as a strategic orchestrator, not just offering incentives but co-shaping industrial policy, research priorities, and innovation hubs that align government, academia, and enterprise into a shared ecosystem vision.

Ecosystem leadership is also about **adaptive governance**. In fast-moving contexts, the ability to revise rules, pilot reforms, and respond to emergent challenges is essential. Static regulation can choke innovation; dynamic governance enables experimentation while still maintaining public interest. Successful ecosystems often institutionalize this adaptiveness—through sandbox policies, feedback loops, and collaborative councils that guide strategic decisions.

Ultimately, institutional quality and ecosystem leadership are not luxuries; they are **strategic levers**. They determine whether an ecosystem becomes fertile ground for innovation or a wasteland of unrealized potential. Strategy, in this sense, is not about competing in markets alone. It is about shaping the very system in which competition and innovation unfold.

10.2 Measuring Ecosystem Vitality and Strategic Fit

Entrepreneurial ecosystems are living systems, not static environments, and like any living system, their health must be measured and monitored. However, measuring ecosystem vitality is far more complex than counting startups or investment rounds. It requires multidimensional indicators that capture the **depth, diversity, dynamism, and directionality** of entrepreneurial activity. These indicators help strategists, policymakers, and investors assess not only how an ecosystem is performing but also whether it aligns with broader strategic goals, its **strategic fit** within a firm, region, or national innovation trajectory.

10.2.1 Beyond Vanity Metrics: What Really Matters

Many ecosystem assessments rely on easily accessible, but often misleading, metrics: number of new ventures, amount of venture capital raised, or accelerator counts. While these offer a surface view, they miss the underlying health of the system. A highly active ecosystem could be structurally fragile if it lacks quality mentorship, inclusive participation, or institutional resilience.

Instead, ecosystem vitality should be measured through **core dimensions** that reflect system-wide performance:

- **Entrepreneurial Diversity**: Are ventures emerging across multiple sectors and demographics, or is the ecosystem narrowly concentrated?
- **Knowledge Flow**: How fluid is the movement of ideas, talent, and capital across actors?
- **Institutional Strength**: Are governance structures adaptive, fair, and trusted?
- **Ecosystem Density and Connectivity**: What's the ratio of support institutions to startups? How often do actors collaborate across silos?
- **Resilience and Renewal**: Can the ecosystem absorb shocks (e.g., COVID-19, economic crises) and regenerate entrepreneurial activity?

Understanding the health of an entrepreneurial ecosystem requires more than anecdotal success stories or funding totals. It demands **multi-dimensional, data-informed measurement**. To capture this complexity, organizations increasingly rely on **mixed methods,** including surveys, network analyses, qualitative ecosystem mapping, and integration of third-party indices such as the **Global Entrepreneurship Monitor**, **World Bank Ease of Doing Business**, and **SDG-related metrics**.

These deeper metrics offer a **more predictive view of long-term ecosystem vitality**, helping stakeholders identify not just where an ecosystem stands today, but how resilient, inclusive, and scalable it is over time.

For example, **startup density** can signal either vibrant activity or resource saturation, depending on context. **Time to product-market fit** reflects how efficiently ventures are supported, while **diversity indices** reveal the breadth of innovation and inclusivity. Metrics like **exit velocity** and **policy adaptiveness** indicate maturity and responsiveness, and **network centrality** helps identify key orchestrators or potential bottlenecks.

The table below summarizes key indicators and their strategic implications:

Table 10.3: Indicators of Ecosystem Vitality and Their Strategic Implications

Indicator	What It Reveals	Strategic Implication
Startup Density	Activity levels and resource competition	Signals potential for collaboration or oversaturation
Time to Product-Market Fit	Ecosystem efficiency in venture development	High fit speed reflects an effective support infrastructure
Diversity Index (Sector/Gender)	Inclusivity and innovation heterogeneity	Diversity strengthens resilience and opportunity space
Mentorship Density	Quality of informal learning and human capital flow	Enhances adaptive capacity and innovation cycles
Exit Velocity (M&A, IPOs)	Maturity and capital recycling in the ecosystem	Reflects scale-readiness and investor confidence
Policy Adaptiveness	Speed and quality of governance responses	Aligns with dynamic strategy execution and systemic learning
Network Centrality	Influence and connectivity of key ecosystem players	Identifies potential orchestrators or bottlenecks

10.2.2 Strategic Fit: Aligning Ecosystem Context with Organizational Goals

Even the most vibrant ecosystem may not be the right one for every organization. Firms must evaluate **strategic fit,** the degree to which an ecosystem's capabilities, norms, and trajectories align with a company's innovation goals, market entry strategy, and cultural DNA. This is not about choosing the "best" ecosystem globally, but the **most synergistic** one for a specific purpose.

For example, a startup in quantum computing may find stronger alignment in Toronto or Delft, where research institutions, policy incentives, and talent pools converge. A fintech firm might favor Nairobi or São Paulo, where digital inclusion and financial infrastructure are advancing rapidly. Strategic fit means asking: *Does this ecosystem accelerate our goals, or distract from them?*

Key dimensions for assessing fit include:

* **Domain specialization** (e.g., biotech clusters in Boston, ag-tech in Tel Aviv)
* **Cultural compatibility** (entrepreneurial risk norms, governance style)
* **Regulatory alignment** (data policies, tax structures, IP frameworks)
* **Co-creation potential** (availability of partners, testbeds, or sandboxes)

Organizations can model this fit using **ecosystem scoring matrices** or even **AI-based location intelligence tools** that weigh these criteria based on strategic priorities.

Ultimately, vitality tells us how alive an ecosystem is. Strategic fit tells us if we belong there. Both are essential for designing ecosystem engagement strategies that are not only visionary but also viable.

10.3 Strategic Participation in Entrepreneurial Ecosystems

In the past, firms viewed ecosystems as external contexts, geographies to enter, and environments to navigate. Today, however, they must think more actively: not just *existing* within ecosystems, but *participating* in them with strategic intentionality. Strategic participation involves choosing the **right role**, building meaningful **relationships**, and leveraging the ecosystem's dynamics to drive long-term advantage—not just short-term gain.

As ecosystems become more complex and decentralized, success depends on how well an organization can engage across multiple levels: contributing to the health of the system while capturing value from it. Participation is no longer passive—it is **strategically designed**.

10.3.1 From Presence to Participation: Strategic Roles

Firms can adopt multiple roles within an entrepreneurial ecosystem. These roles are not mutually exclusive, and they often evolve:

* **Enabler**: Providing infrastructure, tools, or services that others build upon (e.g., AWS in global startup ecosystems).
* **Orchestrator**: Coordinating activities and setting standards or incentives across actors (e.g., Ping An's OneConnect fintech platform).
* **Investor**: Supplying financial capital and strategic guidance (e.g., corporate venture capital arms like Intel Capital or GV).
* **Testbed Partner**: Opening up real-world environments for pilots, experimentation, and data collection (e.g., telcos partnering with IoT startups).
* **Policy Shaper**: Engaging with regulators to influence innovation policy and governance frameworks.

Each role comes with trade-offs. Enablers must stay modular and neutral; orchestrators risk overreach or backlash; investors require long-term patience and alignment. The strategic question is not just *where to be,* but *who to be* in the ecosystem.

10.3.2 Building Ecosystem Capital: Relationships, Trust, and Reputation

Ecosystem participation depends heavily on **relational capital,** the quality of trust, shared vision, and informal networks a firm cultivates. Unlike traditional supply chains, ecosystems are fluid and emergent. There are no guaranteed outcomes, only **co-created trajectories**.

To build ecosystem capital, organizations must:

- **Show up consistently** at ecosystem events, forums, and collaborative initiatives.
- **Co-develop pilots** or proofs-of-concept with startups, universities, or NGOs.
- **Contribute to open standards**, policy dialogues, or shared datasets that benefit the entire system.
- **Develop "listening infrastructure"**—dedicated roles, teams, or platforms that scan weak signals and emerging ideas from across the ecosystem.

Reputation in an ecosystem spreads quickly. Firms seen as extractive or transactional may be shut out of valuable informal loops. Conversely, firms that offer mentorship, platforms, or resources without immediate extraction often become trusted anchors—gaining early access to partnerships, talent, and intelligence.

10.3.3 Timing and Strategic Entry: First Movers, Fast Followers, and Catalysts

Not all ecosystem entry strategies are the same. Strategic timing matters. Some firms enter early, shaping the ecosystem's architecture. Others wait until models stabilize, then scale rapidly. Both approaches, **first-mover** and **fast-follower,** can succeed, but the strategic intent must match execution capability.

A third role is the **catalyst**: a firm that intervenes not to dominate but to accelerate the evolution of the ecosystem itself. This is often done through funding grand challenges, launching innovation prizes, or creating ecosystem-wide APIs or platforms. Catalysts drive value creation across the system, even if they don't extract all of it.

The table below outlines the strategic trade-offs and ideal contexts for each entry strategy:

Table 10.4: Strategic Timing Approaches in Ecosystem Entry

Entry Strategy	Advantages	Risks	Best Fit Context
First Mover	Shape standards, build early network power	Uncertain returns, immature infrastructure	Nascent ecosystems with high disruption upside
Fast Follower	Lower risk, faster scale-up	Competitive crowding, fewer design options	Mature ecosystems with validated demand
Catalyst	High ecosystem influence, long-term trust	Indirect monetization, diffuse benefits	Fragmented ecosystems need coordination

Strategic participation is ultimately about reciprocity: those who **contribute meaningfully** to the ecosystem gain not just access, but advantage. In this era of decentralized innovation, the firm that knows *how to show up* and *how to lead without controlling* will shape the ecosystem from within.

10.4 Orchestrating Ecosystem Innovation and Collaboration

Innovation no longer happens in isolated R&D labs or corporate think tanks. Instead, it increasingly unfolds across a distributed landscape of universities, startups, accelerators, NGOs, public agencies, and citizen communities. In this environment, success requires a new strategic capability: **orchestration**. Orchestrating ecosystem innovation means coordinating diverse actors, aligning incentives, and enabling shared value creation, without necessarily owning all the assets or making every decision.

Rather than being the "central planner," the orchestrator is the **conductor of a symphony,** making sure each part plays in harmony while encouraging improvisation. This role is central in modern entrepreneurial ecosystems where innovation emerges from **co-creation**, not control.

10.4.1 Platforms, APIs, and Innovation Marketplaces

Technology platforms are one of the most effective orchestration tools. By offering shared infrastructure, standards, or interfaces, platforms enable innovation at scale, often by others. Think of Amazon Web Services (AWS), Google Play, or SAP's business technology platform. These platforms do not dictate how others innovate; they create the **conditions** for scalable, interoperable innovation.

One key orchestration tool is the **Application Programming Interface (API)**. APIs are not just technical tools; they are strategic assets. They allow firms to expose selected capabilities to third parties in a way that enables modular innovation. For instance, Ping An's healthcare division offers APIs that allow startups and partners to build new apps using its diagnostic algorithms, insurance data, or hospital networks.

Innovation marketplaces take this logic further. These are curated spaces, often digital, where ecosystem actors can pitch problems, propose solutions, match needs with capabilities, and co-develop offerings. Examples include open innovation platforms like InnoCentive or government-backed tech challenge portals. These marketplaces don't just exchange goods—they exchange **problems and ideas**.

10.4.2 Managing Intellectual Property in Open Systems

One of the biggest concerns in ecosystem collaboration is **intellectual property (IP)**. If innovation is co-created, who owns it? If data is shared, how is it protected? These are not trivial questions, and they require new models of governance.

In open ecosystems, IP strategies must balance protection and permeability. Too much control chokes collaboration; too little undermines incentives to invest. Leading firms use a range of mechanisms:

* **Tiered IP Access**: Different ecosystem participants receive different rights based on their role and contribution.
* **Data Trusts**: Shared repositories governed by neutral rules, where multiple players can access anonymized datasets under fair use agreements.
* **Joint IP Pools**: Co-created IP is shared within a defined group, often with usage fees or strategic licensing options.

Ping An, for example, engages in joint patent filings with ecosystem startups and academic partners. This not only protects innovation but fosters a sense of shared ownership—turning legal tools into **relationship enablers**.

10.4.3 Balancing Openness and Control: The Governance Challenge

At the heart of orchestration is a paradox: the most innovative ecosystems are open and emergent, but they also need **some degree of coordination**. Governance, therefore, must be **lightweight but effective**—guiding behavior without stifling it.

Key governance principles include:

* **Transparency**: Clear rules for participation, access, data sharing, and conflict resolution.

* **Inclusivity**: Broad participation across sectors, disciplines, and geographies—avoiding elite capture or gatekeeping.
* **Adaptive Rules**: Governance structures that evolve with ecosystem needs rather than being fixed at inception.
* **Trust Anchors**: Respected institutions (e.g., a university, public agency, or foundation) that act as honest brokers.

Orchestration also involves **managing power dynamics**. Large firms must avoid crowding out smaller players or monopolizing value capture. Instead, they must signal a commitment to mutual benefit through fair partnership models, shared infrastructure, or open calls for collaboration.

To operationalize these principles, ecosystem leaders deploy a range of tools that shape participation, enable collaboration, and reduce friction. The table below outlines key orchestration tools and their strategic effects:

Table 10.5: Orchestration Tools and Their Strategic Effects

Tool	Purpose	Strategic Effect
APIs	Enable modular innovation by others	Scales ecosystem innovation without direct control
Innovation Marketplaces	Match problems and capabilities	Accelerates problem-solving and co-creation
Tiered IP Agreements	Balance access and protection	Encourages strategic contribution from partners
Data Trusts	Enable secure data collaboration	Builds trust and compliance in data-driven systems
Governance Frameworks	Guide behavior and roles	Aligns incentives and reduces ecosystem friction

In the orchestration era, strategy is not about defending boundaries; it's about **designing flows**. The most successful organizations will be those that can attract the right partners, enable the right conditions, and shape innovation without owning every piece of it.

10.5 Strategic Modeling of Ecosystems as Dynamic Systems

Ecosystems are not static landscapes; they are living, breathing systems composed of interdependent actors, fluid relationships, nonlinear change, and emergent behavior. Traditional strategy tools, SWOT, Five Forces, and static industry maps, are ill-equipped for this level of complexity. Instead, ecosystems demand **dynamic strategic modeling** tools that can simulate feedback loops, policy shocks, innovation cascades, and co-evolution over time.

Strategic modeling of ecosystems allows decision-makers to move beyond descriptive mapping into **predictive insight and foresight**. It equips them with simulation-based tools to test hypotheses, stress-test interventions, and design resilient strategies under uncertainty.

10.5.1 Why Static Analysis Fails in Ecosystems

Conventional static analysis assumes linear causality and stable structures. But in entrepreneurial ecosystems, **cause and effect are nonlinear**, interactions are recursive, and outcomes often emerge unexpectedly. For example, a policy change intended to boost startup formation might backfire by flooding the market with poorly prepared ventures, straining incubators and investor confidence.

Moreover, static models cannot capture **timing effects** (e.g., delayed consequences), **path dependencies** (e.g., lock-in to early choices), or **adaptive feedback** (e.g., how ecosystem actors learn and react). Without dynamic models, strategy becomes reactive rather than anticipatory.

Recognizing this, advanced firms and policymakers increasingly adopt **systemic modeling tools,** not as substitutes for judgment, but as lenses that expand strategic imagination.

10.5.2 Agent-Based Modeling and System Dynamics

Two powerful modeling approaches, **Agent-Based Modeling (ABM)** and **System Dynamics (SD),** offer complementary ways to understand ecosystem behavior.

Agent-Based Modeling simulates individual actors (agents) with distinct behaviors, goals, and interaction rules. These agents can represent startups, investors, regulators, universities, etc. By modeling how agents interact over time, ABM reveals **emergent patterns,** such as clustering, tipping points, or runaway success-failure spirals.

For example, an ABM could simulate how changes in mentorship availability affect startup survival rates in under-resourced ecosystems. Or how social influence among entrepreneurs accelerates or suppresses risk-taking behaviors.

System Dynamics, on the other hand, uses **causal loop diagrams** and **stock-and-flow models** to represent feedback structures. It is ideal for capturing delays, reinforcements, and accumulations across the ecosystem. For instance, an SD model might show how research funding flows influence innovation outputs, which in turn shape investor behavior, feeding back into university partnerships.

These models can be calibrated with real-world data or run as **policy laboratories** to explore "what-if" scenarios, before decisions are made in the real world.

10.5.3 Scenario Engines and Digital Twins for Ecosystems

A third frontier in ecosystem modeling is the use of **Scenario Simulation Engines** and **Digital Twins**.

Scenario Engines combine qualitative foresight tools with computational models to explore plausible ecosystem futures. These platforms allow strategists to rank key drivers (e.g., regulation, AI adoption, climate shocks), define uncertainty ranges, and simulate multiple trajectories. Scenario planning becomes not a one-time workshop, but an ongoing strategic practice.

Digital Twins go even further. Originally used in engineering, a digital twin is a real-time virtual replica of a physical system. Applied to ecosystems, digital twins simulate the interactions of startups, funding flows, IP creation, talent migration, and policy shifts—using real-time data streams. This enables continuous monitoring, prediction, and experimentation.

Urban innovation hubs like **MIT's City Science Labs** and smart city platforms like **Barcelona's Urban Lab** use digital twins to simulate entrepreneurship outcomes, infrastructure needs, and innovation policy effects. Corporate venture studios are beginning to adopt similar tools to simulate **portfolio dynamics and ecosystem health**.

The table below summarizes the core simulation tools used in ecosystem strategy and their respective inputs, outputs, and use cases:

Table 10.6. Simulation Tools for Ecosystem Strategy

Tool	Inputs	Outputs	Use Case
Agent-Based Modeling	Agent rules, network data	Emergent behavior	Startup survival, behavior contagion
System Dynamics	Feedback loops, causal diagrams	System trajectories	Innovation funding, talent bottlenecks
Scenario Engine	Driver rankings, uncertainty axes	Strategic narratives, scenarios	Ecosystem future exploration
Digital Twin	Real-time data, API integrations	Predictive insights, alerts	Policy testing, infrastructure readiness

Strategic modeling transforms how we design, govern, and adapt ecosystems. It moves us from *describing* complexity to *navigating* it. In a world where tomorrow's advantage depends on today's foresight, firms that embrace dynamic modeling will gain an invaluable edge—not just in reacting to the future, but in **rehearsing and shaping it**.

10.6 Designing Ecosystem-Based Strategy for Sustainable Growth

As the competitive landscape shifts from firm-to-firm rivalry to network-versus-network dynamics, strategy must evolve. Leading organizations are no longer just participants in ecosystems—they are becoming **ecosystem architects**. Designing strategy within an ecosystem context involves balancing influence, inclusion, coordination, and adaptability. The goal is not just competitive advantage, but **systemic advantage**: the ability to shape, stabilize, and scale the very environment in which a firm operates.

This section introduces a new strategic paradigm—**ecosystem-based strategy**—that emphasizes shared growth, resilience, and innovation through purposeful ecosystem design. It builds on foundational principles of systems thinking, but brings them into the realm of actionable strategy.

10.6.1 Firms as Ecosystem Architects

Ecosystem participation is no longer sufficient. Strategic leaders must learn to **design and shape the ecosystems** they rely on. This means going beyond exploiting existing platforms to **curating infrastructure, norms, governance, and partnerships** that attract and enable others.

Firms like Ping An, Amazon, and Alibaba exemplify this approach. They do not merely enter markets—they construct interconnected systems of fintechs, logistics players, insurers, health techs, and cloud providers. These ecosystems extend their reach, embed themselves in critical value flows, and generate self-reinforcing advantages.

Becoming an ecosystem architect involves five key capabilities:

1. **Framing the Opportunity Space**: Defining the core problem or opportunity that the ecosystem exists to address.
2. **Attracting Complementors**: Designing incentives, APIs, interfaces, and shared goals to bring others into the fold.
3. **Structuring Value Exchange**: Clarifying who captures what value and under what conditions.
4. **Setting Governance Norms**: Establishing participation rules, dispute resolution, and strategic direction.
5. **Monitoring and Evolving the System**: Continuously adapting to feedback, exit signals, or unintended consequences.

This approach is not about dominance; it's about orchestration at scale.

10.6.2 From Competitive Advantage to Systemic Advantage

Traditional competitive advantage rests on controlling unique resources, IP, talent, and customer data. Systemic advantage arises from a **position within a network of co-creation**. It depends on relational assets, reputation, interoperability, and the ability to shape shared outcomes.

Systemic advantage means:

* **Being a connector**, not just a competitor.
* **Investing in shared infrastructure** that all players use, not just proprietary tools.
* **Shaping standards and norms**, which create lock-in and path dependence.
* **Absorbing and diffusing innovation**, rather than hoarding it.

Ping An's smart city initiative is a case in point. By integrating data from transportation, health, finance, and security into a unified platform, the company isn't just selling services—it's architecting urban ecosystems. Its advantage comes from being the backbone of local value creation, not from direct market share alone.

Systemic advantage is harder to imitate because it emerges from **long-term relationship building, trust, and institutional alignment**. It is also more resilient—because it adapts with the system, rather than being exposed to its volatility.

10.6.3 Co-Creation, Governance, and Resilience

The most effective ecosystem strategies are co-created—not imposed. They require distributed governance, mutual accountability, and mechanisms for dispute resolution and trust formation.

Co-creation means involving diverse stakeholders—from entrepreneurs to NGOs- in shaping solutions. It's a design process, not just a feedback loop. Tools like **design thinking**, **collaborative foresight**, and **participatory platform design** enable this engagement.

Governance must be light-touch but legitimate. Some options include:

- **Multi-stakeholder councils** that rotate leadership
- **Decentralized decision rights**, enabled by digital voting or smart contracts
- **Open audit trails** for decisions, metrics, and impact

Resilience becomes a core outcome of good design. Ecosystems that are too centralized, opaque, or exclusive tend to collapse under stress. Resilient ecosystems, by contrast, exhibit:

- **Redundancy** (multiple pathways to value)
- **Diversity** (of actors, perspectives, and capabilities)
- **Feedback sensitivity** (early warning systems)
- **Distributed adaptation** (local experimentation)

These features aren't accidental—they are designed. To help organizations embed these principles into their strategy, we introduce the **Strategic Ecosystem Design (SED) Model**. This model outlines three foundational dimensions, **Co-Creation**, **Coordination**, and **Adaptability,** each with a distinct strategic role and set of enabling actions, as shown in the table below:

Table 10.7: Strategic Ecosystem Design (SED) Model

Dimension	Strategic Role	Key Actions
Co-Creation	Shared problem framing and solution generation	Stakeholder mapping, design sprints, innovation labs
Coordination	Alignment of flows, interfaces, and contributions	API standards, incentive structures, shared metrics
Adaptability	Systemic learning and response to shocks	Feedback loops, modularity, scenario rehearsals

This model helps firms evaluate their current posture and **intentionally design ecosystems** that can evolve, scale, and sustain collective advantage.

In conclusion, ecosystem-based strategy marks a shift from power *over* others to power *with* others. It requires a new mindset—one that embraces uncertainty, designs for emergence, and views value creation as a **team sport across organizational boundaries**.

10.7. Chapter 10 Wrap-Up: Ecosystem Thinking as Strategic Advantage

As we close this chapter, one truth becomes clear: strategy can no longer be confined to firms, markets, or even industries—it must extend to **ecosystems**. In today's world, innovation, resilience, and value creation increasingly emerge from dynamic constellations of actors, not individual organizations. Strategic success now depends on understanding and shaping the **system-level forces** that define entrepreneurial ecosystems.

Throughout this chapter, we explored how STEM-powered strategy reframes the way we view ecosystems, not as background context but as **living systems**. We challenged the linear, siloed thinking that dominates many traditional models and replaced it with a **systemic view**: one that accounts for interdependencies, feedback loops, emergence, and evolution.

We examined how **institutional quality, coordination infrastructure, and leadership** function as enablers, or bottlenecks, for ecosystem vitality. We emphasized the need to design **adaptive strategy metrics,** tools that reveal not just firm-level performance but ecosystem-wide health, velocity, and resilience.

From systems modeling to digital twins, from co-creation networks to scenario engines, the chapter introduced an advanced toolkit for ecosystem strategists. We saw that the future of competitive advantage lies not in isolated positioning but in mastering **ecosystem orchestration**: the ability to coordinate complexity, catalyze partnerships, and model long-term system behavior.

We also studied why **static analysis fails** in dynamic ecosystems and how agent-based models, feedback diagrams, and simulation engines help us visualize strategy as a process of continual co-evolution. Competitive advantage, in this new paradigm, is no longer static, durable, or purely internal; it is **systemic**, **interactive**, and **perishable**.

As you progress in your strategic journey, remember: the most powerful strategies will no longer ask, "How can my firm win?" but rather, "How can we design and sustain a system in which everyone thrives, and innovation becomes continuous?"

10.7.1. Discussion Questions

1. How does adopting a systems-thinking lens transform traditional strategic analysis within entrepreneurial ecosystems?
2. In what ways can governments and institutions shape the trajectory of an entrepreneurial ecosystem—and how should firms strategically respond to these forces?
3. Compare the difference between ecosystem participation and ecosystem orchestration. What strategic capabilities are required for firms to act as ecosystem architects?
4. How do interdependencies and feedback loops within an ecosystem create both opportunities and risks for emerging ventures? Can you think of a real-world example?
5. Why does static analysis fail to capture the dynamics of modern entrepreneurial ecosystems? What modeling tools can help overcome this limitation?
6. Discuss the role of data density, connectivity, and trust in shaping the health and resilience of an innovation ecosystem. How can leaders strengthen these dimensions?
7. How does the concept of 'systemic advantage' challenge the traditional view of 'competitive advantage'? What does this shift imply for long-term strategy?
8. How might agent-based modeling or digital twins enhance a policymaker's or entrepreneur's ability to forecast and guide ecosystem evolution?
9. In complex ecosystems, what governance structures or co-creation mechanisms are most effective in balancing control with innovation?
10. Zara (the strategist, not the brand) is dropped into a fragmented, underperforming startup ecosystem in a midsize city. What ecosystem elements should she prioritize rebuilding first, and why?

10.7.2. Micro Case: Zara as an Ecosystem Architect

Zara has long been admired for its vertically integrated supply chain, agile product cycles, and fashion-forward market sensing. But behind this operational excellence lies something deeper: **ecosystem design thinking**. Zara operates as more than a retailer—it's a **strategic orchestrator** of a high-velocity ecosystem.

By co-locating suppliers, digitizing feedback loops, and embedding real-time analytics across its network, Zara has created a system that learns, adapts, and scales. Its strategic partnerships—whether with logistics providers, local manufacturers, or in-house AI teams—are not transactional; they are co-evolutionary.

When global shocks such as pandemics or shipping crises disrupt markets, Zara's resilience stems from its **ecosystem density, connectivity, and modularity**. New product designs are tested quickly through digital channels, and supply responses are recalibrated within days. This is not accidental—it is the result of strategic modeling, feedback-informed governance, and ecosystem foresight.

Zara's ability to act as a **systemic strategist**—not just a fast fashion brand—offers a powerful lesson: in turbulent environments, resilience and innovation come from **systems that are intentionally designed, strategically led, and continuously adapted**.

Discussion Questions

1. How does Zara's strategy reflect ecosystem coordination and orchestration rather than traditional supply chain management?
2. What systemic features of Zara's ecosystem enable it to adapt in real-time?
3. How might digital twins or agent-based models improve Zara's ecosystem forecasting?
4. What would ecosystem decay look like for Zara—and how could it be prevented?
5. What indicators should Zara use to monitor the health of its strategic ecosystem?

10.7.3. Interactive Exercise: Designing a Resilient Ecosystem Strategy

Scenario:
You are part of a policy-innovation team tasked with designing a startup ecosystem in a secondary city aiming to become a sustainability hub. Your team includes representatives from local government, a technical university, two anchor corporations, and a climate-tech startup.

Tools Available:

- Agent-based simulation engine
- Ecosystem density and connectivity dashboards
- Institutional trust index

Your Task:
Using the frameworks from Chapter 9:

- Propose a systemic strategy to seed and scale the ecosystem.
- Identify at least three feedback loops that need to be reinforced.
- Suggest a hybrid governance model for coordination.
- Design a resilience dashboard with five real-time indicators.

10.7.4. Recommended HBR Case Studies

❖ The Journey from Corporate Professional to InsurTech Startup Founder: Vivien Le, Woman Entrepreneur in Vietnam

- **Overview**: Published on December 13, 2024, and authored by Cheng Boon Koh and Joo Seng Tan, this 13-page case follows Vivien Le, co-founder and CEO of Saladin, Vietnam's first tech-enabled insurance brokerage. The case traces her transformation from a corporate professional to a pioneering entrepreneur in Vietnam's emerging InsurTech sector. Saladin achieved rapid growth, serving over 300,000 customers with a 95% claim success rate, by leveraging technology to address inefficiencies in Vietnam's insurance market. The case explores Vivien's adaptive leadership, her navigation of regulatory and digital infrastructure challenges, and her role as a female founder in a traditionally male-dominated industry.
    - Students are invited to analyze:
        - Her strategic decision-making in scaling the business
        - The cultural and gender dynamics of entrepreneurship in Vietnam
        - The trade-offs between regional expansion, product innovation, and consolidating core operations

  This case is ideal for discussions on entrepreneurship, leadership, gender in business, and emerging market innovation.
- **Citation**: Koh, C. B., & Tan, J. S. (2024). *The journey from corporate professional to InsurTech startup founder: Vivien Le, woman entrepreneur in Vietnam* (Case No. NTU376-PDF-ENG). ABCC at Nanyang Technological University.

❖ CARBON MASTERS INDIA LIMITED

- **Overview**: Published by the Indian Institute of Management Bangalore on November 1, 2024, this 13-page case study, authored by Sabarinathan G. and Ramakrishna S. Velamuri, explores the strategic and operational challenges faced by Carbon Masters India Limited, a startup focused on sustainable waste management and clean energy. The company addresses two major environmental issues in India: the disposal of municipal solid waste and the replacement of fossil fuels and chemical fertilizers with biogas and organic fertilizer. Its flagship product, Carbonlites®, is a bottled biogas brand derived from organic waste. As the company seeks to scale, it faces a growth dilemma—balancing social impact with financial sustainability. The case highlights the difficulties of working with public institutions, the fragmented regulatory environment across Indian states, and the limited interest from investors in early-stage, impact-driven ventures.
- **Citation**: Sabarinathan, G., & Velamuri, R. S. (2024). *CARBON MASTERS INDIA LIMITED* (Case No. IM035B-PDF-ENG). Indian Institute of Management Bangalore.

❖ Duolingo: On a "Streak"

* **Overview**: Published on January 24, 2025, and authored by Jeffrey Rayport, Nicole Tempest Keller, and Nicole Luo, this 25-page case explores the strategic evolution of Duolingo, the world's most downloaded education app. By late 2024, Duolingo had over 100 million monthly active users, 8 million paid subscribers, and a market capitalization nearing $15 billion. Its success was driven by a gamified learning experience, adaptive AI, and a powerful behavioral incentive: the daily streak. This feature not only helped users stay engaged but also fueled subscription growth and reduced churn. The case focuses on a pivotal moment when Severin Hacker, Duolingo's Co-Founder and CTO, considered expanding beyond language learning into new verticals like math and music, powered by generative AI. This strategic shift raised key questions:
    - Does Duolingo have the brand credibility to expand into adjacent educational domains?
    - What are the risks and rewards of integrating new subjects into the flagship app?
    - How should the company balance focus vs. diversification in a competitive edtech landscape?

This case is ideal for discussions on platform strategy, AI-driven innovation, product diversification, and scaling in digital education.

* **Citation**: Rayport, J., Keller, N. T., & Luo, N. (2025). *Duolingo: On a "Streak"* (Case No. 825097-PDF-ENG). Harvard Business School.

10.7.5. Strategy-in-Film: Learn by Watching

Strategy today extends beyond individual organizations; it is deeply shaped by the ecosystems they inhabit. The films below showcase how entrepreneurial environments foster innovation, competition, and collaboration across industries and sectors. As you explore entrepreneurial ecosystems in this chapter, use these narratives to reflect on how system-level dynamics influence opportunity creation, value capture, and long-term strategic success.

Movie / Documentary	Strategic Theme Highlighted
The Founder	Ecosystem franchising, supply-side integration
Coded Bias	Institutional governance, algorithmic oversight
Social Dilemma	Platform ecosystems, unintended feedback effects
Ford v Ferrari	System competition, innovation under constraint

Movie / Documentary	Strategic Theme Highlighted
The Biggest Little Farm	Ecosystem adaptation, feedback loops in regenerative systems
Startup.com	ecosystem rise and fall
Pandora's Promise	Energy ecosystems, policy conflict, long-term foresight
Citizenfour	Strategic systems risk and decentralized actors
Inside Job	Institutional failure, feedback collapse in financial systems
Our Planet	Interdependence, resilience in natural ecosystems

References

Abootorabi, H., Wiklund, J., Johnson, A. R., & Miller, C. D. (2021). A holistic approach to the evolution of an entrepreneurial ecosystem: An exploratory study of academic spin-offs. *Journal of Business Venturing*, *36*(5), 106143.

Acs, Z. J., Stam, E., Audretsch, D. B., & O'Connor, A. (2017). The lineages of the entrepreneurial ecosystem approach. *Small Business Economics*, *49*, 1-10.

Alvedalen, J., & Boschma, R. (2017). A critical review of entrepreneurial ecosystems research: Towards a future research agenda. *European Planning Studies*, *25*(6), 887-903.

Audretsch, D. B., Belitski, M., Eichler, G. M., & Schwarz, E. (2024). Entrepreneurial ecosystems, institutional quality, and the unexpected role of the sustainability orientation of entrepreneurs. *Small Business Economics*, *62*(2), 503-522.

Bergman, B. J., & McMullen, J. S. (2022). Helping entrepreneurs help themselves: A review and relational research agenda on entrepreneurial support organizations. *Entrepreneurship Theory and Practice*, *46*(3), 688-728.

Bourai, S., Arora, R., & Yadav, N. (2024). Structure-conduct-performance (SCP) paradigm in digital platform competition: a conceptual framework. *Journal of Strategy and Management*, *17*(2), 322-347.

Brown, R., & Mason, C. (2017). Looking inside the spiky bits: a critical review and conceptualisation of entrepreneurial ecosystems. *Small Business Economics*, *49*, 11-30.

Candeias, J. C., & Sarkar, S. (2024). Entrepreneurial ecosystems policy formulation: A conceptual framework. *Academy of Management Perspectives*, *38*(1), 77-105.

Cao, Z., & Shi, X. (2021). A systematic literature review of entrepreneurial ecosystems in advanced and emerging economies. *Small Business Economics*, *57*, 75-110.

Cantner, U., Cunningham, J. A., Lehmann, E. E., & Menter, M. (2021). Entrepreneurial ecosystems: a dynamic lifecycle model. *Small Business Economics*, *57*, 407-423.

Cennamo, C. (2021). Competing in digital markets: A platform-based perspective. *Academy of Management Perspectives*, *35*(2), 265-291.

Chandra, B., & Rahman, Z. (2024). Artificial intelligence and value co-creation: a review, conceptual framework and directions for future research. *Journal of Service Theory and Practice*, *34*(1), 7-32.

Cho, D. S., Ryan, P., & Buciuni, G. (2022). Evolutionary entrepreneurial ecosystems: A research pathway. *Small Business Economics*, *58*(4), 1865-1883.

Cobben, D., Ooms, W., Roijakkers, N., & Radziwon, A. (2022). Ecosystem types: A systematic review on boundaries and goals. *Journal of Business Research*, *142*, 138-164.

Cohen, M. C., & Zhang, R. (2022). Competition and coopetition for two-sided platforms. *Production and Operations Management*, *31*(5), 1997-2014.

Daniel, L. J., de Villiers Scheepers, M. J., Miles, M. P., & de Klerk, S. (2022). Understanding entrepreneurial ecosystems using complex adaptive systems theory: Getting the big picture for economic development, practice, and policy. *Entrepreneurship & Regional Development*, *34*(9-10), 911-934.

De Bernard, M., Comunian, R., & Gross, J. (2022). Cultural and creative ecosystems: a review of theories and methods, towards a new research agenda. *Cultural Trends*, *31*(4), 332-353.

de Vasconcelos Gomes, L. A., Castillo-Ospina, D. A., Facin, A. L. F., dos Santos Ferreira, C., & Ometto, A. R. (2023). Circular ecosystem innovation portfolio management. *Technovation*, *124*, 102745.

Fernandes, A. J., & Ferreira, J. J. (2022). Entrepreneurial ecosystems and networks: a literature review and research agenda. *Review of Managerial Science*, *16*(1), 189-247.

Fischer, B., Meissner, D., Vonortas, N., & Guerrero, M. (2022). Spatial features of entrepreneurial ecosystems. *Journal of Business Research*, *147*, 27-36.

Garcia Martin, P. C., Sjödin, D., Nair, S., & Parida, V. (2024). Managing start-up–incumbent digital solution co-creation: a four-phase process for intermediation in innovative contexts. *Industry and Innovation*, *31*(5), 579-605.

Harima, A., Harima, J., & Freiling, J. (2024). Ecosystem orchestration: Unpacking the leadership capabilities of anchor organizations in nascent entrepreneurial ecosystems. *Entrepreneurship Theory and Practice*, *48*(6), 1404-1450.

Hruskova, M. (2024). Ecosystem pipelines: Collective action in entrepreneurial ecosystems. *International Small Business Journal*, *42*(1), 39-66.

Jacobides, M. G., Cennamo, C., & Gawer, A. (2018). Towards a theory of ecosystems. *Strategic management journal*, *39*(8), 2255-2276.

Juma, N., Olabisi, J., & Griffin-EL, E. (2023). External enablers and entrepreneurial ecosystems: The brokering role of the anchor tenant in capacitating grassroots ecopreneurs. *Strategic Entrepreneurship Journal*, *17*(2), 372-407.

Khurana, I., Dutta, D. K., & Ghura, A. S. (2022). SMEs and digital transformation during a crisis: The emergence of resilience as a second-order dynamic capability in an entrepreneurial ecosystem. *Journal of Business Research*, *150*, 623-641.

Klimas, P., & Czakon, W. (2022). Species in the wild: a typology of innovation ecosystems. *Review of Managerial Science*, *16*(1), 249-282.

Kretschmer, T., Leiponen, A., Schilling, M., & Vasudeva, G. (2022). Platform ecosystems as meta-organizations: Implications for platform strategies. *Strategic Management Journal*, *43*(3), 405-424.

Leendertse, J., Schrijvers, M., & Stam, E. (2022). Measure twice, cut once: Entrepreneurial ecosystem metrics. *Research Policy*, *51*(9), 104336.

Liang, L., & Li, Y. (2023). How does government support promote digital economy development in China? The mediating role of regional innovation ecosystem resilience. *Technological Forecasting and Social Change*, *188*, 122328.

Linde, L., Sjödin, D., Parida, V., & Wincent, J. (2021). Dynamic capabilities for ecosystem orchestration: A capability-based framework for smart city innovation initiatives. *Technological Forecasting and Social Change*, *166*, 120614.

Marcon, A., Ribeiro, J. L. D., Olteanu, Y., & Fichter, K. (2024). How the interplay between innovation ecosystems and market contingency factors impacts startup innovation. *Technology in Society*, *76*, 102424.

Mele, C., & Russo-Spena, T. (2025). Agencement of onlife and phygital: smart tech–enabled value co-creation practices. *Journal of Service Management*, *36*(2), 217-240.

Miah, M. T., Aiupova, N., Erdei-Gally, S., & Fekete-Farkas, M. (2025). Digital entrepreneurship ecosystems: Then vs. now-a future perspectives. *Digital Business*, *5*(1), 100110.

Neto, J. R., Figueiredo, C., Gabriel, B. C., & Valente, R. (2024). Factors for innovation ecosystem frameworks: Comprehensive organizational aspects for evolution. *Technological Forecasting and Social Change*, *203*, 123383.

O'Connor, A., & Audretsch, D. (2023). Regional entrepreneurial ecosystems: learning from forest ecosystems. *Small Business Economics*, *60*(3), 1051-1079.

Priyono, A., & Hidayat, A. (2024). Fostering innovation through learning from digital business ecosystem: A dynamic capability perspective. *Journal of Open Innovation: Technology, Market, and Complexity*, *10*(1), 100196.

Rashid, S., & Ratten, V. (2021). Entrepreneurial ecosystems during COVID-19: the survival of small businesses using dynamic capabilities. *World Journal of Entrepreneurship, Management and Sustainable Development*, *17*(3), 457-476.

Rietveld, J., & Schilling, M. A. (2021). Platform competition: a systematic and interdisciplinary review of the literature. *Journal of Management*, *47*(6), 1528-1563.

Roundy, P. T., Bradshaw, M., & Brockman, B. K. (2018). The emergence of entrepreneurial ecosystems: A complex adaptive systems approach. *Journal of Business Research*, *86*, 1-10.

Silva, L. E. N., de Vasconcelos Gomes, L. A., de Faria, A. M., & Borini, F. M. (2024). Innovation processes in ecosystem settings: An integrative framework and future directions. *Technovation*, *132*, 102984.

Snihur, Y., & Bocken, N. (2022). A call for action: The impact of business model innovation on business ecosystems, society, and planet. *Long Range Planning*, *55*(6), 102182.

Solakis, K., Katsoni, V., Mahmoud, A. B., & Grigoriou, N. (2024). Factors affecting value co-creation through artificial intelligence in tourism: a general literature review. *Journal of Tourism Futures*, *10*(1), 116-130.

Stahl, B. C. (2022). Responsible innovation ecosystems: Ethical implications of the application of the ecosystem concept to artificial intelligence. *International Journal of Information Management*, *62*, 102441.

Stam, E., & Spigel, B. (2016). *Entrepreneurial ecosystems* (Vol. 16, No. 13, pp. 1-15). USE Discussion paper series.

Theodoraki, C. (2024). Building entrepreneurial ecosystems sustainably. *Foundations and Trends® in Entrepreneurship*, *20*(4), 384-480.

Theodoraki, C., Dana, L. P., & Caputo, A. (2022). Building sustainable entrepreneurial ecosystems: A holistic approach. *Journal of Business Research*, *140*, 346-360.

Torres, P., & Godinho, P. (2022). Levels of necessity of entrepreneurial ecosystems elements. *Small Business Economics*, *59*(1), 29-45.

Wurth, B., Stam, E., & Spigel, B. (2022). Toward an entrepreneurial ecosystem research program. *Entrepreneurship Theory and Practice*, *46*(3), 729-778.

Yuan, N., & Li, M. (2024). Research on collaborative innovation behavior of enterprise innovation ecosystem under evolutionary game. *Technological Forecasting and Social Change*, *206*, 123508.

Zhang, X., Yang, L., Gao, T., & Zhou, W. (2024). The coordination mechanism of value co-creation between developers and users in digital innovation ecosystems. *Electronic Markets*, *34*(1), 1.

Zhao, X., Xu, Y., Vasa, L., & Shahzad, U. (2023). Entrepreneurial ecosystem and urban innovation: Contextual findings in the lens of sustainable development from China. *Technological Forecasting and Social Change*, *191*, 122526.

CHAPTER 11

Strategic Sustainability Analytics

"What gets measured gets managed but in sustainability what gets modeled shapes the future"

by MAK

Learning Outcomes

By the end of this chapter, readers will be able to:

Sustainability is not a sprint—it's a memory, a model, and a movement. Like the elephant, it shapes the future by walking with wisdom, purpose, and presence.

- ❖ Explain the evolution of ESG (Environmental, Social, and Governance) as a strategic priority, moving beyond compliance into systemic value creation.
- ❖ Differentiate between CSR and CSV, and apply stakeholder theory in data-intensive ESG planning.
- ❖ Interpret and design ESG dashboards and impact analytics to inform strategic decisions.
- ❖ Evaluate scoring models and assess the limitations and biases of ESG quantification.
- ❖ Use scenario modeling and multi-objective optimization to manage ESG trade-offs in strategic planning.
- ❖ Analyze case examples of climate, circular economy, and social equity strategies using STEM tools such as lifecycle assessment and decision intelligence.
- ❖ Formulate governance recommendations for ESG strategy implementation and accountability using predictive analytics and board foresight.

Environmental, Social, and Governance (ESG) concerns have evolved from boardroom buzzwords into defining pillars of long-term competitiveness. What once existed as parallel reporting structures and "do-no-harm" initiatives now sits at the heart of strategy. ESG is no longer an optional add-on; it is a core performance dimension and a lens through which risk, innovation, and value are assessed.

This chapter explores the intersection of ESG and STEM. As sustainability concerns grow more complex and global in scope, STEM tools, ranging from dynamic dashboards and AI-based risk models to lifecycle assessments and optimization algorithms, offer new ways of visualizing impact, managing trade-offs, and designing resilient futures.

The goal is not simply to help practitioners understand ESG definitions. Rather, it is to equip them with the tools, frameworks, and strategic mindset to navigate ESG in real time. We'll move from foundational ideas such as CSR and stakeholder theory to advanced methods like multi-objective modeling and circular design. Along the way, we will illustrate how ESG strategy is already transforming industries like finance, manufacturing, retail, and tech, and why the future belongs to firms that embed sustainability into every decision.

Opening Case: Ping An's ESG Transformation, From Green Intentions to Strategic Intelligence

Ping An, one of the world's largest and most tech-forward insurance and financial service firms, is redefining ESG strategy for the data age. Faced with tightening regulatory frameworks, shifting investor expectations, and growing stakeholder activism, Ping An chose not just to report ESG performance, but to model and simulate it using cutting-edge STEM tools.

Their transformation began with a central question: Can sustainability be algorithmically managed at scale? The firm's ESG Intelligence Engine says yes. It combines satellite imagery, emissions databases, market behavior models, and AI-based scenario planning to dynamically track carbon intensity, social equity investments, and board governance risks across hundreds of subsidiaries and portfolios.

Internally, Ping An restructured its ESG operations around a central data hub. Real-time dashboards allowed decision-makers to see not just current ESG scores but project future trajectories under different scenarios, such as a 1.5°C policy environment or a carbon tax shock. These projections informed capital allocation, strategic partnerships, and innovation sprints.

Moreover, Ping An used ESG metrics not only for external reporting but to evaluate internal performance incentives. For example, product development teams earned bonuses not just for market growth but for measurable impact in financial inclusion and green finance KPIs. This alignment of strategy and ESG made sustainability a competitive asset, not just a reporting burden.

The case of Ping An reveals a new playbook for ESG strategy, one where ethics meets engineering, and where firms use STEM to transition from reactive reporting to proactive sustainability leadership.

Play the Strategist

1. How does Ping An's approach to ESG differ from traditional sustainability reporting? What strategic advantages does this create?
2. In what ways did Ping An use STEM tools (e.g., AI, data dashboards, scenario modeling) to enhance ESG integration into decision-making?
3. How does the alignment of performance incentives with ESG metrics influence organizational behavior and culture at Ping An?
4. What ethical considerations should guide the use of AI in ESG scoring, especially when it affects external reporting and internal incentives?

11.1 ESG Strategy in a STEM World

In today's business landscape, Environmental, Social, and Governance (ESG) strategy is no longer a side agenda; it's a core dimension of corporate competitiveness and long-term value creation. Investors, regulators, customers, and employees increasingly demand that organizations demonstrate measurable commitment to sustainability, equity, and ethical governance. Yet the challenge is not just about commitment, it's about execution.

Traditional ESG approaches have struggled with vague metrics, compliance-driven reporting, and the risk of greenwashing. What's changing now is the infusion of **STEM-based thinking,** data science, modeling, AI, engineering simulations, and systems design into ESG decision-making. These tools allow firms to quantify trade-offs, track performance in real time, model unintended consequences, and ensure transparency at scale.

This section explores how ESG is evolving from aspirational values to an analytically grounded strategy. We examine foundational concepts like the **Triple Bottom Line** and **Stakeholder Theory**, and show how STEM tools are enabling firms to move beyond compliance toward **strategic sustainability**. By integrating dashboards, scenario modeling, and multi-dimensional scoring, organizations can embed ESG into their core decision engines.

11.1.1 From Greenwashing to Strategic Sustainability

Greenwashing, the superficial promotion of environmental credentials without real impact, has undermined trust in corporate ESG claims. Consumers and investors are increasingly skeptical of vague sustainability reports and generic pledges. What distinguishes today's leaders is not ESG rhetoric, but the ability to measure, model, and act on it strategically.

Strategic sustainability means integrating ESG into long-range planning, capital allocation, product development, and performance incentives. Firms like **Unilever**, **Ping An**, and **Microsoft** now embed carbon pricing into financial models, use AI to track supply chain emissions, and tie executive bonuses to sustainability outcomes.

What's driving this shift is the recognition that ESG isn't just ethical, it's financially material. Data now links ESG performance to risk mitigation, resilience, talent retention, and long-term returns. In a STEM-driven world, the question is no longer "Should we care about ESG?" but "How do we quantify it, simulate its effects, and embed it into strategic choices?"

11.1.2 Triple Bottom Line and ESG as Strategic Data

The **Triple Bottom Line (TBL)** framework, **People, Planet, and Profit,** was once seen as a moral compass for business. Now, it's becoming a data architecture. ESG metrics are being used to structure not only sustainability reports, but also strategic models.

For example:

- **People (Social)** metrics track employee well-being, diversity, supply chain labor practices, and community impact.
- **Planet (Environmental)** metrics include emissions, water use, biodiversity risk, and circularity ratios.
- **Profit (Governance + Financial)** includes ethical compliance, board accountability, risk disclosure, and anti-corruption systems.

What's changing is how these metrics are **quantified, visualized, and embedded into dashboards**. ESG data is now being captured in real time via IoT devices, satellite imagery, employee platforms, and regulatory APIs. This transformation turns ESG into **strategic data**—usable for scenario planning, investment decisions, and even algorithmic trading.

11.1.3 CSR vs. CSV: Beyond Compliance

Michael Porter and Mark Kramer's distinction between **Corporate Social Responsibility (CSR)** and **Creating Shared Value (CSV)** offers a powerful lens for rethinking ESG strategy. While CSR often focuses on mitigating harm or meeting stakeholder expectations, CSV reframes ESG as a **platform for innovation, growth, and competitive advantage**.

Traditional CSR is typically treated as an **add-on activity,** reactive, cost-centered, and often disconnected from core business value. In contrast, CSV is **embedded in the business model**, proactively seeking opportunities where social and environmental progress align with commercial success.

This strategic shift is summarized in the table below:

Table 11.1: Strategic Comparison — CSR vs. Creating Shared Value (CSV)

Traditional CSR	Creating Shared Value (CSV)
Add-on activity	Core to competitive strategy
Reactive (responding to pressure)	Proactive (seeking opportunity)
Cost center	Growth and innovation engine
Often disconnected from value creation	Fully aligned with profit and purpose

CSV reframes ESG not as a compliance burden, but as a platform for innovation and advantage. For example, Nestlé improved agricultural practices in its supply chain, benefiting both farmers and its product quality. Similarly, Ping An's green finance initiatives reduce systemic financial risk while creating new market opportunities.

Using **STEM tools**, firms can map where shared value creation is most viable, e.g., where emissions reduction also lowers costs, or where inclusive hiring boosts innovation.

11.1.4 Stakeholder Theory Reimagined with STEM Tools

Stakeholder Theory holds that firms should consider the interests of all parties affected by their actions, not just shareholders. While this theory has guided ESG efforts for decades, it often lacked operational precision. STEM tools are now changing that.

Today's strategy teams can:

- **Simulate stakeholder outcomes** using agent-based models.
- Use **dashboards** to balance KPIs across groups (employees, regulators, investors).
- Integrate **real-time feedback** from social media, sentiment analysis, and digital communities.
- Apply **decision intelligence** to optimize trade-offs (e.g., wage increases vs. automation).

In short, firms can now **visualize stakeholder networks**, quantify externalities, and model ripple effects. This allows ESG to move from fuzzy goals to actionable insights, embedding stakeholder logic into algorithms, investment decisions, and board reporting.

The table below illustrates how traditional ESG concepts are being reframed through STEM tools to deliver measurable strategic value:

Table 11.2. Evolving ESG from Principle to Practice with STEM Tools

Traditional Concept	Strategic Reframing	STEM Tool or Method	Strategic Benefit
Greenwashing	Strategic Sustainability	Carbon pricing models, AI emissions tracking	Aligns ESG actions with core business strategy
Triple Bottom Line (People, Planet, Profit)	ESG as Strategic Data	ESG dashboards, real-time sensors, analytics engines	Enables quantifiable, real-time impact assessment
Corporate Social Responsibility (CSR)	Creating Shared Value (CSV)	Optimization algorithms, social ROI modeling	Converts compliance into innovation and growth opportunities
Stakeholder Theory	Stakeholder Simulation & Feedback	Agent-based modeling, sentiment analytics, decision intelligence	Quantifies stakeholder impact and trade-offs

11.2 ESG Dashboards and Impact Analytics

In the age of strategic transparency, ESG (Environmental, Social, and Governance) dashboards are no longer window dressing; they are operational nerve centers. For firms navigating sustainability, inclusion, ethics, and governance, ESG dashboards provide a real-time lens into performance, exposure, and strategic alignment. But simply tracking metrics is not enough. The real challenge lies in transforming raw ESG data into actionable insight.

This section explores how leading organizations leverage ESG dashboards and analytics not just to report performance, but to drive it. Firms like BlackRock use sophisticated ESG indices to shape investment portfolios. Patagonia visualizes supply chain sustainability risks in real-time. Ping An integrates ESG scores into its lending decisions, modeling societal outcomes alongside creditworthiness. These aren't isolated initiatives; they reflect a growing shift from qualitative aspiration to quantitative accountability.

As ESG moves from voluntary disclosure to regulatory expectation, through frameworks like the EU Taxonomy and TCFD, firms need more than spreadsheets. They need impact intelligence. This section outlines how ESG dashboards are built, the anatomy of scoring models, and how visualization tools like heat maps, time series, and risk zones guide strategic ESG decisions. It also addresses the critical issues of auditability, bias, and the quantification limits of inherently value-driven concepts like equity and justice.

By the end of this section, readers will understand how ESG dashboards can evolve from compliance checklists to strategy engines, turning values into data, and data into competitive foresight.

11.2.1 Anatomy of an ESG Dashboard: From Metrics to Meaning

An ESG dashboard is a decision-support interface that aggregates, visualizes, and interprets environmental, social, and governance data in real time. But the best ESG dashboards are more than repositories of metrics; they are strategic storytelling devices. They make patterns visible, trade-offs transparent, and performance understandable for a wide range of stakeholders, from C-suite executives to regulators, investors, and the public.

A typical ESG dashboard includes performance indicators across E, S, and G dimensions. On the environmental side, you may find carbon intensity, energy use, waste generation, and water impact. Social metrics could include gender diversity, labor standards compliance, and community investment. Governance indicators often feature board independence, anti-corruption policies, and executive pay ratios.

Leading firms build ESG dashboards that integrate both backward-looking KPIs and forward-looking risk alerts. For instance, BlackRock's Aladdin Climate platform incorporates scenario-based stress tests for climate risk exposure across portfolios. Ping An's ESG dashboard for corporate clients evaluates not just emissions but health insurance inclusion and education access—linking social impact with financial performance.

The key is **contextualization**: data points must be embedded within strategic narratives. It's not enough to know that Scope 3 emissions are rising—what matters is whether they are rising due to faster growth, supply chain shifts, or lack of policy compliance. A great ESG dashboard turns data into a diagnosis.

11.2.2 Scoring Models: Environmental, Social, and Governance Dimensions

To simplify complex ESG profiles, many dashboards include scoring systems. These scores, often on a 0–100 or AAA–CCC scale, summarize ESG performance using weighted models. However, these models are not neutral; they reflect value judgments and assumptions about what matters most.

For instance:

- **Environmental scores** may emphasize carbon emissions (e.g., CO_2/Revenue), energy transition plans, or biodiversity impact.
- **Social scores** could prioritize labor practices, health and safety, community engagement, or product accessibility.
- **Governance scores** often capture board diversity, ownership structures, audit integrity, and whistleblower protection mechanisms.

Each dimension must balance **quantitative rigor** with **qualitative nuance**. Patagonia, for example, integrates worker feedback loops into its social scoring, while the EU Taxonomy requires scientifically validated environmental disclosures across six climate objectives.

To ensure transparency and comparability, some firms rely on third-party ratings (e.g., MSCI, Sustainalytics), while others, like Ping An, build proprietary ESG scoring engines aligned with internal strategy and national development priorities.

11.2.3 Visualizing Impact: Heat Maps, Time Series, and Risk Zones

How ESG data is visualized determines whether it sparks insight or confusion. Effective dashboards use visual tools to highlight both hotspots and progress zones.

- **Heat Maps**: Useful for comparing ESG performance across departments, geographies, or suppliers. A heat map might show low diversity scores in overseas offices or carbon spikes in a specific product line.
- **Time Series Charts**: Reveal performance trends. Are emissions decreasing over time? Is board diversity improving annually?
- **Risk Zones**: These overlays signal where ESG underperformance intersects with financial or reputational risk, e.g., a region where labor violations coincide with customer boycotts.

Ping An's impact maps, for example, track green finance disbursements against air quality indices in urban regions, creating spatial feedback loops that link financing to measurable social value.

Visualization not only makes ESG actionable, but it also democratizes it. Employees, investors, and regulators don't need to parse spreadsheets; they can see, at a glance, where the system is working and where intervention is needed.

11.2.4 Auditability, Bias, and the Limits of ESG Quantification

While ESG dashboards promise clarity, they also raise serious questions about bias, accountability, and ethical framing.

- **Auditability**: How are ESG scores calculated? Can firms trace the origin of data? Do internal systems match third-party ratings? For ESG to become strategic, firms must ensure traceability and audit readiness, especially under emerging global disclosure regimes (e.g., CSRD in Europe).
- **Bias**: Algorithms used to generate ESG scores can inherit systemic biases. For instance, over-weighting governance practices common in Western firms may penalize high-performing emerging market companies. Similarly, under-representation of informal economies can distort social impact assessments.

- **Quantification Limits**: Not all ESG values are easily measurable. Inclusion, fairness, well-being, or dignity resist full numerical expression. Over-reliance on dashboards risks ignoring the lived realities behind the numbers.

Thus, leaders must treat ESG dashboards as decision aids, not decision makers. A dashboard should be a tool for dialogue and deliberation, not a final verdict. To help navigate this complexity, the table below categorizes common ESG metrics by their **strategic relevance** and **data maturity**, offering a more nuanced view of what can—and cannot—be reliably measured:

Table 11.3: ESG Metrics Categorized by Strategic Relevance and Data Maturity

Metric	ESG Category	Strategic Relevance	Data Maturity	Example
Scope 1–3 Carbon Emissions	Environmental	High – Regulatory and reputational	Mature	EU Carbon Reporting / TCFD
Water Stress in the Supply Chain	Environmental	Medium – Operational risk	Emerging	Textile firms in arid regions
Gender Pay Ratio	Social	High – Cultural reputation, lawsuits	Moderate	UK Gender Pay Gap reporting
Community Investment (% of Profits)	Social	Low – Branding, license to operate	Low	Voluntary CSR disclosures
Board Diversity (% by gender/race)	Governance	High – Investor and ESG index appeal	Moderate	Nasdaq listing rules
Whistleblower Protection Policies	Governance	Medium – Ethical risk management	Mature	Governance audits
ESG Controversies Score (media scans)	Composite	High – Crisis management	Emerging	AI-based sentiment tools (e.g., RepRisk)

In summary, ESG dashboards and impact analytics are no longer optional; they are strategic imperatives. But building them well requires more than data; it requires clarity of values, rigor in modeling, and humility in interpretation. Done right, these tools can align stakeholder expectations, guide innovation, and make firms not just more transparent but more transformative.

11.3 Modeling ESG Trade-Offs in Strategic Planning

Sustainability strategy is full of good intentions, but it is also full of **difficult trade-offs**. Reducing emissions might increase short-term costs. Improving worker benefits could slow automation efforts. Expanding into underserved markets may carry higher ESG risk exposure. Traditional strategic planning often avoids these tensions, treating ESG as a checklist rather than a decision domain. But in the STEM-enabled strategy landscape, we don't avoid trade-offs; we **model them**.

This section explores how advanced modeling techniques, especially those grounded in **multi-objective optimization**, **scenario simulation**, and **stakeholder analytics,** allow firms to navigate competing ESG priorities with transparency and precision. By using these tools, strategy becomes a laboratory of futures, where choices are stress-tested, stakeholder perspectives simulated, and sustainability becomes not just a principle, but a practiced discipline.

11.3.1 Sustainability Is Full of Trade-Offs—So Model Them

Every ESG goal has opportunity costs. Reducing emissions may mean switching suppliers, slowing delivery, or raising prices. Improving equity may mean challenging legacy power structures or rethinking hiring pipelines. The point is not to avoid these tensions, but to make them visible and manageable.

In this context, **STEM tools** offer clarity. For example:

- **Carbon abatement cost curves** help firms identify which emissions cuts provide the best ROI.
- **Supply chain modeling** shows the environmental vs. resilience trade-offs of local versus global sourcing.
- **HR impact models** quantify how different DEI policies affect retention, innovation, or legal exposure.

By mapping these tensions, strategy becomes a system of **informed prioritization,** not ideological purity. The key is to build structured methods that allow competing values to **coexist within decision frameworks**.

11.3.2 Multi-Objective Optimization: Carbon vs. Cost vs. Inclusion

Traditional strategy often seeks a single "optimal" outcome, to maximize profit or minimize cost. But ESG strategy lives in a **multi-objective world**. Success may require **balancing** three or more performance criteria simultaneously.

This is where **multi-objective optimization (MOO)** comes in. Techniques like **Pareto analysis** or **goal programming** help firms explore trade-off frontiers, combinations of outcomes where no single objective can improve without sacrificing another.

For example, a sustainability leader might model the following:

- **Objective A**: Reduce carbon footprint by 30% by 2030
- **Objective B**: Maintain cost increase under 5%
- **Objective C**: Improve workforce inclusion metrics by 15%

Using MOO, the firm can generate a **solution space** that reveals which strategies meet all three goals, and where hard choices must be made. More importantly, it can show how small changes in one objective (e.g., budget) can unlock gains in others (e.g., inclusion).

Ping An's ESG portfolio modeling does exactly this: it simulates investment strategies that optimize for green finance returns, social outreach, and regulatory alignment, all while maintaining long-term profitability.

11.3.3 Scenario Modeling for ESG Futures

Sustainability operates under uncertainty. Climate regulations, consumer expectations, and supply shocks shift constantly. That's why ESG planning must incorporate **scenario modeling,** structured narratives that explore different plausible futures.

Firms can use ESG scenarios such as:

- **"Green Leap"**: Aggressive policy shifts make carbon expensive and transparency mandatory.
- **"Social Backlash"**: Rising inequality leads to stakeholder activism and boycotts.
- **"Tech Fix"**: AI and clean energy tech dramatically reduce ESG costs, but widen data inequality.

Each scenario reveals different strategic vulnerabilities and opportunities. Using modeling platforms, firms can simulate how product lines, geographic expansions, or supply chains perform under these futures, and develop **contingency strategies**.

For example, Unilever uses ESG scenarios to stress-test its supply chain resilience against droughts and carbon pricing shocks. Ping An's financial ESG models incorporate health and aging trends to forecast future social impact risk zones.

11.3.4 Decision Intelligence and Stakeholder Simulation

One of the most promising frontiers in ESG strategy is **stakeholder simulation,** a method for modeling how different stakeholder groups will **react to ESG decisions**. Using behavioral data, sentiment analysis, and historical patterns, firms can anticipate likely reactions to key moves.

For instance:

- What happens if a company closes a high-emissions plant in a low-income community?
- How will investors respond to a drop in short-term profit caused by new DEI initiatives?
- How do customers respond to carbon labeling or gender pay gap disclosures?

Firms can use **agent-based models**, **digital twins**, and **decision trees** to simulate responses across customers, regulators, employees, NGOs, and shareholders. This transforms ESG planning into **multi-stakeholder strategy gaming**—with feedback loops and rapid iteration.

A leading example is Ping An's ESG dashboard, which includes a stakeholder reaction layer that estimates policy adoption risk based on media sentiment, community health scores, and political climate.

To support this approach, the following simulation grid outlines how firms can structure ESG trade-off modeling, defining inputs, constraints, key dilemmas, and measurable outcomes:

Table 11.4. Trade-Off Simulation Grid – Inputs, Constraints, Outcomes

Scenario	Input Variables	Constraints	Key Trade-Offs	Outcome Metrics
Green Supply Chain Upgrade	Supplier switch, transport distance, cost	<5% cost increase	Carbon reduction vs. delivery delays	Scope 3 Emissions, On-time Delivery
Inclusive Hiring Strategy	DEI training budget, new recruitment policies	6-month launch window	Inclusion gains vs. hiring timeline impact	Diversity %, Time to Hire, Turnover
Circular Product Line Launch	Material inputs, recycling logistics	Maintain gross margin	Eco-design vs. production complexity	Circularity Score, Margin Retention
Ethical Sourcing Disclosure	Transparency level, audit depth	Reputational risk threshold	Brand trust vs. supplier retaliation risk	Sentiment Index, Supplier Retention Rate
Climate Scenario Stress Test	Carbon tax levels, insurance costs	10-year strategic plan horizon	Emissions path vs. asset risk exposure	VaR (Value at Risk), ESG Ratings

In summary, **modeling ESG trade-offs** doesn't mean solving every problem, but it does mean making dilemmas visible, quantifiable, and ethically conscious. In a STEM-enabled strategy framework, ESG becomes not a constraint but a dynamic arena for innovation, simulation, and systemic value creation.

11.4 ESG Strategy in Climate, Circular Economy, and Social Impact

While ESG strategy often begins with data dashboards and compliance frameworks, its deepest value lies in how it shapes **real-world transformations**—from decarbonization to product redesign to inclusive growth. This section explores how ESG strategy can be applied to three of the most critical domains of systemic impact: **climate action, circular economy, and social equity**.

Each domain demands not only vision but also **STEM-enabled capabilities**. Advanced analytics help map emissions trajectories. Lifecycle assessments uncover waste and regeneration potential. Scenario modeling helps firms navigate just transitions. And board structures increasingly reflect the need for governance that aligns strategy with societal expectations.

By the end of this section, you'll understand how ESG strategy moves beyond reporting to **reshape business models, product lifecycles, and governance design**—creating value that is environmental, social, and enduring.

11.4.1 Climate Analytics: Emissions, Offsets, and Decarbonization Paths

Climate strategy is no longer optional. Investors, regulators, and consumers demand credible, science-based targets. STEM tools are essential here, enabling firms to measure their **carbon footprint**, simulate **transition scenarios**, and plan **decarbonization paths** aligned with global targets.

Key tools include:

- **Carbon Accounting Platforms** (e.g., Scope 1–3 emissions calculators)
- **Carbon Forecasting Models** to evaluate future emissions based on strategy changes
- **Marginal Abatement Cost Curves** to prioritize emissions cuts by ROI
- **Offset Validation Systems** to ensure purchased credits meet high-integrity standards

For example, Ping An uses AI to track carbon emissions across its loan portfolios, identifying high-risk sectors and simulating investment reallocation impacts. Unilever uses carbon modeling to adjust packaging materials and transportation methods to align with Net Zero targets.

These tools allow firms to create not only targets, but **credible transition maps**.

11.4.2 Circular Strategy: Designing for Regeneration

The circular economy moves beyond "reduce, reuse, recycle" to **designing products and systems that regenerate rather than deplete**. ESG strategy in this space requires lifecycle thinking and engineering-enabled design.

Key elements of circular strategy include:

- **Lifecycle Assessment (LCA):** End-to-end environmental impact analysis
- **Circular Design Principles:** Products that can be disassembled, repaired, or remanufactured
- **Reverse Logistics Modeling:** Efficient pathways for product return and reuse
- **Material Ledgering:** Digital tracking of components and recycled inputs

IKEA's materials ledger system, for instance, tracks inputs through a circular loop, from raw materials to stores to customer returns and recycling partners. Unilever's "loop" partnerships redesign packaging for refill and reuse, reducing both cost and waste.

Circular strategy is a natural fit for STEM-enabled leadership; it demands **systems thinking**, **data integration**, and **supply chain redesign** grounded in analytics.

11.4.3 Just Transition: From Job Risk to Social Equity

As firms decarbonize and automate, not all impacts are positive—**workers, communities, and regions may be left behind**. A just transition strategy ensures ESG action aligns with equity, inclusion, and long-term social resilience.

Strategic levers for just transition include:

- **Workforce Reskilling Models** to move labor from sunset industries to green growth areas
- **Geospatial Mapping** of ESG risks across regions and populations
- **Stakeholder Co-Creation Platforms** to develop equitable implementation strategies
- **ESG Impact Bonds** tied to employment and social targets

Ping An, for example, funds rural health and digital literacy initiatives to bridge social gaps created by fintech expansion. Firms like Ørsted pair renewable energy investments with local hiring and job transition funds.

STEM tools help firms **map social risks, model equity outcomes**, and simulate the distributional effects of ESG investments—transforming social responsibility into proactive resilience design.

11.4.4 Governance Foresight: Board Design and ESG Accountability

ESG success depends not only on models and strategy, but on **governance structures that embed ESG into leadership itself**. Boards must evolve from compliance monitors to **stewards of systemic value**.

Key governance trends include:

- **Board ESG Committees** with direct oversight of climate, DEI, and ethics
- **ESG Skills Audits** to ensure director readiness
- **Executive Incentive Design** tied to ESG metrics
- **AI-enabled Board Analytics** to surface ESG risks and stakeholder sentiment

Ping An, for example, integrates ESG metrics into board-level scorecards and has pioneered AI tools to monitor real-time reputational risk from ESG issues across its financial holdings.

This approach reflects a deeper shift: governance must **anticipate, simulate, and respond** to sustainability risks and opportunities, not merely report on them.

STEM Tool Highlight

- **Lifecycle Assessment (LCA):** Evaluates the full environmental impact from raw material to disposal.
- **Carbon Forecasting Models:** Predict emission pathways under different strategies.
- **AI-Driven ESG Risk Detection:** Scans unstructured data for signals of compliance gaps, reputational threats, or stakeholder backlash.

11.5. Chapter 11 Wrap-Up: ESG Strategy in a STEM World

This chapter reframed ESG from a reporting function into a strategic lever—deeply connected to value creation, risk management, and innovation. We began by tracing ESG's evolution from greenwashing fears to **strategic sustainability**, revealing how STEM tools enable firms to move beyond intention to **measurement, modeling, and transformation**.

We introduced the **Triple Bottom Line,** people, planet, and profit, not as a philosophical idea but as a strategic architecture that can now be **quantified, visualized, and optimized**. We differentiated **Corporate Social Responsibility (CSR)** from **Creating Shared Value (CSV)**, then showed how stakeholder theory, when augmented by analytics and simulation, allows leaders to engage complexity with confidence.

You explored how ESG dashboards transform scattered metrics into **strategic dashboards**—complete with scoring models, impact maps, and bias audits. From Ping An's green finance platforms to Patagonia's mission-aligned decisions, the message was clear: ESG is now a **design space for competitive and systemic advantage**.

We also addressed ESG trade-offs—such as carbon reduction versus employment—and demonstrated how **multi-objective modeling, decision intelligence, and stakeholder simulations** help firms navigate them with rigor. This culminated in the integration of ESG into three critical domains: climate, circular economy, and social equity. ESG became not only a risk lens but a canvas for **governance innovation and resilient strategy**.

As STEM strategists, your role is not to choose between ethics and economics—but to **design strategies that deliver both**. With the right tools, ESG is not a constraint. It is a blueprint for long-term impact, trust, and adaptive leadership.

11.5.1. Discussion Questions

1. How does ESG strategy differ when framed as a data problem versus a values problem?
2. In what ways do ESG dashboards change how executives make trade-off decisions?
3. What are the biggest risks of relying solely on ESG scores or ratings for strategy?
4. How can STEM tools like lifecycle assessment or carbon forecasting enhance ESG credibility?

5. What are the limitations of current ESG quantification frameworks—and how might AI improve them?
6. In modeling trade-offs between carbon and employment, what ethical considerations arise?
7. How can scenario modeling support more equitable ESG outcomes?
8. Why is board design increasingly critical in ESG strategy—and what STEM capabilities do boards need?
9. What does a "just transition" require from business beyond good intentions?
10. How does Ping An illustrate ESG strategy as a system-level innovation challenge?

11.5.2. Micro-Case: Zara's ESG Sprint

Zara, now an executive at a large multinational apparel firm, is asked to lead the development of a **circular design strategy** to meet new EU sustainable textile mandates. Her task: design a plan that cuts lifecycle emissions by 40%, reduces landfill waste, and improves labor transparency—**within 6 months**.

She convenes a STEM-enabled task force including material engineers, data scientists, and circular economy specialists. Using **lifecycle assessment**, the team maps emissions hotspots across the supply chain—identifying fabric sourcing and transport as the biggest contributors.

Zara also builds an **ESG dashboard** that allows factory partners to track their environmental performance and receive green incentives. A digital "materials passport" system is piloted—using blockchain to trace sustainable inputs across the entire product lifecycle.

When cost models show a short-term hit to margins, Zara runs a **multi-objective simulation** to model reputational benefit and customer retention under different ESG profiles. She convinces the CFO that customers with high ESG values show **2.5x loyalty**, and her model links long-term gains with near-term redesign.

As production nears, Zara collaborates with HR and NGOs to co-design a **labor equity audit** using geospatial tools and anonymized worker surveys. The result: the firm not only meets EU mandates but is recognized for ESG innovation at a global summit.

Her sprint demonstrates how ESG, when paired with STEM intelligence, becomes a strategy accelerator—not a drag.

Discussion Questions

1. How did Zara integrate ESG principles into design, supply chain, and labor practices using STEM tools?
2. What challenges did the team face in balancing short-term financial pressures with long-term ESG goals?

3. How did Zara use simulation and scenario modeling to influence executive decision-making?
4. What lessons can other companies draw from Zara's approach to circular economy strategy and stakeholder alignment?

11.5.3. Interactive Exercise: ESG Trade-Off Simulation

Scenario:
You are advising a global consumer electronics firm preparing to launch a new smartphone line. The product must meet ESG performance expectations across carbon footprint, labor standards, and supply chain transparency.

Your Task:

Using at least two tools from this chapter (e.g., ESG dashboard, multi-objective modeling, lifecycle assessment):

- Model three trade-off scenarios (e.g., low-carbon + high cost; low labor risk + longer launch; fast launch + medium ESG risk).
- Recommend the optimal balance of ESG and financial performance.
- Identify which stakeholders must be engaged and what data signals they require.

Deliverable:
Prepare a one-page strategy brief including trade-off visualizations, KPI proposals, and ethical risks.

11.5.4. Recommended HBR Case Studies for Chapter 10

- **ESG at George Weston Ltd.: Building Generational Value in Loblaw and Choice Properties**
  - **Overview:** Published on March 4, 2025, and authored by Michael R. King (University of Victoria) and Murray J. Bryant (Ivey Business School), this 22-page case explores how George Weston Limited (GWL), a 142-year-old, family-controlled Canadian holding company, approaches environmental, social, and governance (ESG) strategy across its subsidiaries: Loblaw Companies Limited and Choice Properties REIT. The case centers on the company's CEO, a fourth-generation family leader, who is reviewing ESG reports and evaluating whether the organization is meeting its sustainability commitments and has the governance structures in place to support long-term, stakeholder-driven value creation. It highlights the challenges of aligning ESG principles with business performance, managing stakeholder expectations, and embedding sustainability

into corporate governance. This case is ideal for discussions on corporate sustainability, family business strategy, stakeholder capitalism, and ESG reporting.

* **Citation:** King, M. R., & Bryant, M. J. (2025). *ESG at George Weston Ltd.: Building Generational Value in Loblaw and Choice Properties* (Case No. W42919-PDF-ENG). Ivey Publishing.

❖ ESG integration at Prescient Investment Management: Becoming a quantitatively responsible investor

* **Overview:** Published on February 28, 2025, and authored by Stephanie Giamporcaro and David Harrison, this 25-page case follows Prescient Investment Management (PIM), a leading South African asset manager, as it develops and integrates a responsible investing (RI) strategy into its core operations. Set in September 2022, the case explores how PIM responds to growing investor pressure to incorporate environmental, social, and governance (ESG) factors into its investment decisions. In the absence of a universal ESG standard, especially one tailored to the South African context, PIM takes the bold step of creating its own quantitative ESG scorecard and prepares to launch the ESG Enhanced Equity Index, a new product aimed at sustainability-conscious investors. The case highlights the technical, organizational, and cultural challenges of embedding ESG into a traditionally quantitative investment process. It also explores the roles of key leaders, Cheree Dyers (CEO), Michelle Green (ESG Committee Chair), and Bastian Teichgreeber (CIO), as they navigate data limitations, third-party partnerships, and internal change management.
* **Citation:** Giamporcaro, S., & Harrison, D. (2025). *ESG integration at Prescient Investment Management: Becoming a quantitative responsible investor* (Case No. UCT023-PDF-ENG). University of Cape Town Graduate School of Business.

❖ Tractor Supply: Steering ESG Strategy as Consumer Beliefs Shift

- **Overview**: Published in 2025, this case explores how Tractor Supply Company, a major U.S. retailer serving rural and agricultural communities, navigated the sociopolitical backlash to its ESG (Environmental, Social, and Governance) and DEI (Diversity, Equity, and Inclusion) initiatives. In 2021, the company launched ambitious ESG goals, including reducing carbon emissions, increasing diversity in leadership, and expanding spending with diverse suppliers. However, in 2024, conservative influencer Robby Starbuck publicly criticized these initiatives, sparking a viral backlash that began to affect customer sentiment and sales. In response, Tractor Supply announced a strategic pivot: it would withdraw its DEI and emissions goals and refocus on "core rural American values" such as agricultural

education and animal welfare. This move, however, triggered criticism from progressive stakeholders, creating a new reputational challenge.
  - The case provides a rich context for discussing:
    - The risks of politicized consumer activism
    - The tension between brand values and stakeholder expectations
    - How companies can navigate ESG strategy in a polarized environment

  It is well-suited for courses in brand management, leadership communication, marketing strategy, and corporate governance.
- **Citation**: Whitler, K., Firebaugh, H., Grossman, J. (2025). *Tractor Supply: Steering ESG Strategy as Consumer Beliefs Shift* (Case No. UV9179-PDF-ENG).

11.5.5. Strategy-in-Film: Learn by Watching

Sustainability has become a defining challenge—and opportunity—for modern organizations. The films below portray how companies confront environmental and social pressures, balance competing priorities, and integrate sustainability into strategic decision-making. As you explore sustainability analytics in this chapter, use these narratives to reflect on how data, systems thinking, and ethical considerations shape sustainable competitive advantage.

Movie / Documentary	Strategic ESG Theme Highlighted
Before the Flood	Global climate impact and systemic business response
The True Cost	Ethical trade-offs in global supply chains and circularity
Our Planet	Data visualization for environmental awareness
Dark Waters	Corporate accountability and regulatory failure
Wall-E	Long-term consequences of unsustainable systems
The Social Dilemma	Tech ethics, bias, and stakeholder transparency
An Inconvenient Truth	Climate modeling and foresight in strategic advocacy
RiverBlue	Fashion industry impact and sustainable design
Cowspiracy: The Sustainability Secret	Emissions, systems thinking, and stakeholder denial
Don't Look Up	Strategic communication failure in the face of global risk

References

Barnes, S. J., Guo, Y., & Chan, J. (2022). Big Data analytics for sustainability: Insight through technological innovation. *Information & Management*, *59*(5), 103627.

Bhardwaj, E., & Khaiter, P. A. (2023). What data analytics can or cannot do for climate change studies: An inventory of interactive visual tools. *Ecological Informatics*, *73*, 101918.

Blomsma, F., Pieroni, M., Kravchenko, M., Pigosso, D. C., Hildenbrand, J., Kristinsdottir, A. R., ... & McAloone, T. C. (2019). Developing a circular strategies framework for manufacturing companies to support circular economy-oriented innovation. *Journal of Cleaner Production*, *241*, 118271.

Bridoux, F., & Stoelhorst, J. W. (2022). Stakeholder theory, strategy, and organization: Past, present, and future. *Strategic Organization*, *20*(4), 797-809.

Bryson, J. M., Crosby, B. C., & Bloomberg, L. (2014). Public value governance: Moving beyond traditional public administration and the new public management. *Public Administration Review*, *74*(4), 445-456.

Dmytriyev, S. D., Freeman, R. E., & Hörisch, J. (2021). The relationship between stakeholder theory and corporate social responsibility: Differences, similarities, and implications for social issues in management. *Journal of Management Studies*, *58*(6), 1441-1470.

Drempetic, S., Klein, C., & Zwergel, B. (2020). The influence of firm size on the ESG score: Corporate sustainability ratings under review. *Journal of Business Ethics*, *167*(2), 333-360.

Eccles, R. G., Ioannou, I., & Serafeim, G. (2014). The impact of corporate sustainability on organizational processes and performance. *Management science*, *60*(11), 2835-2857.

Fiedler, T., Pitman, A. J., Mackenzie, K., Wood, N., Jakob, C., & Perkins-Kirkpatrick, S. E. (2021). Business risk and the emergence of climate analytics. *Nature Climate Change*, *11*(2), 87-94.

Freeman, R. E., Dmytriyev, S. D., & Phillips, R. A. (2021). Stakeholder theory and the resource-based view of the firm. *Journal of Management*, *47*(7), 1757-1770.

Hamdan, R. K., & Buallay, A. (Eds.). (2024). *Artificial Intelligence (AI) and Customer Social Responsibility (CSR)*. Springer.

Karpoff, J. M. (2021). On a stakeholder model of corporate governance. *Financial Management*, *50*(2), 321-343.

Kayikci, Y., Kazancoglu, Y., Gozacan-Chase, N., & Lafci, C. (2022). Analyzing the drivers of smart sustainable circular supply chain for sustainable development goals through stakeholder theory. *Business Strategy and the Environment*, *31*(7), 3335-3353.

Khan, S. A. R., Yu, Z., & Farooq, K. (2023). Green capabilities, green purchasing, and triple bottom line performance: Leading toward environmental sustainability. *Business strategy and the Environment*, *32*(4), 2022-2034.

Khan, I. S., Ahmad, M. O., & Majava, J. (2021). Industry 4.0 and sustainable development: A systematic mapping of triple bottom line, Circular Economy, and Sustainable Business Models perspectives. *Journal of Cleaner Production*, *297*, 126655.

Khurshid, H., & Snell, R. S. (2022). Examining distinctions and relationships between creating shared value (CSV) and corporate social responsibility (CSR) in eight Asia-based firms. *Asian Journal of Business Ethics*, *11*(2), 327-357.

Kortetmäki, T., Heikkinen, A., & Jokinen, A. (2023). Particularizing nonhuman nature in stakeholder theory: The recognition approach. *Journal of Business Ethics*, *185*(1), 17-31.

Kotsantonis, S., Pinney, C., & Serafeim, G. (2016). ESG integration in investment management: Myths and realities. *Journal of Applied Corporate Finance*, *28*(2), 10-16.

Kristoffersen, E., Blomsma, F., Mikalef, P., & Li, J. (2020). The smart circular economy: A digital-enabled circular strategies framework for manufacturing companies. *Journal of Business Research*, *120*, 241-261.

Kumar, S., Sharma, D., Rao, S., Lim, W. M., & Mangla, S. K. (2025). Past, present, and future of sustainable finance: insights from big data analytics through machine learning of scholarly research. *Annals of Operations Research*, *345*(2), 1061-1104.

Li, Q., Zou, G., Zeng, W., Gao, J., He, F., & Zhang, Y. (2024). ESG guidance and artificial intelligence support for power systems analytics in the energy industry. *Scientific Reports*, *14*(1), 11347.

Liute, A., & De Giacomo, M. R. (2022). The environmental performance of UK-based B Corp companies: An analysis based on the triple bottom line approach. *Business Strategy and the Environment*, *31*(3), 810-827.

Pozzi, F. A., & Dwivedi, D. (2023, September). ESG and IoT: ensuring sustainability and social responsibility in the digital age. In *International Conference on Artificial Intelligence: Towards Sustainable Intelligence* (pp. 12-23). Cham: Springer Nature Switzerland.

Sain, S. R. (2023). Data science and climate risk analytics. *Environmetrics*, *34*(2), e2749.

Salvador, R., Barros, M. V., Freire, F., Halog, A., Piekarski, C. M., & De Francisco, A. C. (2021). Circular economy strategies on business modelling: Identifying the greatest influences. *Journal of Cleaner Production*, *299*, 126918.

Sariyer, G., Mangla, S. K., Chowdhury, S., Sozen, M. E., & Kazancoglu, Y. (2024). Predictive and prescriptive analytics for ESG performance evaluation: A case of Fortune 500 companies. *Journal of Business Research*, *181*, 114742.

Schnase, J. L. (2016). Climate analytics as a service. In *Cloud Computing in Ocean and Atmospheric Sciences* (pp. 187-219). Academic Press.

Stoelhorst, J. W., & Vishwanathan, P. (2024). Beyond primacy: A stakeholder theory of corporate governance. *Academy of Management Review*, *49*(1), 107-134.

Chapter 12

Strategy Implementation and Execution With STEM

"Execution is where strategy becomes reality. In a world fueled by data, algorithms, and real time feedback, execution isn't the final step—it's the continuous engine that drives competitive advantage."

by MAK

Learning Outcomes

After completing this chapter, future strategists will be able to:

Execution isn't about following orders—it's about sensing, adjusting, and delivering in real time. Like the Border Collie, the modern strategist must move with purpose, precision, and perpetual awareness.

- Translate strategic objectives into measurable KPIs using strategy maps and Balanced Scorecard principles.
- Align cross-functional execution using digital governance systems, OKRs, and STEM-enabled dashboards.
- Apply agile frameworks to monitor, adjust, and accelerate strategic execution in real time.
- Evaluate and integrate digital tools, including dashboards, predictive analytics, and sprint analytics, into execution workflows.
- Diagnose and manage strategic change using data-driven techniques such as sentiment analysis and digital simulations.
- Design end-to-end execution systems that connect strategic intent to frontline action through smart, adaptive technologies.
- Develop the mindset and toolkit of a STEM-literate strategy executor capable of navigating dynamic business environments.

Strategy may begin in boardrooms and whitepapers, but it either triumphs or fails in execution. A well-crafted vision loses all value if it cannot be translated into visible, measurable, and sustainable results. In a STEM-enabled world, where algorithms predict disruptions before they occur, sensors monitor performance in real time, and agile teams iterate faster than plans can be printed, execution has evolved from an operational afterthought into a strategic imperative.

Gone are the days when execution was treated as a linear rollout or the final phase of a quarterly plan. Today, strategic execution is an integrated, live system. Powered by data dashboards, AI-driven KPIs, cloud collaboration tools, and digital governance platforms, it allows organizations to sense performance deviations, respond to real-time feedback, and recalibrate tactics continuously. In this environment, strategy is not complete when the PowerPoint ends; it is only beginning.

Consider the case of Moderna, which during the COVID-19 crisis operationalized its bold strategy using digital twins, cloud-based monitoring, and real-time decision dashboards. It wasn't just the science that saved lives; it was the precision and agility of execution. This example underscores a fundamental truth: in volatile, tech-driven markets, the ability to execute with speed, intelligence, and adaptability often matters more than the brilliance of the original plan.

In this chapter, we reframe execution as a strategic capability, one that blends classical performance management with STEM logic and tools. You'll learn how to design KPIs that reflect strategic goals, implement agile frameworks that support dynamic change, build dashboards that connect data to action, and lead change with measurable precision. Whether you're revamping a product line or launching a sustainability initiative, this chapter equips you to implement strategy with the same analytical rigor used in strategy design.

Execution is no longer a handoff. It is a system, measured in metrics, shaped by feedback, and steered through dashboards. In this new paradigm, strategic leaders must become execution architects. Let's begin.

Opening Case: From Planning to Performance—Ping An's Execution Engine

Ping An, a financial services giant headquartered in Shenzhen, China, redefined the rules of execution, not through cost-cutting or better planning, but by turning strategy into a digital operating system. Originally a traditional insurance firm, Ping An transformed itself into a technology-driven conglomerate, using AI, cloud infrastructure, and digital twins not only to shape strategy, but to execute it at scale and speed.

Instead of treating execution as a downstream operational task, Ping An embedded it into the strategic core. Performance dashboards integrated with every unit, from customer service to

R&D, allowed real-time monitoring of strategic KPIs like customer churn risk, claims cycle time, and cross-product engagement. The company developed a modular IT architecture where execution feedback loops constantly informed strategic priorities, turning each department into a sensor in the broader system.

For example, Ping An's "Smart Insurance Brain" uses AI to process claims, predict fraud, and recommend next actions, automatically adjusting resource allocation and operational flows. Its digital twin systems simulate policy portfolios and regulatory responses, while decision dashboards alert leadership to deviations from performance targets before they become risks.

The result? Strategy is no longer reviewed annually; it evolves continuously. Performance gaps are not analyzed months later; they are flagged and addressed within hours. Execution is no longer a lagging function; it's a living, learning process that drives strategic advantage in real time.

Play the Strategist

1. How did Ping An blur the line between strategy formulation and execution using STEM tools?
2. What role did real-time dashboards and digital twins play in Ping An's execution system?
3. How can organizations in less tech-intensive industries apply similar execution logic?
4. What challenges might a firm face when embedding live feedback loops into its execution system, and how could they be addressed?

12.1 Translating Strategy into Actionable Systems

Strategy, no matter how compelling on paper, remains inert until it is translated into action. The real test of a strategic plan is whether it leads to meaningful performance on the ground. In STEM-enabled organizations, this translation is no longer manual or delayed; it is systemic, measurable, and often real-time. This section explores how to convert abstract strategic intent into actionable systems by using tools like strategy maps, Balanced Scorecards (BSC), and advanced KPI architectures.

By applying these tools, strategists ensure that goals don't remain aspirational; they become operational. You will learn how to design key performance indicators (KPIs) that reflect strategic objectives, how to automate measurement through digital dashboards, and how to align execution systems across teams, departments, and organizational layers.

Execution begins with translation, and that translation requires measurement. Strategy maps, when paired with analytics, become the bridge between vision and velocity.

12.1.1 From Strategy Maps to Execution KPIs

Strategic objectives often begin as high-level goals: expand markets, lead in sustainability, improve customer retention. But without translation into KPIs, measurable markers of progress, these goals remain conceptual. The first step in execution is designing a performance system that gives strategy visibility, accountability, and momentum.

The Balanced Scorecard (BSC), introduced by Kaplan and Norton, remains a foundational tool for this translation. It organizes strategy into four perspectives:

- **Financial** (e.g., revenue growth, cost efficiency)
- **Customer** (e.g., satisfaction, retention, NPS)
- **Internal Processes** (e.g., cycle time, error rate)
- **Learning and Growth** (e.g., innovation rate, employee engagement)

What makes the BSC powerful in a STEM context is its adaptability. Each strategic objective becomes a node in a **strategy map**, which visually shows causal links across departments and time. These maps can be enhanced with real-time data pipelines, predictive indicators, and simulation layers. For example, a strategy to increase digital engagement might start with staff training KPIs (learning), drive app improvements (internal), boost app usage (customer), and increase upsells (financial).

Importantly, KPIs should not be limited to end results. High-performing execution systems distinguish between:

- **Input KPIs**: resources invested (e.g., hours of training)
- **Process KPIs**: operational actions (e.g., claims processed per day)
- **Output KPIs**: outcomes achieved (e.g., reduction in service response time)
- **Outcome KPIs**: strategic effects (e.g., customer loyalty increase)

A well-structured KPI system allows leaders to track not just *what* is happening but *why*. In STEM-powered organizations, this is the foundation for digital governance, agile adaptation, and intelligent execution.

12.1.2 KPI Automation and Real-Time Monitoring

In the past, tracking KPIs meant compiling monthly reports or quarterly dashboards, often static, backward-looking, and prone to delay. But today's execution environments demand faster insight. In STEM-enabled organizations, KPIs are no longer just snapshots; they are dynamic signals flowing through live systems. This shift transforms performance measurement from reactive reporting into active steering.

Automation is the core enabler. With modern data architectures, KPIs can be continuously collected, calculated, and visualized from multiple data streams, ERP systems, CRM platforms,

IoT sensors, cloud apps, and even AI models. Tools like **Power BI**, **Tableau**, or custom **Python/R dashboards** connect strategic metrics to live operations, giving leaders a real-time view of execution health.

For example, in a logistics firm, KPIs such as "on-time delivery rate," "warehouse throughput," and "fleet utilization" can be monitored by integrating GPS, inventory databases, and workforce management software. These data flows update execution dashboards minute by minute—allowing instant course correction if a metric slips below target.

Another advantage of KPI automation is **early warning.** By embedding predictive analytics into the KPI framework, systems can alert managers not only when something has gone wrong, but when it *might*. Churn risk, production delays, and budget overruns—these can all be forecasted before they surface, allowing proactive adjustment.

Automated KPIs also enhance transparency across the organization. When every department sees its metrics in real time—and understands how it contributes to the overall strategy—alignment improves. It becomes easier to reward progress, spot performance gaps, and maintain execution momentum.

Ultimately, automated KPIs are more than just digital scoreboards. They are execution triggers, strategic barometers, and feedback mechanisms, fueling continuous adaptation and learning.

12.1.3 Aligning KPIs Across Functions and Levels

Even the most sophisticated KPI systems can fail if they aren't aligned across the organization. Strategy loses traction when metrics live in silos, when the finance team tracks profitability, operations tracks efficiency, and customer success focuses only on satisfaction, with no shared view of what success means. True execution power comes from cascading alignment—when every function and every role can see how their performance connects to strategic goals.

This is where **KPI alignment** becomes essential. Using frameworks like the **Balanced Scorecard** and **OKRs (Objectives and Key Results)**, organizations can create a structured logic from boardroom objectives down to team and individual actions. A clear KPI chain helps maintain focus, eliminate redundancy, and ensure that progress at the edge of the organization reinforces, not conflicts with, the overall strategy.

For example, consider a strategic objective such as "reduce carbon emissions by 30% within three years." This high-level goal can be translated into aligned KPIs across organizational levels, as shown in the table below:

Table 12.1: Cascading KPI Alignment from Strategy to Execution

Strategic Objective	Executive KPI	Team KPI	Individual KPI
Reduce carbon emissions	Total CO_2 output	Emissions per facility	Machine-level energy efficiency
Improve customer experience	Net Promoter Score	Average support resolution time	First-call resolution rate
Boost digital transformation	% of automated tasks	AI models deployed per quarter	Code quality in automation APIs

To support alignment, real-time **dashboard layering** can be deployed. Senior leaders see aggregate KPIs across units, while managers drill down into team performance, and individuals track their own contributions. These dashboards often integrate with tools like **Salesforce**, **Workday**, or **custom analytics platforms**, creating a unified performance system with role-based access and feedback loops.

In STEM-enabled systems, this KPI architecture is not just about cascading metrics—it's about creating a living model of strategic alignment. Data flows continuously across layers, and updates in one area can trigger adjustments in others. Machine learning can detect misalignment (e.g., when team KPIs improve but strategic outcomes don't) and suggest recalibration.

When alignment is achieved, execution becomes focused, integrated, and measurable. Everyone from interns to executives knows not just what they're doing, but *why it matters strategically*.

12.1.4 Designing KPIs with STEM Intelligence

Designing KPIs is not just an exercise in choosing what to measure; it's a strategic act of engineering. In a STEM-enabled organization, KPIs must go beyond intuition or managerial consensus. They should be designed with statistical rigor, systems logic, and computational awareness. Why? Poorly designed KPIs lead to poor decisions, misaligned behavior, and distorted performance.

To avoid these traps, effective KPIs in the STEM world follow **four key design principles**:

1. **Statistical Validity and Reliability**

   A KPI must consistently measure what it claims to measure. Metrics like customer satisfaction, employee engagement, or sustainability performance should be validated with surveys, behavioral data, or audits. Sampling methods, confidence intervals, and reliability testing become critical in KPI design—especially when decisions depend on them.

2. **Causal Relevance**

   Great KPIs reflect not just what is easy to measure, but what actually drives outcomes. STEM thinkers use **system modeling**—including causal loops, regression analysis, or even Bayesian networks—to trace how operational inputs influence strategic goals. This avoids the common mistake of optimizing the wrong variable (e.g., cutting service time but increasing complaints).

3. **AI and Machine Learning Compatibility**

   In a world of predictive analytics and algorithmic forecasting, KPIs must be **machine-readable**. That means designing metrics that can feed directly into models—such as churn predictors, risk engines, or recommendation systems. Clean data structure, clear labeling, and historical time-series tracking make your KPIs ready for AI integration.

4. **Ethical and Strategic Balance**

   Not all KPIs are purely numeric. Some require **ethical judgment** and **qualitative insight**, especially in areas like diversity, well-being, or community impact. In these cases, KPIs may be blended: combining survey data, behavioral signals, and sentiment analysis to create multi-dimensional metrics. The key is to ensure they support—not undermine—the strategic identity of the organization.

KPI Reengineering in Retail Strategy

A global e-commerce firm wanted to improve its "customer satisfaction score," originally based on post-purchase surveys. However, STEM strategists redesigned the KPI suite to include delivery time variance, chatbot resolution latency, and personalized recommendation accuracy. This not only made satisfaction measurable in real time but also tied it to actionable engineering levers, leading to a 20% drop in churn within a single quarter.

In sum, KPIs are the instruments of execution. Designed well, they create transparency, precision, and adaptability. Designed poorly, they obscure reality. In a STEM-powered execution system, KPIs are no longer static—they are modeled, tested, iterated, and optimized like any intelligent system.

12.2 Strategic Alignment through Digital Governance

Execution doesn't break down because people don't care; it breaks down because systems don't connect. In many organizations, strategy lives in PowerPoint slides while operations run on disconnected spreadsheets and task trackers. Alignment becomes a matter of chance, not design. But in a STEM-enabled enterprise, alignment is a **live process**, sustained through **digital governance systems** that integrate objectives, metrics, tools, and teams in real time.

Digital governance transforms strategy execution from a top-down command structure into a **distributed control system**. It ensures that everyone, from executive leadership to frontline contributors, is not only aware of strategic priorities but actively aligned with them through data-driven systems, automated alerts, and decision protocols.

This section explores how tools like **OKRs**, **role-based dashboards**, and **scenario-triggered workflows** can turn static alignment models into responsive, intelligent frameworks. You'll learn how to engineer alignment through architecture, not memos.

12.2.1 Aligning Across Functions and Levels

Strategic success depends on synchronized movement, teams pulling in the same direction, at the same pace, toward the same goals. But alignment doesn't happen by accident. In complex organizations, each function has its own language, tools, and priorities. Without an intentional architecture, strategy execution fragments into isolated initiatives.

To solve this, STEM-driven organizations implement **alignment by design**. They use frameworks like **OKRs (Objectives and Key Results)** and **digital strategy maps** to ensure every layer of the organization translates strategic goals into measurable responsibilities. When done well, this alignment becomes dynamic and visible, powered by digital dashboards and continuous data flows.

Take, for example, a strategic objective to "lead the market in sustainable logistics." That vision must cascade into:

- **Executive-level OKRs**:
  - Objective: Achieve carbon-neutral delivery
  - Key Result: Reduce fleet emissions by 40% in 12 months
- **Operations Team Metrics**:
  - Transition to electric vehicles by Q3
  - Optimize delivery routing using AI to cut fuel usage
- **Warehouse KPIs**:
  - Improve energy efficiency score by 15%
  - Track real-time emissions from equipment and HVAC systems
- **Driver-Level KPIs**:
  - Maintain average emissions per delivery under the target
  - Adopt eco-driving protocols monitored by IoT trackers

This **cascading logic** ensures that strategy is not just understood, it's **lived**. Each team sees its impact. Each role becomes a contributor to the strategic mission.

In STEM environments, this alignment is managed through **real-time dashboards** tailored by role. Executives monitor aggregate performance. Middle managers see team-level progress. Individuals track their daily contribution. These views are updated automatically via data pipelines, eliminating the delay and distortion of manual reporting.

Furthermore, **cross-functional alignment** becomes easier when everyone operates from the same digital infrastructure. Cloud-based project tools (e.g., Jira, Monday.com), data visualization platforms (e.g., Power BI, Tableau), and shared OKR software (e.g., Perdoo, Gtmhub) allow transparency and accountability across silos.

Finally, advanced organizations go a step further: using **analytics to detect misalignment**. Machine learning models can flag when KPIs trend positively at the team level but fail to impact top-level objectives, signaling a need for recalibration. In this way, alignment is not just structural; it becomes measurable and adaptable.

Alignment in the digital era is no longer about meetings or memos. It is engineered into the system, visible, measurable, and reinforced continuously.

12.2.2 Digital Governance: From Control to Coordination

Traditional governance often conjures images of control, approvals, compliance, and top-down oversight. But in today's fast-moving, STEM-powered organizations, governance must evolve from **rigid control mechanisms** to **adaptive coordination systems**. The goal is no longer to enforce plans; it's to enable continuous strategic alignment in dynamic conditions.

Digital governance refers to the integration of **real-time monitoring**, **automated triggers**, and **data-driven decision protocols** into the fabric of execution. It replaces one-size-fits-all reporting cycles with intelligent systems that know *who* needs to act, *when*, and *why,* based on live performance data.

At the heart of digital governance are four core components:

1. **Live KPI Dashboards**

   These dashboards are role-specific, strategic, and responsive. They display not only current performance but also trend forecasts and risk thresholds. For instance, a dashboard for a VP of Supply Chain may show order fulfillment rates, forecasted stockouts, and carbon footprint per shipment, with color-coded alerts based on thresholds.

2. **Scenario-Triggered Alerts and Escalations**

   When key indicators deviate, such as a drop in conversion rate, a delay in product development, or a spike in churn, the system triggers alerts. These can be routed to the right individuals or teams, often with suggested actions based on predictive models. This ensures **responsiveness over retrospection**.

3. **Automated Workflows and Decision Gates**

   Rather than waiting for weekly stand-ups or email approvals, digital governance systems can launch **automated actions** when certain conditions are met. For example, if customer complaints exceed a threshold, the system might trigger a quality audit, notify the product team, and re-prioritize related backlog items.

4. **Embedded Ethics and Compliance Protocols**

   In areas like ESG, DEI, and data privacy, governance isn't just operational—it's ethical. Digital governance can incorporate **algorithmic guardrails** that block decisions that violate sustainability targets, fairness thresholds, or regulatory policies. These rules are enforced automatically—not just recommended.

Dynamic Governance at Ping An

At Ping An, strategy execution is governed by a **real-time intelligence system**. Every function—from insurance claims to healthcare services—feeds into dashboards powered by AI and predictive analytics. When deviations are detected (e.g., bottlenecks, fraud risk, underperformance), automated alerts are escalated and linked to scenario-specific workflows. Leaders no longer ask, "What happened?"—they ask, "What is about to happen, and how should we respond?"

By embedding coordination into the system, digital governance minimizes lags, reduces friction, and improves execution agility. It turns the organization into a **responsive network**—one that adapts continuously while staying grounded in strategic intent.

In this model, **strategy becomes self-adjusting**, **execution becomes intelligent**, and governance becomes a **real-time choreography of action**.

12.2.3 Dashboards and Role-Based Visibility

A dashboard is more than a digital display—it is the **nervous system** of strategic execution. In a STEM-driven organization, dashboards connect the strategic brain (top-level goals) with the operational limbs (daily actions), allowing data to flow seamlessly in both directions. But for dashboards to truly drive performance, they must be tailored—not generic. This is where **role-based visibility** becomes critical.

Every stakeholder in the organization engages with strategy from a different vantage point. A CFO needs macro-level trends and risk flags. A marketing manager needs campaign conversion rates. A frontline supervisor needs shift performance indicators. If all of them stare at the same dashboard, execution suffers from information overload or blind spots. **Dashboards must be customized by role, relevance, and responsibility.**

Here's how effective role-based dashboards function in execution:

1. **Executive Dashboards: Strategic Oversight**

   These dashboards highlight performance against strategic objectives—such as financial targets, ESG commitments, innovation metrics, and customer retention. They also integrate **predictive insights**, such as trend trajectories or scenario outcomes. Executives need synthesis, not noise.

2. **Manager Dashboards: Operational Alignment**

   Middle managers require a blend of leading and lagging indicators, typically segmented by unit or geography. Their dashboards often include KPI breakdowns, variance analysis, and cross-functional dependencies—allowing them to troubleshoot and adjust in real time.

3. **Team and Individual Dashboards: Execution Clarity**

   At the team level, dashboards offer granular performance views tied to OKRs, sprint goals, or process KPIs. When designed well, they offer **immediate feedback**, gamify progress, and foster ownership. Employees can see exactly how their tasks impact broader objectives.

4. **Governance and Compliance Views: Oversight with Integrity**

   For legal, sustainability, or compliance teams, dashboards must include threshold breaches, incident flags, and policy adherence metrics—enabling real-time governance and ethical oversight.

STEM Tools in Action:

Tools like **Microsoft Power BI**, **Tableau**, **Looker**, or **custom dashboards built in R or Python (e.g., using Shiny)** allow real-time connectivity to data sources and the flexibility to create multiple dashboard views, each aligned with a user's strategic role. Some platforms also offer **AI-driven dashboard assistants**, which surface anomalies, insights, or risks automatically.

Critically, these dashboards are not static. They evolve with strategy. As goals shift or priorities change, dashboards can be reconfigured through agile design, ensuring they remain strategically relevant.

When dashboards reflect the right level of detail, match the user's decision scope, and offer real-time interactivity, they become more than tools—they become **strategic interfaces**. They allow everyone in the organization to see what matters most, act on what's relevant, and align with what's next.

12.2.4 Ethically Aligned Decision Protocols

As organizations automate decision-making and accelerate execution through STEM tools, one essential principle must not be lost: **ethics must travel with speed**. In a digital governance system where data flows freely and dashboards trigger automated actions, the absence of ethical checks can lead to strategic missteps, reputational damage, or even systemic harm. That's why the most advanced execution systems now embed **ethically aligned decision protocols**, ensuring that every automated insight or recommendation also passes through a lens of responsibility.

Unlike compliance-focused controls that intervene late, ethical protocols in STEM-aligned strategy are **built into the decision architecture**. These protocols act as dynamic rules, boundary conditions, and exception alerts, guiding actions without requiring constant human oversight.

Here are three core components of ethically aligned decision systems:

1. **Embedded Thresholds for Social and Environmental Impact**

   Dashboards and workflows can be programmed to **block** or **escalate** decisions that violate pre-set ethical or sustainability limits. For example:

   - A cost-reduction plan that cuts labor hours below ethical labor standards triggers a review.
   - An emissions-reducing KPI that inadvertently worsens working conditions or public health metrics is flagged for rebalancing.
2. **Algorithmic Bias Detection and Transparency Flags**

   When AI is involved in execution—such as in hiring, pricing, or risk scoring—**bias detection tools** should monitor outputs for systemic unfairness. Dashboards can include **bias alerts**, explainable AI outputs (e.g., SHAP values), and **fairness indicators** to guide decision-makers toward inclusive and just outcomes.

3. **Ethics-as-Code: Scenario-Based Decision Protocols**

   Leading organizations are now codifying their values into **decision trees** and **scenario rules**. For example:

    - "If profit margin exceeds 25% but customer churn rises above 10%, pause bonus payouts and re-evaluate value delivery."
    - "If a product recommendation engine increases engagement but lowers diversity exposure, initiate an ethics review."

Case Insight: ESG-Linked Dashboards in Financial Services

A European investment firm implemented ESG-linked execution dashboards. If a proposed portfolio adjustment improved returns but reduced the ESG rating below a threshold, the platform automatically rerouted the proposal to an ethics committee. This protocol ensured that strategy execution remained consistent with both fiduciary duty and societal values—without slowing down day-to-day operations.

In a STEM world, **speed and ethics must coexist**. Strategists must ensure that the systems built for efficiency and precision are also programmed for integrity. Ethically aligned protocols are not obstacles—they are design features that safeguard execution from short-sighted success.

12.3 Agile Execution and Continuous Adaptation

Strategy doesn't fail because it's wrong—it fails because the world changes faster than the plan. In traditional organizations, execution unfolds in sequential phases: plan, execute, report, adjust. But in today's volatile, feedback-rich environments, this linear model is no longer sufficient. Strategy must adapt continuously, and execution must keep pace.

This is where **agile execution** enters the frame. Originally developed for software teams, agile principles now underpin how high-performing organizations execute strategy across marketing, operations, sustainability, and finance. In a STEM-enabled context, agility is more than flexibility; it's a **structured system of feedback loops**, analytics tracking, sprint cycles, and data-informed pivots.

Agile execution allows organizations to:

- Break strategy into testable increments.
- Track delivery and impact in real time.
- Respond to performance signals with speed and precision.
- Prioritize continuously based on shifting insights.

In this section, you'll explore how agile frameworks enable execution to become **adaptive**, **intelligent**, and **aligned with evolving strategy**. You'll also learn how analytics, dashboards, and role-based metrics fuel a culture of continuous learning.

Execution is no longer about hitting fixed milestones—it's about moving in rhythm with your environment. Agile is that rhythm.

12.3.1 Agile Frameworks for Strategic Projects

Traditional execution frameworks rely on predictability, milestones planned months in advance, progress tracked through Gantt charts, and success measured at the end. But strategy, especially in fast-moving industries, rarely enjoys such stability. Markets shift, user needs evolve, and technologies disrupt assumptions. That's why strategic execution now demands **agile frameworks**—flexible, iterative models that turn execution into a continuous, adaptive process.

In a STEM-enabled organization, agile execution means breaking down big strategic goals into **short, measurable cycles** (often called sprints), enabling cross-functional teams to rapidly deliver value, evaluate performance, and adjust based on data.

Here are three widely used agile frameworks that can be adapted for strategic projects:

1. Scrum for Strategic Initiatives

- **Structure**: Projects are divided into 2–4 week sprints, each ending with a review and retrospective.
- **Roles**: Strategy lead (product owner), sprint facilitator (scrum master), cross-functional team members (data science, marketing, ops).
- **Use Case**: Ideal for innovation projects, ESG rollouts, or digital product launches.

For example, A healthcare provider launching a digital patient experience strategy used Scrum to test and refine each touchpoint—e.g., appointment booking, symptom checkers, feedback channels—one sprint at a time, with KPIs tracked live on Power BI.

2. SAFe (Scaled Agile Framework) for Enterprise Strategy

- **Structure**: SAFe aligns multiple agile teams to enterprise-wide strategy using synchronized planning, shared goals, and tiered backlogs.
- **Use Case**: Suitable for large organizations executing transformation programs across departments.

For example, A telecom firm digitizing its service model used SAFe to coordinate efforts between IT, marketing, HR, and customer experience. Strategic objectives were linked to program-level OKRs and tracked via dashboard portfolios.

3. OKRs with Agile Cadence

- **Structure**: Objectives and Key Results (OKRs) are set quarterly, reviewed in short cycles, and adjusted based on performance insights.
- **Use Case**: Best for organizations that want lightweight agility without full Scrum adoption.

For example, a sustainability-focused retail brand used OKRs with agile cadences to roll out its circular fashion initiative. Objectives were adjusted every four weeks based on supply chain feedback, customer uptake, and regulatory developments.

Agile frameworks do more than improve project velocity—they **embed learning into execution**. Each cycle becomes an experiment. Each metric becomes a guidepost. Teams aren't just delivering—they're **adapting**, **iterating**, and **advancing strategy in real time**.

In STEM-aligned execution, agile frameworks are the operating system. They keep strategy alive, responsive, and resilient.

12.3.2 Analytics-Driven Sprint Management

Agile execution delivers speed—but **analytics delivers direction**. Without data, agile teams may move quickly but blindly, iterating without insight. In STEM-powered strategy execution, sprint management is not just fast—it's **data-driven, performance-tracked, and impact-optimized**.

Each sprint becomes a mini-laboratory, where hypotheses about strategy are tested through short cycles of implementation, measurement, and adjustment. This transformation—moving from project updates to real-time metrics—turns every sprint into a strategic sensor.

Here's how analytics transforms agile sprint management:

1. Sprint-Level KPIs

Each sprint should have defined performance indicators, such as:

- **Completion rate**: % of planned tasks completed
- **Deployment frequency**: number of updates or releases
- **Velocity**: effort delivered per sprint (measured via story points or hours)
- **Defect ratio**: issues found per unit of output

In strategic initiatives, these metrics are complemented by **impact KPIs**, such as:

- Customer adoption rate
- Revenue contribution per feature
- Change in net promoter score (NPS)

Insight: In a financial services firm redesigning its customer onboarding, each sprint was evaluated not just by deliverables, but by downstream effect—e.g., reduction in time-to-first-transaction and increase in onboarding completion rates.

2. Continuous Dashboarding

Live dashboards (built in Power BI, Tableau, or Notion + Jira integrations) display sprint performance metrics, flag bottlenecks, and visualize progress toward strategic OKRs. These dashboards are **visible to all teams**, enabling transparency and joint problem-solving.

For example, a green tech startup used a sprint analytics dashboard to monitor both progress (task completion, code quality) and outcomes (reduction in energy usage by beta customers). This dual focus helped teams prioritize features that mattered most.

3. Retrospective Data Reviews

At the end of each sprint, analytics should drive the retrospective discussion:

- Which metrics improved? Which stalled?
- What user signals or market data emerged?
- What will we change next sprint—and why?

Retrospectives become **mini strategy labs**, where decisions are based on pattern recognition, not opinion.

4. Predictive Sprint Planning

Advanced teams use AI or statistical modeling to **forecast sprint performance**. For example:

- Expected backlog velocity based on team history
- Likelihood of delay based on dependency map
- User behavior models to prioritize next backlog items

This makes sprint planning **forward-looking**, not just capacity-based.

In short, analytics transforms agile from a **method** into a **strategy engine**. It ensures that execution is not just iterative, but intelligent, moving the strategy forward based on evidence, not assumption.

12.3.3 Agile Tools and Platform Integration

Agile execution is not just a mindset—it's an infrastructure. To operationalize agile strategy at scale, organizations must connect people, tasks, data, and decisions through **integrated digital platforms**. These tools don't merely support execution—they shape how strategy flows through the organization. In a STEM-enabled firm, agile tools serve as **strategic operating systems**, unifying workflows, tracking performance, and enabling continuous adaptation.

Here are the four key categories of agile tools that fuel seamless execution:

1. Task and Sprint Management Platforms

These tools help break down strategic goals into sprints, assign responsibilities, and visualize progress:

- **Jira** – Ideal for complex, cross-functional teams. Integrates with codebases, product roadmaps, and backlog grooming.
- **Asana / Trello** – Lightweight and visual for marketing, HR, and non-tech teams.
- **ClickUp / Monday.com** – All-in-one platforms that combine tasks, timelines, docs, and dashboards.

Strategic Use Case: A sustainability team rolling out a waste reduction strategy used Monday.com to assign initiatives by region, track sprint progress, and monitor pilot results, all within a unified view.

2. Performance and Analytics Dashboards

Agile tools are powerful when combined with real-time performance analytics:

- **Power BI / Tableau / Looker** – Visualize sprint KPIs, team velocity, backlog trends, and customer impact.
- **Jira Advanced Roadmaps or GitLab Insights** – Built-in analytics for sprint forecasting, deployment metrics, and blockers.

Insight: A global logistics firm integrated Jira with Power BI to create a strategy execution dashboard—showing OKR alignment, team delivery rates, and ESG performance in one real-time panel.

3. Communication and Collaboration Layers

Agility requires fast, fluid communication:

- **Slack / Microsoft Teams** – Real-time messaging and alert routing tied to sprint progress.
- **Miro / Figma** – Collaborative whiteboarding for strategy ideation, backlog prioritization, and systems mapping.

Tip: Connect tools through APIs—e.g., trigger Slack alerts when KPIs drop below target or backlog items are blocked for more than 48 hours.

4. Strategic Integration Platforms

For agility to scale, tools must talk to each other. Integration is no longer optional—it's a core capability.

- **Zapier / Make / Workato** – No-code automation across platforms (e.g., update CRM when customer-facing sprint completes).
- **Custom APIs and Python/R scripts** – For STEM teams with data science capacity, integrations enable advanced analytics, machine learning pipelines, or digital twin simulations feeding into agile workflows.

In summary, agile execution is accelerated by **tool orchestration**. Strategy is no longer trapped in slides or quarterly meetings—it's lived in sprint boards, dashboards, and message threads. The more tightly integrated the tools, the more responsive the organization becomes.

Tools don't make strategy agile. But they **make agility possible** at the speed, scale, and complexity the STEM world demands.

12.3.4 Learning Loops and Adaptive Strategy Cycles

Agile execution is more than a project rhythm—it's a learning system. In high-performing, STEM-enabled organizations, strategy is not delivered once and reviewed annually. Instead, it is treated as a **living hypothesis**, tested and adjusted continuously through **learning loops** and **adaptive strategy cycles**.

A learning loop is the core mechanism that turns execution data into strategic insight. When sprints generate real-time metrics, retrospectives uncover causes, and feedback loops inform the next cycle, the organization becomes **self-correcting**. Execution doesn't just follow strategy—it improves it.

Here's how learning loops function in adaptive strategy:

1. Micro-Cycles: Sprint Retrospectives as Strategic Reviews

At the end of every agile sprint, teams reflect on what worked, what didn't, and what to change. In STEM-enabled settings, these retrospectives aren't anecdotal—they're data-driven:

- Sprint-level KPIs are reviewed for trends and anomalies.
- Root causes are mapped using visual tools or causal models.
- Adjustments are proposed not just to tactics, but to strategic assumptions.

Example: A city government piloting smart waste systems found, through sprint retrospectives, that data transmission lag—not collection logistics—was the primary bottleneck. The strategy was updated mid-cycle to prioritize IoT optimization.

2. Meso-Cycles: Quarterly OKR and KPI Recalibration

Every few sprints, organizations hold **strategy syncs** to revisit and recalibrate OKRs based on:

- Real-world learning from execution
- External signals (e.g., customer behavior, competitive shifts, market data)
- Emergent risks or opportunities

This mid-level cycle ensures that agile execution doesn't drift away from strategy—or that outdated strategy doesn't constrain high-potential pivots.

Practice Tip: Pair quarterly OKR reviews with "KPI retirement" discussions—remove metrics that no longer serve evolving goals.

3. Macro-Cycles: Strategic Hypothesis Refinement

At the highest level, firms now treat their strategic plans as **testable models**. Like a scientist revising a theory, leadership continuously refines its understanding of:

- What creates value
- What drives differentiation
- What unlocks resilience

These macro-cycles often involve simulation, forecasting, and scenario planning tools—powered by machine learning or system dynamics—to test updated hypotheses.

Case Insight: A fintech company used machine learning to model which customer engagement strategies actually predicted retention. Quarterly insights from learning loops led to a strategic pivot—reallocating resources from outreach to in-app personalization.

"Agile is not just about how fast you move. It's about how quickly you learn—and how strategically you respond."

12.4 Managing Strategic Change in Tech-Enabled Firms

Even the most innovative strategies and well-aligned execution systems can fail—if the organization resists change. In tech-enabled firms, the pace of transformation is relentless: AI tools reshape workflows, new platforms replace legacy systems, and cross-functional teams must adapt quickly to strategic pivots. Yet human systems often lag behind digital ones. Habits, hierarchies, and uncertainty can slow or derail even the best-executed plans.

That's why managing change is not a side function—it's a **strategic capability**. In a STEM-powered execution environment, change management must evolve from training manuals and top-down announcements to **data-informed, feedback-driven, and psychologically attuned systems**.

This section explores how to diagnose change readiness using data, manage organizational transformation with agility, and embed adaptive behaviors into the culture. You'll also learn how STEM tools—like sentiment analysis, organizational network mapping, and simulation models—are transforming how we lead strategic change.

Execution is not just about *what* people do—it's about how they adapt to doing it differently.

12.4.1 Diagnosing Change Readiness with Data

Most change initiatives fail not because the strategy was wrong, but because the organization wasn't ready. Resistance is rarely loud—it's silent. It shows up in delayed adoption, reduced engagement, or quiet workarounds. That's why leading firms now use **data to diagnose change readiness** before launching major transformations. In STEM-enabled strategy execution, readiness is no longer guessed—it's measured.

Here are four data-driven approaches to assess and predict organizational readiness for change:

1. Organizational Sentiment Analysis

By analyzing employee feedback—from surveys, performance reviews, open-ended comments, and internal communication channels—leaders can detect readiness patterns:

- Confidence in leadership and vision
- Perceived psychological safety
- Trust in new tools or systems
- Hidden anxieties or skepticism

Advanced platforms apply **natural language processing (NLP)** to mine this feedback in real time, revealing patterns across departments or teams. A sudden drop in engagement scores after a strategy announcement, for example, can signal resistance early—before behaviors surface.

Example: A global bank used sentiment dashboards to track reactions to its digital transformation strategy. Negative sentiment clusters in regional offices led to targeted leadership coaching and clearer communication.

2. Network Analysis and Change Influence Mapping

Change doesn't spread evenly. It travels through **informal networks**, where influence and trust reside. Using tools like **organizational network analysis (ONA)**, firms can map internal relationships to:

- Identify early adopters and hidden influencers
- Locate isolated or resistant nodes
- Target communication and training more effectively

Rather than pushing change through formal roles, ONA allows organizations to **mobilize the social architecture** of change.

Tip: Pair influencers with visible ownership roles in the change initiative. Let influence work *with* formal structure, not around it.

3. Readiness Heatmaps by Role, Region, or Function

Using survey data, system usage logs, and engagement metrics, leaders can create **change readiness heatmaps** across units. These maps visualize which areas:

- Are enthusiastic and already experimenting
- Are cautious but open with support
- Are actively resistant or disengaged

Insight: A tech company implementing AI decision tools used heatmaps to see where resistance correlated with low data literacy. It launched targeted learning modules before full deployment, increasing adoption by 30%.

4. Predictive Modeling of Adoption Risk

With enough historical data, machine learning models can predict:

- Which users are likely to adopt or resist new systems
- Which teams might miss implementation milestones
- Where communication gaps will cause delays

These models feed into **preemptive change strategies**, guiding where to invest time, coaching, or communication before problems arise.

Change readiness is no longer a soft estimate—it's a **strategic signal**. By applying STEM tools, firms can replace intuition with insight, build tailored change interventions, and give their strategy a better chance to succeed.

In a data-driven world, **successful execution starts with readiness.**

12.4.2 STEM Tools for Change Management

In tech-enabled organizations, change is no longer episodic—it's constant. Traditional change management, focused on static communication plans and annual training cycles, can't keep up. Instead, modern firms deploy **STEM-powered tools** that enable **real-time sensing, adaptive learning, and system-level coordination** throughout the change process.

Here are four STEM-integrated tools that transform how change is managed:

1. Digital Twin Models for Change Simulation

Before launching a strategic initiative, firms can create **digital twins**—simulated replicas of the organization or its subsystems—to test the potential impacts of change. These models help leaders:

- Forecast operational bottlenecks
- Visualize interdependencies between teams
- Test scenarios for resource shifts, role changes, or tech adoption

Example: A manufacturing firm used a digital twin of its production system to simulate a shift to automation. It uncovered that the real risk wasn't technical—it was a retraining bottleneck in middle management. The change plan was revised accordingly.

2. Real-Time Adoption Dashboards

Change leaders can track adoption metrics live, using dashboards connected to learning platforms, communication tools, and system usage logs. These dashboards visualize:

- Login frequency or usage rates of new tools
- Course completion for training programs
- Click-through or response rates for change communications
- Variations by region, team, or role

Insight: A healthcare provider used a Power BI dashboard to monitor physician engagement with a new diagnostics platform. When one department lagged, leadership offered optional micro-coaching sessions—which doubled adoption within a month.

3. Automated Nudges and Behavioral Reinforcement

Instead of relying on one-time announcements, smart systems now deliver **personalized nudges**—short, targeted prompts delivered via email, Slack, or dashboards—to reinforce behavior aligned with the change.

Nudges might include:

- "You haven't used the new reporting tool this week—want a 2-minute refresher?"
- "Your team is 80% aligned with the new workflow. Here's what's left."

These can be triggered by **behavioral rules** or **machine learning models** that detect risk patterns.

Tool Examples: Microsoft Viva, Pendo, or custom nudge bots built into Slack ecosystems.

4. Continuous Learning Systems with Feedback Loops

Instead of separating change from learning, modern platforms integrate both. Tools like Coursera for Business, Degreed, or even in-house LMS systems track what employees are learning, offer adaptive pathways, and feed data back into the change dashboard.

Learning becomes **real-time and responsive**, not just instructional.

Practice Tip: Use completion rates and quiz performance to adapt the content dynamically—e.g., simplify modules if users struggle with key concepts.

Strategy in Action: Change as a Dynamic System

A multinational tech firm rolled out a strategy to embed AI into all business units. Rather than rely on memos and one-time town halls, it launched a STEM-powered change platform:

- Simulated risks with digital twins
- Used dashboards to monitor AI usage and skill-building
- Delivered nudges when engagement dropped
- Adapted content weekly based on learning analytics

The result? Full rollout in six months with minimal disruption and high employee satisfaction.

In the STEM era, change is not managed with posters and patience. It's **measured, modeled, and adapted**—just like any other strategic system. With the right tools, change becomes less of a shock—and more of a capability.

12.5 Designing Execution Systems in STEM Organizations

In traditional organizations, execution is often a chain of tools: a project plan here, a KPI spreadsheet there, a few dashboards patched together. But in STEM-driven firms, execution is **not a toolkit—it's a system**: integrated, intelligent, and constantly learning. Strategy doesn't just flow down from leadership—it moves across platforms, through algorithms, and into every corner of the organization in real time.

This section explores how to architect complete **strategic execution systems**—built not just for performance, but for precision, coordination, and adaptability. You'll learn how to structure a technology-enabled execution stack, connect strategic planning to frontline behavior, and create platforms that reinforce strategy continuously.

In a STEM world, execution systems must do more than manage tasks—they must **sense, simulate, respond, and evolve.** This is execution by design.

12.5.1 Building the Strategic Execution Stack

Think of a STEM-enabled organization as a living system. Strategy is the brain, execution is the nervous system, and the **execution stack** is the infrastructure that keeps them connected. In traditional organizations, execution is fragmented—tools don't talk, data isn't shared, and feedback loops are slow. In contrast, modern firms build **strategic execution stacks** that are integrated from top to bottom—turning goals into action with speed, intelligence, and precision.

A strategic execution stack is composed of five tightly integrated layers:

1. Strategy Layer

This is where enterprise goals, OKRs, and high-level plans are defined.

- Tools: Strategic planning platforms (e.g., WorkBoard, Cascade), cloud-based strategy maps
- Key Functions: Priority setting, goal alignment, scenario modeling
- STEM Impact: Models like Monte Carlo simulations and digital twins feed into the planning layer to test assumptions dynamically.

2. Translation Layer

Here, strategy is broken into sprints, programs, and cross-functional objectives.

- Tools: Jira, Asana, ClickUp, Monday.com
- Key Functions: Backlog creation, dependency mapping, sprint planning
- STEM Tip: Use AI assistants (e.g., Jira Smart Suggestions) to optimize backlog prioritization or detect bottlenecks in resource allocation.

3. Execution Layer

The engine room of delivery—where tasks, workflows, and decisions unfold.

- Tools: Workflow engines, robotic process automation (RPA), ERP platforms
- Key Functions: Task assignment, approvals, cross-team collaboration
- Integration: Connect execution tools with dashboards and performance platforms to sync real-time status with higher-level KPIs.

4. Monitoring & Feedback Layer

This layer senses performance, identifies deviations, and sends real-time signals.

- Tools: Power BI, Tableau, custom Python dashboards, telemetry data feeds

- Key Functions: Performance tracking, KPI alerting, anomaly detection
- STEM Bonus: Add predictive analytics and SHAP models to flag outcome risks before they happen.

5. Learning & Adaptation Layer

The final—and often neglected—layer is what gives the system its intelligence. Here, data from the previous layers is used to adjust the entire stack.

- Tools: Learning management systems, data lakes, strategic analytics platforms
- Key Functions: OKR recalibration, behavior tracking, training personalization
- Example: Machine learning tools analyze adoption patterns and recommend targeted coaching or reallocation of resources.

Execution Stack in Action

A logistics company deploying a digital route optimization strategy built its execution stack using:

- Cascade for OKRs (strategy layer)
- Jira for sprint execution (translation + execution)
- Power BI for real-time delivery performance (monitoring)
- An AI-driven analytics platform to reforecast delivery timelines weekly (adaptation)

The result: A 25% reduction in delivery variance and improved customer satisfaction within one quarter.

Strategic execution stacks aren't built overnight. But as each layer is added and integrated, the organization gains not just efficiency—but **strategic responsiveness**. In a STEM world, that responsiveness is no longer optional—it's a competitive necessity.

12.5.2 Connecting Strategy, Technology, and People

A well-designed execution stack is only powerful if it connects three core elements: **strategy, technology, and people**. In many organizations, these elements exist in silos. Strategy is defined by leadership, technology is managed by IT, and people operate in workflows shaped by neither. But in STEM-enabled firms, **alignment is engineered**—not assumed.

This section explores how to create that alignment through **interface design**, **behavioral data**, and **feedback mechanisms** that ensure strategy lives in both systems and people.

1. Strategy Translated into Interfaces

A strategy that lives only in a PowerPoint presentation is already dead. For execution to succeed, strategic priorities must be **visible and actionable** at the interface level. This means:

- Embedding strategic KPIs into dashboards that employees use daily.
- Linking OKRs to task management views, so contributors see how their work drives goals.
- Making progress toward objectives as easy to view as unread emails.

Example: A consumer goods firm connected its sustainability strategy to role-based dashboards. Factory workers could see CO_2 savings from their process improvements—making impact visible, not abstract.

2. Human-Centered Tech Configuration

Technology must **enable people**, not confuse them. Execution systems should be designed with behavioral friction in mind. That means:

- Reducing clicks and complexity in task systems.
- Using nudges to guide action (e.g., deadline alerts, priority highlights).
- Personalizing tools based on role, experience level, or performance trends.

Insight: Use heatmaps and behavior logs to identify where users drop off or delay tasks. Then revise the UI or automate the flow.

3. Feedback Loops Between Tools and Teams

Strategy lives or dies by how fast it **learns**. People need feedback on performance; tools need feedback from users. Strong execution systems use:

- In-app feedback (e.g., thumbs-up/down on dashboard insights).
- Weekly "pulse" surveys tied to OKRs.
- Embedded analytics that show progress—and flag stalling efforts.

Practice Tip: Pair dashboards with reflection prompts. For example: "You're 80% to your Q2 impact target. What's helping you succeed?"

4. Culture-Technology Synchronization

No system can fix a culture that punishes failure or hoards information. Execution systems should **reinforce behaviors** that the strategy requires, such as:

- Cross-functional collaboration
- Transparency of progress
- Safe failure and rapid iteration

Case Insight: A biotech firm redesigned its execution system to include "celebrated failure" tags—highlighting lessons from ideas that didn't work. This normalized experimentation in high-stakes teams.

"Execution isn't just what gets done—it's how strategy is wired into systems, screens, and minds."

12.5.3 Feedback Systems and Behavioral Engineering

In a STEM-enabled execution environment, feedback is not an afterthought—it's the engine. Strategy, once set in motion, must be **shaped continuously by real-world signals**. But not all feedback is equal. To drive meaningful behavior change and organizational learning, feedback must be **engineered—intentionally designed, well-timed, and role-sensitive**.

This section explores how feedback systems, powered by behavioral insights and STEM tools, can shape execution from the inside out.

1. Designing Feedback That Drives Action

Effective feedback is:

- **Timely**: Delivered at or near the moment of action.
- **Contextual**: Linked to specific goals, roles, or metrics.
- **Constructive**: Focused on improvement, not just measurement.

In execution platforms, this translates to:

- Micro-feedback on dashboard performance metrics
- Real-time nudges ("Your response rate dropped 15% this week—here's what to check.")
- Outcome-linked alerts ("This sprint fell short of KPI—review root cause?")

Tip: Use A/B testing to refine feedback formats—does a visual indicator outperform a text message? Which prompt boosts engagement?

2. Gamification and Behavioral Nudging

Behavioral science shows that people respond better to **reinforcement than reprimand**. Execution systems can incorporate elements such as:

- Progress bars and achievement streaks
- Peer benchmarks (e.g., "Top 10% of teams by response rate")
- Positive reinforcement nudges ("You're ahead of target—nice work!")

Example: A global retailer used weekly leaderboard dashboards for regional teams implementing a new supply chain tool. Engagement rose 28% quarter-over-quarter, with improved consistency across markets.

3. Multichannel Feedback Loops

Effective feedback doesn't only come from managers or dashboards. Modern systems collect and return feedback across multiple channels:

- **System-to-user**: Live dashboards, alerts, AI-generated suggestions
- **User-to-system**: Surveys, feedback buttons, qualitative notes
- **Peer-to-peer**: Collaboration platforms (Slack, MS Teams) that allow recognition or shared learning

Practice Tip: Embed two-way feedback in tools—allow users to annotate reports or suggest improvements within the execution platform.

4. Engineering Behavior at Scale

Behavioral engineering uses analytics and design principles to **shape habits** across the organization. This includes:

- Structuring workflows to reduce cognitive load (e.g., pre-filled templates, smart defaults)
- Using analytics to detect fatigue, disengagement, or overload
- Automating escalation or support triggers based on behavioral patterns

Application: A tech firm integrated an SHAP-based model into its HR platform to identify predictors of task completion delays. The system then proactively assigned light coaching resources to teams with high predicted risk.

"What gets measured gets managed—but what gets engineered gets done."

In STEM organizations, feedback is no longer episodic or reactive. It is **designed, personalized, and predictive**—creating a strategic execution culture where behaviors align with goals by default, not just intention.

12.5.4 Platforms That Reinforce Strategic Behavior

Technology platforms don't just support execution—they shape it. In STEM-enabled firms, platforms are no longer neutral tools. They act as **behavioral environments** that reinforce, redirect, or erode strategic intent based on how they're designed, deployed, and integrated. When platforms reflect the priorities of a firm's strategy, they help encode desired behaviors into the daily rhythm of work.

This section explores how to build and use platforms that **actively reinforce strategic behaviors**—by making the right actions easier, more visible, and more rewarding.

1. Strategic Friction and Flow

Well-designed platforms reduce **friction** for high-priority behaviors and increase it for low-value ones. This includes:

- Pre-configured templates aligned with strategic goals (e.g., sustainability reports auto-aligned with ESG standards)
- Automations for repeatable tasks, so more effort goes to strategic thinking
- Blockers for non-strategic work (e.g., alerts if task deviates from project scope)

Example: A media company embedded strategic prompts into its project tracker—teams couldn't close a campaign without tagging how it aligned with one of the firm's top OKRs.

2. Embedded Strategy Prompts

Execution platforms can remind users of the **"why"** behind their tasks through:

- Strategy banners or OKR callouts embedded in dashboards
- Role-based prompts (e.g., "This task contributes to ESG Goal 3: Waste Reduction")
- Real-time milestone tracking toward strategic themes

Tip: Link every task, campaign, or sprint to a visible strategy goal or metric. This transforms platforms into **strategy mirrors**—reflecting intent in every action.

3. Integrated Recognition and Incentive Loops

Behavior is reinforced when people see progress and receive recognition. Strategic execution platforms can:

- Show progress toward OKRs at individual, team, and org levels
- Highlight milestone completions with visual cues or badges
- Auto-generate recognition emails or Slack messages when key behaviors occur

Practice Insight: A health-tech firm used an integrated dashboard that awarded monthly badges to teams contributing most to patient onboarding KPIs—reinforcing both performance and alignment.

4. Adaptive Workflows and Smart Suggestions

Modern platforms use AI to learn from usage patterns and recommend next steps, priorities, or resources. These **intelligent systems** ensure that execution is both dynamic and aligned:

- Suggest re-prioritization based on lagging KPIs
- Recommend collaborators based on skill profiles or past task success
- Offer micro-training or tips based on recent performance

STEM Tool Example: Platforms like Notion AI, Microsoft Viva Insights, and WorkBoard AI coach users toward more strategic execution habits by learning and adapting to their behavior.

Final Insight:

"In high-performing firms, the platform isn't just where work happens—it's how strategy happens."

By designing execution platforms that **reflect, reinforce, and reward** strategic behavior, organizations transform tools from passive utilities into **active strategic engines**. In a STEM-driven environment, platforms become culture—digitally embedded, always aligned, and built to scale impact.

12.6. Chapter 12 Wrap-Up: Strategy Execution in a STEM World

Execution has always been essential to strategy. But in a world shaped by intelligent systems, real-time data, and algorithmic interfaces, execution is no longer just a phase—it is the **ongoing system** through which strategy lives, adapts, and scales.

This chapter reframed execution not as the endpoint of strategy but as its **living infrastructure**. Execution systems today are embedded in dashboards, nudges, digital workflows, AI-driven alerts, and feedback loops. In the STEM-powered firm, strategic behavior doesn't happen by accident—it's designed, reinforced, and measured with precision.

We began by challenging the myth that execution is merely about "doing." Section 11.1 argued that strategy must be engineered into workflows and decision systems. When strategy exists only in PowerPoint slides or executive memos, it vanishes on contact with reality. Execution excellence means that strategy is **always visible, always aligned, and always actionable**.

In Section 12.2, we explored the anatomy of STEM-powered execution—from OKR systems and role-specific dashboards to ethically aligned nudging. Strategy-aware interfaces can elevate behavior or degrade it, depending on how they're structured. The rise of behavioral engineering in strategy execution shows that **how we shape choices determines what people choose**.

Section 12.3 introduced agile strategy loops. Rather than static, top-down rollouts, modern firms use iterative frameworks where feedback, retrospectives, and real-time analytics recalibrate strategy continuously. Strategy becomes a learning loop—not a one-way signal.

In Section 12.4, we examined change as a systemic design problem. Rather than waiting for resistance, STEM-enabled change management anticipates friction using sentiment analysis, readiness diagnostics, and simulation modeling. Execution systems must not only carry change—they must absorb it, adapt to it, and evolve through it.

Finally, Section 12.5 explored execution platforms as the **behavioral fabric of the organization**. From strategic prompts to smart suggestion engines, platforms now shape the daily decisions that determine whether strategy succeeds or stalls. Designing these systems intentionally ensures that the **strategy is reflected in every interface**.

As we transition to future chapters, remember: execution is not just about performance metrics. It is about **strategic coherence**. The most successful organizations don't just plan better. They **design execution systems** that make the right behaviors easier, faster, and more rewarding—every single day.

12.6.1. Discussion Questions

1. In what ways is execution a design problem rather than just an implementation problem?
2. How do behavioral nudges and role-based dashboards improve alignment between individual actions and organizational strategy?
3. Why are feedback loops critical to strategic execution in dynamic environments?
4. What ethical concerns arise from embedding nudges and AI-driven decision protocols into execution systems?
5. How do STEM-enabled firms manage resistance to change differently from traditional models?
6. What does it mean to say that a platform can "reinforce" or "distort" strategy? Can you give an example from your own work experience?
7. How does strategic agility differ from flexibility? How do tools like retrospectives and real-time dashboards support agility?
8. What lessons from this chapter would you apply in designing an execution system for a cross-functional digital transformation?

12.6.2. Micro-Case: Zara's Execution Engine

Making Speed Work—How Zara Executes Strategy in Real Time

Zara, one of the world's most agile fashion retailers, is not just known for its design cycle—it's known for its execution system. Behind the brand's rapid response to fashion trends is a tightly engineered, data-powered execution engine. But this wasn't always the case.

Facing growing competition from digital-first brands, Zara's leadership launched a strategy to reduce product-to-shelf time by 40%—without increasing cost or compromising sustainability. The challenge was not just technical—it was organizational. Multiple teams, legacy systems, and cultural inertia all stood in the way.

Zara's solution was not to push strategy harder, but to **build execution smarter**:

- Every store became a sensor, feeding real-time sales and style feedback into centralized dashboards powered by predictive analytics.
- Design, procurement, and logistics were linked through an AI-enabled platform that nudged decision-makers toward faster, sustainability-aligned options.
- Visual dashboards tracked not just outputs but alignment with broader strategic goals like circular design adoption and emissions reduction.
- Resistance to the new execution system was anticipated early through sentiment analytics, and tailored onboarding sessions were delivered via micro-learning platforms based on user behavior.

Within six months, the execution gap closed. Lead times dropped, decision cycles accelerated, and strategic alignment became measurable. Execution wasn't managed—it was engineered.

Discussion Questions:

1. How did Zara turn strategy into a real-time system of action rather than a static rollout plan?
2. What role did dashboards and nudges play in shaping daily behaviors across departments?
3. How could similar execution systems be adapted to industries outside of fashion?
4. What risks might arise from over-automating execution behavior? Where should human judgment remain central?

12.6.3. Interactive Exercise: Execution by Design – The Strategic Systems Challenge

Scenario:
You've just joined a fast-growing med-tech firm, **VitalSync**, as VP of Strategic Execution. The company is rolling out a new home-based diagnostic product across three continents. The strategy is sound—but past initiatives have failed in execution due to silos, unclear accountability, and outdated change processes.

Your Task:
Design a next-generation execution system to bring this strategy to life across the organization. Your design must integrate platforms, roles, analytics, and change mechanisms.

Deliverables (one-page strategy brief):

1. **Execution Stack Design**
    - Outline your core platforms, dashboards, and decision support tools.

- o Who sees what, when, and why?

2. **Strategic Alignment Tactics**
   - o How will role-specific nudges and OKRs ensure behavioral alignment?
   - o What feedback loops or alerts will track progress?
3. **Change Management Layer**
   - o How will you simulate resistance?
   - o What tools will you use to increase adaptability and engagement?
4. **Ethical Guardrails**
   - o How will you ensure nudges, feedback systems, or automation remain fair and transparent?
   - o Where should human oversight be emphasized?

Optional Bonus:
Use diagrams or mock interface sketches to visualize your system. You'll present your system in a strategy design sprint next class.

12.6.4. Recommended HBR Case Studies

❖ The Hyderabad Metro from Idea to Execution: The World's Largest Metro Rail Project under a Public-Private Partnership

* Overview: Published on February 20, 2025, and authored by Vijaya Sunder M, Ramakrishna Nidumolu, and Sridhar Bhagavatula, this 15-page case study examines the ambitious and complex development of the Hyderabad Metro Rail project, the world's largest metro rail initiative executed under a Public-Private Partnership (PPP) model. The case follows Reddy and his team as they navigate a range of challenges, including resource constraints, a volatile political environment, and the need to align diverse stakeholders. Through a combination of innovative financial structuring, advanced engineering, and strategic stakeholder management, the team successfully completed the first phase of the project. This case is ideal for discussions on infrastructure strategy, project management, PPP models, and urban development. It offers valuable insights into how large-scale public infrastructure projects can be executed effectively in emerging markets.
* Citation: Sunder, V. M., Nidumolu, R., & Bhagavatula, S. (2025). *The Hyderabad Metro from Idea to Execution: The World's Largest Metro Rail Project under a Public Private Partnership* (Case No. ISB477-PDF-ENG). Indian School of Business.

❖ Satya Nadella at Microsoft: Leading the next transformation into AI

* **Overview***:* Published on July 16, 2024, and authored by Kannan Ramaswamy and William E. Youngdahl from the Thunderbird School of Global Management, this 8-page case explores the strategic and cultural transformation of Microsoft under CEO Satya Nadella. The case traces Nadella's leadership journey

over the past decade, focusing on two intertwined themes: strategy and leadership. It begins with Microsoft's pivot to a cloud-first, customer-centric model, which required dismantling legacy structures and reshaping the company's culture. Nadella's emphasis on a growth mindset, collaboration, and innovation helped reenergize the organization. The case culminates in Microsoft's bold move into generative AI, particularly through its partnership with OpenAI, raising questions about how the company can sustain its momentum in an increasingly competitive and fast-evolving tech landscape. This case is ideal for discussions on digital transformation, strategic leadership, organizational change, and AI-driven innovation. Explores how Netflix scaled a strategy of radical transparency and trust via digitally enabled performance systems.

* **Citation:** Ramaswamy, K., & Youngdahl, W. E. (2024). *Satya Nadella at Microsoft: Leading the next transformation into AI* (Case No. TB0720-PDF-ENG). Thunderbird School of Global Management.

❖ Four Visions' Plant Medicines: Overcoming Western Cultural Challenges Through Disruptive Marketing

- **Overview:** Published by Temple University and authored by Dennis Paris, Jennifer Sundstrom-Fitzgerald, Sheri Lambert, and Denise Donaghue, this case explores the journey of Four Visions Market LLC, a company founded by Mariah Gannessa to bring Indigenous plant medicines to Western markets. With over a decade of experience living among Indigenous communities in South America, Gannessa built Four Visions on principles of ethical sourcing, fair trade, and cultural respect. However, despite a strong mission and loyal customer base, the company hit a financial plateau. To reignite growth, Gannessa partnered with Nicole Bezinski, a marketing strategist, to develop a disruptive marketing strategy that could overcome Western skepticism, entrenched medical norms, and cultural biases.

    - The case challenges students to consider how to:
        - Position a culturally rooted wellness brand in a competitive and skeptical Western market
        - Balance authenticity with scalability
        - Use marketing to bridge cultural divides and build trust

    It's ideal for discussions on brand strategy, cultural entrepreneurship, ethical marketing, and social enterprise growth.

- **Citation:** Paris, D., Sundstrom-Fitzgerald, J., Lambert, S., & Donaghue, D. (2024). *Four Visions' Plant Medicines: Overcoming Western Cultural Challenges Through Disruptive Marketing* (Case No. W41813-PDF-ENG). Temple University.

12.6.5. Strategy-in-Film: Learn by Watching

Even the best strategy fails without disciplined execution. The films below capture how organizations translate strategic intent into action, navigate operational complexities, and adapt execution in real time. As you explore implementation and execution in this chapter, use these narratives to reflect on how data-driven systems, feedback loops, and agile processes drive successful strategy delivery.

Film / Documentary	Execution Insight Highlighted
Ford v Ferrari	Platform speed, execution under pressure, system-level alignment
The Founder	Strategic replication and process engineering (McDonald's systemization)
Apollo 13	Adaptive execution, feedback loops, cross-functional crisis response
The Social Network	Speed-to-execution, scaling platforms, and system fragmentation risks
Moneyball	Real-time decision modeling and execution alignment in high-stakes teams
The Intern	Organizational culture vs execution systems in startups
Enron: The Smartest Guys in the Room	Ethics failure in performance-driven systems
The Playbook (Netflix)	Coaches as system designers—strategy meets execution on the field

References

Atkinson, H. (2006). Strategy implementation: a role for the balanced scorecard?. *Management decision*, *44*(10), 1441-1460.

Bartosova, V., Drobyazko, S., Bielialov, T., Nechyporuk, L., & Dzhyhora, O. (2023). Company strategic change management in the open innovation system. *Journal of Open Innovation: Technology, Market, and Complexity*, *9*(2), 100087.

Bellis, P., Magnanini, S., & Verganti, R. (2023). Dialogue for strategy implementation: how framing processes enable the evolution of new opportunities. *Journal of Knowledge Management*, *28*(11), 1-32.

Beste, T., & Klakegg, O. J. (2022). Strategic change towards cost-efficient public construction projects. *International Journal of Project Management*, *40*(4), 372-384.

Cândido, C. J., & Santos, S. P. (2015). Strategy implementation: What is the failure rate?. *Journal of Management & Organization*, *21*(2), 237-262.

Christie, A., & Tippmann, E. (2024). Intended or unintended strategy? The activities of middle managers in strategy implementation. *Long range planning*, *57*(1), 102410.

Elbanna, S., Kamel, H., Fatima, T., & Eid, R. (2022). An investigation of the causality links in the balanced scorecard: The case of the Gulf Cooperation Council hospitality industry. *Tourism Management Perspectives*, *41*, 100934.

Frigo, M. L., & Krumwiede, K. R. (2000). The Balanced Scorecard. *Strategic Finance*, *81*(7), 50-50.

Hanelt, A., Bohnsack, R., Marz, D., & Antunes Marante, C. (2021). A systematic review of the literature on digital transformation: Insights and implications for strategy and organizational change. *Journal of Management Studies*, *58*(5), 1159-1197.

Heesen, B. (2015). *Effective strategy execution: improving performance with business intelligence*. Springer.

Hitt, M. A., Jackson, S. E., Carmona, S., Bierman, L., Shalley, C. E., & Wright, M. (2017). *The Oxford handbook of strategy implementation*. Oxford University Press.

Holm, C. G., Kringelum, L., & Anand, A. (2025). Creating effective strategy implementation: a systematic review of managerial and organizational levers. *Review of Managerial Science*, 1-33.

Ishaq Bhatti, M., Awan, H. M., & Razaq, Z. (2014). The key performance indicators (KPIs) and their impact on overall organizational performance. *Quality & Quantity*, *48*, 3127-3143.

Kaplan, R. S., & Norton, D. P. (2006). *Alignment: Using the balanced scorecard to create corporate synergies*. Harvard Business Press.

Kirtley, J., & O'Mahony, S. (2023). What is a pivot? Explaining when and how entrepreneurial firms decide to make strategic change and pivot. *Strategic management journal*, *44*(1), 197-230.

Kumar, S., Lim, W. M., Sureka, R., Jabbour, C. J. C., & Bamel, U. (2024). Balanced scorecard: trends, developments, and future directions. *Review of Managerial Science*, *18*(8), 2397-2439.

Li, H., An, N., & Liu, J. (2021). Strategic planning, strategy map, and management control: a case study. *Nankai Business Review International*, *12*(3), 386-408.

López-Torres, J. F., Sánchez-García, J. Y., Núñez-Ríos, J. E., & López-Hernández, C. (2023). Prioritizing factors for effective strategy implementation in small and medium-size organizations. *European Business Review*, *35*(5), 694-712.

Massicotte, S., & Henri, J. F. (2021). The use of management accounting information by boards of directors to oversee strategy implementation. *The British Accounting Review*, *53*(3), 100953.

Mio, C., Costantini, A., & Panfilo, S. (2022). Performance measurement tools for sustainable business: A systematic literature review on the sustainability balanced scorecard use. *Corporate social responsibility and environmental management*, *29*(2), 367-384.

Noble, C. H. (1999). Building the strategy implementation network. *Business Horizons*, *42*(6), 19-19.

Parmenter, D. (2015). *Key performance indicators: developing, implementing, and using winning KPIs.* John Wiley & Sons.

Perkins, M., Grey, A., & Remmers, H. (2014). What do we really mean by "Balanced Scorecard"?. *International Journal of Productivity and Performance Management*, *63*(2), 148-169.

Ritala, P., Baiyere, A., Hughes, M., & Kraus, S. (2021). Digital strategy implementation: The role of individual entrepreneurial orientation and relational capital. *Technological Forecasting and Social Change*, *171*, 120961.

Schuler, B. A., Orr, K., & Hughes, J. (2023). My colleagues (do not) think the same: Middle managers' shared and separate realities in strategy implementation. *Journal of Business Research*, *160*, 113782.

Soderberg, M., Kalagnanam, S., Sheehan, N. T., & Vaidyanathan, G. (2011). When is a balanced scorecard a balanced scorecard?. *International Journal of Productivity and Performance Management*, *60*(7), 688-708.

Sull, D., Homkes, R., & Sull, C. (2015). Why strategy execution unravels—and what to do about it. *Harvard Business Review*, *93*(3), 57-66.

Tawse, A., & Tabesh, P. (2023). Thirty years with the balanced scorecard: What we have learned. *Business Horizons*, *66*(1), 123-132.

Tawse, A., & Tabesh, P. (2021). Strategy implementation: A review and an introductory framework. *European Management Journal*, *39*(1), 22-33.

Tichy, N. (1983). The essentials of strategic change management. *Journal of Business Strategy*, *3*(4), 55-67.

Verweire, K. (2014). *Strategy implementation.* Routledge.

Williams, D. A., & Wade-Golden, K. C. (2023). *The chief diversity officer: Strategy, structure, and change management.* Taylor & Francis.

CHAPTER 13

Strategic Leadership and STEM Capabilities

"In a STEM driven world leadership is no longer about having the answers—it s about building the systems that ask better questions detect blind spots and adapt with integrity"

by MAK

Learning Outcomes

In a world of algorithms and uncertainty, leadership is not about control—it's about building systems that learn, adapt, and lead with integrity, like the orangutan in the canopy.

By the end of this chapter, you will be able to:

- Define the evolving role of strategic leadership in STEM-intensive organizations and explain how it differs from traditional leadership paradigms.
- Analyze how systems thinking, data literacy, and ethical intelligence are reshaping leadership competencies in the age of AI and digital transformation.
- Evaluate behavioral and AI-augmented diagnostic tools for leadership development and decision-making calibration.
- Apply frameworks to identify cognitive biases and decision traps common in high-stakes strategic environments.
- Design a leadership development approach that integrates analytics, feedback systems, and ethical decision protocols.
- Critically reflect on bounded rationality, algorithmic delegation, and the future of leadership accountability in data-driven ecosystems.

Strategic leadership is being redefined in real time. No longer confined to vision statements and quarterly reports, today's leaders are increasingly expected to **navigate uncertainty using data, shape systems that learn, and guide decisions that adapt in real time**. This is not just a shift in tools; it's a shift in what leadership means.

In STEM-driven organizations, leadership is not about knowing everything; it's about architecting environments where **evidence guides action**, where **algorithms augment judgment**, and where **ethical foresight keeps pace with technical speed**. A strategic leader must now be a systems integrator, a translator between humans and machines, and a custodian of values embedded in digital logic.

Traditional models of leadership emphasized charisma, vision, and authority. While these still matter, they are no longer sufficient. Leaders must now **understand feedback loops**, **interpret machine learning outputs**, and **anticipate the second-order effects of automated decisions**. They must work across disciplines, integrating insights from engineering, behavioral science, and ethics—all while steering their organizations toward long-term strategic goals.

This chapter reframes leadership for the age of intelligent systems. It explores the **new competencies**, **diagnostic tools**, and **design principles** required to lead in environments shaped by AI, data analytics, and algorithmic interdependence. Along the way, we'll surface cognitive traps like overconfidence and confirmation bias, explore how feedback tools can support adaptive leadership, and investigate how ethics and decision integrity must be embedded—not added—as organizations scale their STEM capabilities.

Ultimately, the question is no longer: *What makes a great leader?* It's: *What kind of systems do great leaders now build—and how do those systems lead us in return?*

Opening Case: Leading the Algorithmic Enterprise – Ping An's Leadership Reinvention

When Ping An Insurance began its digital transformation, it didn't just invest in technology—it reengineered how leadership worked. Once a traditional financial conglomerate rooted in hierarchical command structures, Ping An is now recognized as one of the most AI-integrated companies in the world. But the transformation wasn't just about deploying AI tools—it was about **designing leadership systems that could scale with data, adapt through feedback, and lead through complexity.**

Rather than rely solely on executive decisions from the top, Ping An embedded **AI-powered feedback loops** across its business units. Managers at all levels began using real-time dashboards not only to track performance but to recalibrate strategies dynamically. Predictive algorithms flagged emerging risks in customer behavior, while **digital twins** of operations allowed leaders to run simulations before making major decisions.

Leadership development also changed. Ping An launched internal platforms where rising leaders were assessed not only by performance outcomes, but by **how well they collaborated with intelligent systems**, interpreted predictive models, and adapted to feedback. The company no longer promoted based on seniority alone—but on how fluently a leader could work in a **data-governed environment**.

This shift was not without challenges. Some leaders resisted ceding decision authority to "black box" algorithms. Others struggled with the ethical tensions of automating decisions that directly affected people's insurance coverage or financial access. To address these concerns, Ping An created an **AI ethics task force** and integrated explainability frameworks (like SHAP) into leadership dashboards, ensuring leaders understood not just *what* the model said—but *why*.

Ping An's transformation is a window into the future of strategic leadership: not heroic decision-makers, but designers of systems that can reason, adapt, and learn.

Play the Strategist

1. How did Ping An's shift from human-centered decision-making to AI-augmented systems reshape its leadership model?
2. What leadership competencies became more important in Ping An's new operating model? Which became less relevant?
3. How did Ping An attempt to balance algorithmic efficiency with ethical responsibility in its leadership design?
4. If you were designing a leadership development program at Ping An today, what new evaluation metrics or training modules would you include?

13.1 STEM-Enabled Leadership Models

In the age of intelligent systems, leadership is no longer defined by proximity to power—it's defined by proximity to insight. As organizations grow more technologically integrated and data-intensive, leadership must evolve from personality-centered models to system-aware frameworks. This is the age of **STEM-enabled leadership**, where strategic influence flows not just from vision, but from an ability to design, interpret, and ethically govern complex decision environments.

STEM-enabled leaders must be fluent in **data, systems, and ethics**. They lead not by controlling every decision, but by building platforms where others can make better decisions. Their value is not in memorizing answers, but in shaping the conditions that surface better questions. They are as concerned with feedback loops as they are with revenue; as attuned to model transparency as they are to market share.

This section explores the evolving architecture of leadership in STEM-driven enterprises. From cognitive bias detection to AI-augmented coaching, from cross-functional intelligence to system-integrated ethics, you'll discover how the very definition of leadership is being redesigned—by leaders who can both guide and adapt within dynamic, data-rich environments.

The future doesn't just need strong leaders. It needs **strategically humble**, **ethically anchored**, and **system-savvy** ones.

13.1.1 From Command to Collaboration: Redefining Leadership Norms

Leadership was once synonymous with control—clear chains of command, rigid hierarchies, and top-down strategy. But as organizations embrace STEM capabilities, the role of the leader is being redefined. Today, effective leadership is less about commanding authority and more about **orchestrating systems, enabling collaboration, and co-creating decisions with intelligent tools**.

In traditional settings, decisions often relied on gut instinct or anecdotal experience. But in STEM organizations, strategic choices are shaped by **simulations, feedback loops, predictive analytics, and ethically aligned algorithms**. Leaders no longer have the luxury of intuition without explanation—they are accountable not just for what they decide, but for how transparently those decisions are made.

This shift demands a different type of leader—one who is data fluent, ethically grounded, and systems-aware. It also demands new cultural norms. Instead of silos, STEM leaders foster cross-disciplinary teams. Instead of guarding information, they promote **data democratization** through shared dashboards and open performance metrics. To visualize this transformation, consider the following comparative table:

As this table illustrates, **collaboration replaces command**, and leadership becomes a function of **designing smart systems** rather than issuing directives. The best STEM leaders act as translators—bridging human insights and machine logic, aligning teams around strategic intent while respecting the nuances of algorithmic input.

The future of leadership is not less human—it is **more intentionally human**, precisely because machines are now involved. STEM leaders must bring empathy to data, context to models, and reflection to every automated insight.

Table 13.1. Comparing Traditional vs. STEM-Enabled Strategic Leadership

Dimension	Traditional Leadership	STEM-Enabled Leadership
Decision-Making Approach	Intuition, experience-based	Data-informed, model-assisted
Core Strength	Authority, charisma	Systems thinking, analytics fluency
Strategic Focus	Long-term vision and hierarchy	Adaptive experimentation and continuous feedback
Use of Technology	Delegated to technical teams	Integrated into daily leadership tasks
Handling Uncertainty	Confidence-driven	Scenario-driven, simulation-informed
Ethics & Governance	Case-by-case judgment	Embedded bias detection and algorithmic scrutiny

13.1.2 Cognitive Biases and the STEM-Aware Leader

Even in data-rich environments, decisions can be skewed by deeply rooted mental shortcuts. Cognitive biases—those subtle, often unconscious distortions in how we process information—pose a persistent threat to strategic clarity. For STEM-enabled leaders, awareness of these biases isn't optional; it's foundational to ethical, data-driven leadership.

Traditional leadership models emphasized decisiveness and confidence. But STEM-aware leaders must **embrace uncertainty**, **interrogate assumptions**, and **validate insights through replicable evidence**. A bias-aware leader doesn't blindly trust data or intuition. Instead, they question both.

STEM tools such as SHAP (for model explainability), Monte Carlo simulations (for probabilistic foresight), and digital dashboards (for performance transparency) offer partial protection. But these tools only work when leaders are also willing to recognize the **limits of their perception**.

To illustrate how bias creeps into even the most sophisticated decision systems, consider the following:

Table 13.2. Common Cognitive Biases in Strategic Decision-Making

Bias	Description	Leadership Risk	STEM Mitigation Strategy
Confirmation Bias	Tendency to seek or prioritize information that confirms pre-existing beliefs.	Reinforces flawed assumptions, blocks innovation.	Use SHAP or PDP to surface surprising model drivers.
Overconfidence Bias	Overestimating the accuracy of one's own judgments or models.	Leads to underestimating risk or ignoring weak signals.	Monte Carlo simulations to expose risk variability.
Anchoring Bias	Relying too heavily on initial information when making decisions.	Limits flexibility and responsiveness to new data.	Real-time dashboards to challenge static assumptions.
Availability Heuristic	Giving weight to information that is recent or emotionally vivid.	Overreacting to anomalies or ignoring structural trends.	Deploy rolling analytics and ensemble forecasts.
Status Quo Bias	Preference for the current state even when a change is beneficial.	Resistance to transformation or innovation.	Scenario modeling to simulate the benefits of change.

Bias awareness is not about eliminating intuition—it's about **disciplining it through design**. The most forward-looking leaders don't just recognize cognitive traps; they embed **bias-disrupting mechanisms** into their decision infrastructure. This might mean using algorithmic explainability tools in boardrooms, or it might mean requiring all strategic proposals to include counterfactual scenarios and dissenting data points.

Bias thrives in silence. STEM-aware leadership replaces silence with structure—transparent, testable, and iteratively improved.

13.1.3 Feedback Loops and Adaptive Leadership

In fast-changing environments, the strength of a leader is not in knowing all the answers—it's in **adapting faster than the complexity evolves**. Adaptive leadership depends on a system's ability to learn, correct, and recalibrate in real time. At the heart of this responsiveness lies a fundamental principle: **the feedback loop**.

In biology, engineering, and computing, feedback loops are essential for maintaining system stability and improving performance. The same applies to organizations. However, while traditional leadership often treats feedback as episodic—annual reviews, quarterly KPIs—STEM-enabled leadership builds **continuous, multi-modal feedback structures** into the decision fabric itself.

This isn't just about listening. It's about sensing. It's about equipping organizations with real-time metrics, machine-generated insights, and behavioral analytics that keep the system aware of its own performance and blind spots. And it's about ensuring that **leaders themselves are part of the loop**, not above it.

The types of feedback loops vary depending on the purpose, scope, and data source. The table below outlines common forms used in STEM-enabled strategy systems:

Table 13.3. Types of Feedback Loops in Strategic Leadership

Feedback Loop Type	Primary Function	Example Tool or Mechanism	Strategic Leadership Benefit
Performance Loops	Track and adjust based on operational outcomes.	Real-time dashboards, KPIs, OKRs	Improves accountability, highlights friction points.
Learning Loops	Capture user/employee input to refine processes.	Pulse surveys, digital suggestion systems	Embeds organizational learning into workflows.
Predictive Loops	Anticipate change and simulate possible outcomes.	Digital twins, Monte Carlo simulations	Enhances readiness, informs risk-based decision-making.
Behavioral Loops	Monitor leadership impact or team dynamics.	Organizational network analysis, NLP tools	Supports culture alignment and ethical behavior tracking.
Ethical Loops	Flag emerging fairness, privacy, or bias issues.	SHAP analysis, AI audit trails	Sustains trust and regulatory alignment.

Adaptive leadership thrives not on perfection but on iteration. STEM-empowered leaders use these feedback systems to sense weak signals, close performance gaps, and enable strategic pivots before disruption strikes. They see feedback not as critique, but as **strategic radar**.

But building feedback loops isn't enough. Leaders must also normalize them—create cultures where **reacting to data is a reflex**, not a burden. They must ask: *Are we learning faster than the environment is changing?*

If the answer is no, then the feedback loop isn't just a support tool—it's a survival imperative.

13.1.4 STEM Skills for Strategic Influence

In today's data-driven enterprise, authority alone no longer guarantees influence. The most impactful leaders are those who can **navigate complex systems, interpret data responsibly, and communicate insights in actionable ways**. These aren't just technical skills—they're strategic assets.

Strategic influence in a STEM-enabled world demands a new fusion of capabilities. It's no longer sufficient for executives to "understand the business" and leave the technical modeling to analysts. To guide organizations through uncertainty and transformation, leaders must be **fluent in the language of data, simulation, and systems dynamics**—and know when to deploy which tool for which type of decision.

But STEM fluency isn't about coding or running regression models. It's about understanding what good data looks like, how models are trained and interpreted, and how to engage ethically with automated systems. It's about **knowing which questions to ask, which assumptions to challenge, and how to co-create insight with data scientists**. The table below outlines the core STEM skill domains that today's strategic leaders must build—not in isolation, but as part of an integrated leadership toolkit:

Table 13.4. STEM Skillsets for Strategic Leaders

STEM Skill Domain	Strategic Leadership Application	Example Tools / Techniques
Data Literacy	Understand, question, and apply insights from structured data.	Dashboards, SQL basics, data visualization
Model Interpretability	Evaluate how predictive models derive insights and detect risk.	SHAP values, PDPs, scenario trees
Systems Thinking	Analyze interactions across complex strategic environments.	Causal loop diagrams, feedback loops
Ethical Intelligence	Assess fairness, transparency, and unintended consequences in models.	Bias audits, AI governance frameworks
Simulation Proficiency	Forecast outcomes under uncertainty to support strategic pivots.	Monte Carlo simulation, digital twin modeling
Collaborative Fluency	Co-create decisions with analysts, engineers, and AI systems.	Agile leadership, cross-functional modeling teams

Leaders don't need to master all of these domains—but they must understand how they **connect**. For example, without ethical intelligence, data literacy becomes dangerous. Without interpretability, simulation can mislead. And without collaboration, even the smartest model becomes a black box that no one acts upon.

STEM fluency empowers strategic influence not by replacing judgment, but by **amplifying it with clarity, transparency, and shared language**. Leaders who embrace these skillsets not only stay relevant—they help their organizations thrive amid accelerating complexity.

13.2 Intelligent Diagnostics for Leadership Development

Leadership development has traditionally relied on self-report assessments, 360-degree feedback, or subjective evaluations—useful tools, but often static, slow, and vulnerable to bias. In the era of intelligent systems, these tools are no longer enough. Leadership is dynamic, situational, and deeply contextual. As such, our diagnostics must evolve to reflect the **fluidity and feedback-rich nature** of real leadership performance.

Enter **intelligent diagnostics**: systems that use real-time behavioral data, predictive analytics, AI-driven sentiment analysis, and dynamic simulations to assess and shape leadership capabilities. These technologies don't just evaluate; they learn and adapt. They don't just measure personality traits; they reveal **decision patterns**, **ethical blind spots**, and **team-level influence dynamics**.

For STEM-enabled organizations, leadership development is no longer an HR side initiative—it's a **core component of competitive intelligence**. Just as we optimize operations with dashboards and digital twins, we must now optimize leadership with the same rigor, speed, and systemic thinking.

This section introduces emerging diagnostic frameworks and technologies that help organizations **identify, coach, and accelerate leadership potential** at scale. You'll explore how platforms integrate data from collaboration tools, performance systems, and even biometrics to generate more nuanced and predictive insights about who leads well—and how.

We begin with a reimagining of the leadership pipeline itself, before diving into the tools, dashboards, and ethical questions surrounding real-time leadership measurement.

13.2.1 Leadership Pipeline 2.0

In the classic model, the leadership pipeline is a slow-moving funnel: individuals move upward through predefined roles over time, based on performance reviews, tenure, and visibility. This approach was built for a stable world—where leadership traits were presumed to be static, and potential could be measured in annual cycles.

But in today's volatile, complex, and data-saturated environment, leadership potential doesn't unfold linearly. It **emerges in context**, often in sudden bursts of insight, action, or influence. Waiting 12 months to recognize a rising leader—or worse, promoting someone based on outdated data—can be fatal for innovation.

The STEM-enabled pipeline doesn't just track **who is moving up**. It constantly senses **how individuals are responding to dynamic challenges, who is building trust, who is adapting under pressure**, and **who is driving systems-level learning**.

Digital tools now allow us to build **Leadership Pipeline 2.0**—a real-time, adaptive, evidence-based view of leadership emergence. This new model integrates cross-functional feedback, AI-based behavioral analysis, and scenario simulations to identify leadership readiness not by role, but by response.

The table below compares the traditional and STEM-enabled approaches to leadership development:

Table 13.5. Comparing Traditional vs. STEM-Enabled Leadership Pipelines

Dimension	Traditional Pipeline	STEM-Enabled Pipeline
Evaluation Cycle	Annual or semi-annual reviews	Continuous, real-time diagnostics
Data Source	Self-report, supervisor rating	Behavioral analytics, collaboration platforms, and AI pattern detection
Focus	Role readiness, performance history	Situational influence, ethical judgment, system thinking
Decision Basis	Tenure, KPIs, subjective endorsement	Predictive modeling, dynamic simulations
Leadership Visibility	Hierarchy-bound	Network-aware (who influences whom, across silos)
Coaching Trigger	After failure or promotion	Adaptive nudges, in-the-moment feedback

This shift isn't about replacing people with algorithms—it's about recognizing that **leadership potential is often visible in micro-decisions and adaptive responses**, long before formal promotions occur. By embedding intelligent diagnostics into collaborative systems, STEM organizations can surface these moments and nurture them with precision.

Leadership Pipeline 2.0 repositions development as a **daily feedback loop**, not an annual gate. And for those aspiring to lead in complexity, that loop is the true proving ground.

13.2.2 Behavioral Analytics for Strategic Insight

Modern leaders are not evaluated solely by what they say—but increasingly by how they behave, collaborate, and respond under pressure. Behavioral analytics brings objectivity to this process by transforming workplace interactions—emails, meetings, decisions, delays—into rich, analyzable signals.

This isn't surveillance; it's **strategic sensing**. By ethically capturing and interpreting digital traces, behavioral analytics allows organizations to identify patterns of influence, trust, alignment, burnout risk, and more. It becomes possible to answer questions that used to rely on gut instinct:

- Who is the actual go-to leader on this project?
- Where are communication bottlenecks forming?
- How are strategy directives cascading—or not cascading—through the system?

STEM-enabled leadership development platforms now use Natural Language Processing (NLP), Organizational Network Analysis (ONA), and sentiment recognition to map these behaviors in real time. When aggregated and anonymized, these insights help **identify hidden leadership, surface weak signals**, and shape strategic coaching interventions. The table below outlines key behavioral signals, the analytic methods used to extract them, and the leadership insights they generate:

Table 13.6. Strategic Signals Extracted Through Behavioral Analytics

Behavioral Signal	Analytic Method	Leadership Insight Generated
Response latency in decisions	Workflow analytics	Decisiveness under pressure, delegation efficiency
Cross-functional engagement	Organizational Network Analysis (ONA)	Influence across silos, bridging roles
Tone and language patterns	NLP / sentiment analysis	Empathy, emotional intelligence, and potential burnout
Decision pattern consistency	Machine learning classification	Strategic alignment vs. reactive behavior
Participation in critical tasks	Workflow & collaboration mining	Strategic ownership, commitment to high-stakes outcomes

The goal isn't to quantify humans into scores, but to **illuminate leadership as it unfolds**—in patterns, signals, and choices. When leaders receive this data as private coaching inputs, they can grow in self-awareness and strategic impact without waiting for performance reviews.

In a STEM world, behavioral analytics doesn't just monitor—it empowers. It transforms raw interaction data into fuel for ethical, agile, and evidence-based leadership development.

13.2.3 Predictive Modeling of Leadership Potential

Can leadership potential be predicted? In a world where machine learning forecasts consumer demand and system failures, it's no surprise that similar approaches are now being used to identify future leaders. Predictive modeling in leadership development moves us from reactive evaluations to **proactive discovery**—helping organizations nurture the right individuals before crises or promotions force the decision.

These models typically integrate multiple data streams—behavioral signals, performance metrics, 360-degree feedback, cognitive assessments, and even personality trait indicators—to find patterns associated with successful leadership transitions. The goal isn't to automate promotions but to **support talent decisions with evidence**, mitigate bias, and surface high-potential individuals who might otherwise be overlooked.

By using tools like random forests, logistic regression, and XGBoost, organizations can not only predict leadership emergence but also simulate **scenario-specific fit**—for example, who is most likely to succeed in a turnaround, innovation launch, or global expansion role?

The key, however, is **interpretability**. Leaders and HR professionals must be able to explain why someone was flagged as high potential—making SHAP values, PDPs, and visual dashboards essential components of this process.

The table below outlines common predictor variables used in leadership modeling, along with their data sources and strategic relevance:

When designed and deployed ethically, predictive models create **leadership equity**—ensuring that potential is spotted based on evidence, not just visibility or likability. They also help de-risk promotions, reduce talent gaps, and focus coaching efforts where they'll have the most strategic return.

In STEM-enabled organizations, these models become a living mirror—one that helps leaders see both their shadow and their light, with clarity and context.

Table 13.7. Variables Commonly Used to Predict Leadership Success

Predictor Variable	Data Source	Why It Matters
Cross-functional influence	Email/Slack metadata, ONA	Indicates reach and trust beyond direct teams
Strategic thinking behaviors	Decision logs, project narratives	Reflects long-term orientation and risk tradeoff skills
Adaptability under stress	Workflow volatility, time-to-decision	Shows cognitive flexibility in high-pressure environments
Ethical decision patterns	Scenario simulations, behavioral coding	Reveals integrity, transparency, and rule adherence
Feedback responsiveness	360 reviews, coaching logs	Tracks growth mindset and openness to development
Initiative-taking	Task initiation rate, innovation logs	Predicts entrepreneurial behavior and proactive leadership

13.2.4 Ethical Boundaries in Leadership Data

As leadership diagnostics become increasingly data-driven, the ethical stakes rise. Behavioral insights, predictive scores, and real-time assessments offer tremendous potential—but also **profound responsibility**. In the wrong hands, these tools can reinforce bias, undermine trust, and turn human development into surveillance.

The challenge isn't just technical; it's **philosophical**. What are the boundaries of acceptable monitoring? Who owns the data that reflects a person's leadership potential? And how do we ensure that data intended for development doesn't become a weapon of judgment?

STEM-enabled organizations must establish **ethical guardrails** for leadership data. These include transparent consent protocols, opt-in models for developmental analytics, anonymized benchmarking, and firewalls between coaching data and promotion decisions. Just because we *can* measure doesn't mean we *should*—unless the purpose is clear, the governance is fair, and the impact is constructive.

Tools like SHAP, fairness audits, and explainable AI provide a technical foundation for transparency, but **organizational culture must carry the weight of ethical use**. That means framing diagnostics as **coaching assets**, not compliance tools. It means training leaders to interpret patterns without reducing people to numbers.

This also requires answering deeper strategic questions:

- Should employees have visibility into how their leadership scores are generated?
- What happens when predictive models conflict with human intuition?
- How do we correct models trained on biased organizational histories?

Ethically aligned leadership diagnostics don't remove humans from the loop—they **expand the loop**, bringing in more signals, more reflection, and more accountability.

In short, the future of leadership development will not be defined by how much data we can collect, but by how wisely and fairly we choose to use it.

13.3 AI-Augmented Leadership Coaching

Traditional leadership coaching has long relied on reflection, dialogue, and qualitative feedback. While still essential, this model faces limitations in a STEM-shaped world: it is time-intensive, often subjective, and difficult to scale across large organizations. Enter **AI-augmented coaching**—a hybrid approach that combines human empathy with machine precision to support leader growth in real time.

AI-based coaching platforms now offer personalized development plans, feedback simulations, and nudges tailored to a leader's unique behavioral patterns. Drawing on data from performance metrics, digital interactions, and 360-degree assessments, these platforms detect learning needs and recommend targeted exercises—such as decision sprints, scenario training, or ethical reflection modules.

Importantly, AI doesn't replace human coaches—it **enhances their reach and precision**. Coaches can access visual dashboards that highlight a leader's growth areas, cognitive biases, stress patterns, and team dynamics. Sessions become more grounded in evidence, enabling deeper and more focused development conversations.

For example:

- An emerging leader struggling with delegation might receive scenario-based prompts to simulate resource allocation under pressure.
- A senior executive with declining cross-functional trust may be guided through NLP-based communication audits and realignment strategies.
- A team manager flagged for ethical ambiguity could undergo simulated ethical dilemmas with explainable AI-generated feedback.

These systems also **close the loop between intention and action**—tracking whether coaching insights are applied in the workplace and how they ripple through teams. Some platforms integrate with email, Slack, and task management tools to observe behavioral changes passively and ethically.

The promise of AI-augmented coaching lies not in automation, but in **hyper-personalization and accountability at scale**. It helps leaders evolve not just through annual workshops, but through daily feedback loops powered by ethically designed algorithms.

However, the same ethical boundaries discussed earlier still apply: transparency, voluntary participation, and clear separation between development and evaluation. When these are respected, AI can democratize leadership coaching—bringing strategic growth opportunities to every layer of the organization.

13.3.1 Micro-Coaching Models and Feedback Loops

In fast-moving strategic environments, leaders often don't need hour-long coaching sessions—they need **brief, high-impact nudges** that align with their daily flow of decisions. This is the promise of **micro-coaching**: short, context-aware interventions that are designed to build leadership capabilities moment by moment.

Micro-coaching doesn't aim to replace deep reflection—it complements it by providing **just-in-time support**, often triggered by data patterns. For example:

- After a tense team meeting, a leader might receive a reflection prompt on emotional tone and listening behavior.
- If analytics show delayed decision-making under ambiguity, a short module might offer decision-tree heuristics to accelerate clarity.
- A spike in multitasking might prompt a self-assessment on focus and prioritization.

These feedback loops are powered by **AI systems trained on behavioral benchmarks** and strategic competencies. They learn from patterns, detect inflection points, and recommend development micro-tasks or self-guided questions. Often delivered via mobile apps, email integrations, or dashboards, they seamlessly embed leadership growth into the workday.

What makes micro-coaching transformative is its **looped design**:

1. **Observation** – Passive sensing of digital behaviors (e.g., collaboration, decision cadence).
2. **Interpretation** – Pattern recognition using AI (e.g., bottlenecks, hesitations, overcontrol).
3. **Intervention** – Delivery of a concise insight, scenario, or question.
4. **Response Tracking** – Monitoring changes or improvements over time.
5. **Model Refinement** – System adapts based on the leader's evolving patterns.

This turns coaching from a static process into a **dynamic, co-evolving system**. And when paired with a human coach who can interpret trends and personalize further, micro-coaching becomes a powerful engine for continuous development—anchored not in vague advice, but in strategic relevance.

13.3.2 AI–Coach Symbiosis

The future of leadership coaching is not AI *versus* humans—it's AI *with* humans. While AI can scale insights, identify patterns, and optimize delivery, human coaches provide the **empathy, narrative framing, and cultural nuance** that machines still lack. The most effective coaching systems operate as **hybrid partnerships**, where AI supports the coach, not supplants them. In this model, AI handles the heavy lifting of data interpretation:

- Analyzing thousands of digital interactions for patterns of influence or conflict.
- Highlighting shifts in motivation or stress based on task flows or sentiment markers.
- Recommending micro-interventions based on leadership archetypes.

Meanwhile, the human coach curates these insights into **meaningful developmental arcs**—connecting data to purpose, behavior to legacy, and leadership style to strategic goals. The coach helps the leader interpret trends, contextualize feedback, and make values-based choices that data alone can't dictate. The table below outlines the **complementary roles** of AI systems and human coaches in this emerging model of leadership development:

Table 13.8. Complementary Roles of AI and Human Coaches

Function	AI Coaching Systems	Human Coaches
Pattern Detection	High-frequency, data-driven pattern recognition	Contextual sensemaking of ambiguous signals
Feedback Delivery	Automated, real-time, scalable	Personalized, narrative-based
Decision Modeling Support	Recommends heuristics, models, and nudges	Challenges assumptions and frames strategic dilemmas
Behavior Change Tracking	Tracks micro-changes over time	Interprets qualitative shifts and resistance
Ethical and Emotional Insight	Limited—based on coded indicators	Deep empathy, cultural and emotional intelligence
Development Planning	Optimized goal tracking and resource suggestions	Long-term visioning and values alignment

The symbiosis is not just functional—it's strategic. By combining computational intelligence with human wisdom, organizations build **scalable, ethical, and high-impact coaching ecosystems**. Leaders get the benefit of both machine precision and human reflection, enabling them to grow faster, decide smarter, and lead with greater self-awareness. As leadership challenges become more complex—especially in STEM fields—this hybrid model isn't a luxury. It's the new baseline for resilient, adaptive development.

13.3.3 Real-Time Coaching via Digital Twins

Digital twins—virtual replicas of physical systems—have traditionally been used in manufacturing, energy, and logistics. But today, **they are emerging as breakthrough tools in leadership development**. By creating digital models of leadership behavior and decision environments, organizations can simulate, observe, and coach in real time.

A **Leadership Digital Twin** integrates data from communication platforms, performance dashboards, project management tools, and feedback systems. It models how a leader's actions ripple across systems—how they allocate resources, influence morale, respond to ambiguity, or resolve conflict. These virtual environments are **testbeds for leadership behavior**. For example:

- A digital twin of a product launch team might simulate how a leader's delegation patterns affect team velocity and emotional tone.
- Ethical dilemmas can be simulated to observe likely responses under pressure, providing preemptive feedback loops.
- Policy shifts, supply chain shocks, or team restructuring can be modeled to assess strategic agility and adaptive leadership behaviors.

Unlike static assessments, these twins evolve with the leader—providing **live coaching insights** based on recent interactions. When integrated with micro-coaching prompts, they enable real-time nudging, allowing the leader to correct course mid-action rather than post-mortem. To illustrate how these systems function in practice, the following table highlights key components of leadership digital twins and the types of insights they generate:

Table 13.9. Real-Time Feedback Elements in Leadership Digital Twins

Feature	Purpose	Coaching Insight
Communication Flow Modeling	Track tone, frequency, and direction of communication	Feedback on influence, inclusion, and clarity
Decision Impact Simulations	Model the consequences of strategic decisions	Preemptive learning from simulated outcomes
Stress and Recovery Index	Monitor workload patterns and recovery behaviors	Nudges for sustainable performance and balance
Role-Based Outcome Forecasting	Simulate team performance based on leadership behaviors	Shows how style affects results across stakeholders
Ethical Tension Heatmaps	Identify zones of potential moral conflict	Scenario coaching to enhance ethical agility

The future of real-time coaching lies in **virtual rehearsal**. Leadership digital twins offer a safe, responsive, and evolving space to practice, fail, learn, and improve—without damaging real teams or outcomes. And by combining these twins with explainable AI and human coaches, leaders receive a **multi-dimensional feedback system** that is truly transformational.

In STEM-driven organizations where speed and consequence are amplified, this is not a futuristic vision—it's a present-day advantage.

13.3.4 Interpretable AI in Personalized Coaching

As AI becomes central to leadership development, the need for transparency and interpretability grows. Leaders must understand not only the insights delivered by AI-based coaching systems but also **how and why** those insights were generated. This is where **interpretable AI** reshapes personalized coaching—turning black-box predictions into actionable, trust-based feedback.

Tools such as SHAP (SHapley Additive exPlanations) and LIME (Local Interpretable Model-Agnostic Explanations) now enable coaching platforms to explain the **key behavioral variables** influencing a leader's performance. For example:

- SHAP can show which communication behaviors most influence team morale scores.
- LIME can highlight how specific decisions affected perceived psychological safety or delegation efficiency.
- PDPs (Partial Dependence Plots) can visualize how leadership agility responds to changes in stress or ambiguity levels.

These insights not only deepen the coach's understanding but also **empower the leader**. By demystifying algorithmic evaluations, leaders feel a greater sense of control over their development path. They can **see the "why" behind each nudge**, enabling more thoughtful responses and ownership of behavioral growth.

The table below outlines how different interpretable AI tools are applied in coaching contexts and the types of insights they deliver:

Table 13.10. Applications of Interpretable AI in Personalized Coaching

AI Tool	Coaching Application	Example Insight Delivered
SHAP	Explains feature importance in behavior-performance links	"Responsiveness during strategic updates drives trust."
LIME	Explains the local impact of a single behavior on an outcome	"Interruptions during ideation lowered innovation score."
PDP	Shows nonlinear relationships between leadership actions and outcomes	"Strategic clarity increases engagement only after 3 days of consistent communication."
Counterfactual AI	Identifies alternate actions that would improve predicted outcomes	"If you had delayed feedback by one day, tension index drops 12%."
Causal Inference	Tests likely cause–and–effect relationships in coaching interventions	"Weekly 1:1s are the main cause of improved team alignment."

Interpretable AI enables not just *coaching at scale*, but **coaching with precision, fairness, and clarity**. It aligns development feedback with the leader's unique patterns and values, while reinforcing accountability and explainability in the coaching process.

In a STEM-powered organization, where models drive decisions at every level, interpretability is not just a technical requirement—it's a leadership imperative. When leaders understand their coaching data, they not only improve—they inspire trust and set new standards for transparent growth.

13.4 Strategic Integration in Practice: Leading Across Systems and Disciplines

As organizations become more complex, distributed, and digitally enabled, strategic leadership can no longer afford to operate in silos. The modern leader must integrate knowledge across functional domains, decision layers, and algorithmic systems. Strategic integration is not about centralized control—it's about intelligent alignment. It requires leaders to recognize where decisions are happening, how data flows between systems, and where human insight is most valuable.

In the STEM-driven landscape, this means actively bridging the divide between technical capabilities and strategic outcomes. Leaders must cultivate the ability to spot decision traps stemming from cognitive biases and outdated assumptions (Section 12.4.1), while learning to deploy algorithms not as crutches, but as strategic scaffolds that elevate human decision-making (Section 12.4.2). Furthermore, as diagnostic tools become embedded within organizational data flows, leaders must also develop the fluency to interpret, refine, and act on dynamic assessments of their own leadership effectiveness (Section 12.4.3).

This section explores how integrated leadership is practiced—not just conceptualized. It offers frameworks and practical insights for navigating cross-disciplinary complexity, ensuring leadership decisions are not just fast but also well-informed, explainable, and systemically aligned. In doing so, we move beyond the lone decision-maker archetype to embrace a more interconnected model of leadership—one that leads not only through vision, but through system design, data transparency, and collaborative intelligence.

13.4.1 Bounded Rationality and Decision Traps

Herbert Simon's concept of *bounded rationality* reminds us that even the most intelligent leaders operate within cognitive, informational, and organizational constraints. In a STEM-augmented environment, these bounds do not disappear—they simply become more subtle and complex. The illusion of "big data" completeness or algorithmic precision can tempt leaders into overconfidence, masking the very biases and traps they aim to eliminate.

Modern strategic environments heighten these risks. The acceleration of information does not equate to better decision-making—unless leaders can recognize when their mental models, data sources, or frameworks are misaligned with the problem at hand. Strategic missteps often emerge not from ignorance, but from *false certainty*: overly simplified assumptions, misplaced confidence in legacy models, or failure to reframe problems across systems.

To navigate these constraints, leaders must internalize three key realities:

1. **All decisions are made under uncertainty**, no matter how advanced the model.
2. **Data does not eliminate bias**—it merely changes its form.
3. **Tools do not make decisions—leaders do.** But tools can help surface blind spots.

The most effective leaders use diagnostic tools not only to assess external threats and opportunities, but also to interrogate their own cognitive patterns. They practice "judgment calibration"—the ability to test their decision weightings against alternative scenarios, diverse perspectives, and systemic feedback. These leaders don't just ask, "What should we do?"—they ask, "What assumptions are shaping this conclusion?" and "How might we be wrong?"

To support this mindset, the following table outlines common leadership decision traps and the STEM-enabled countermeasures that can help mitigate them:

Table 13.11. Common Leadership Decision Traps and Diagnostic Countermeasures

Decision Trap	Description	STEM-Enabled Countermeasure
Overfitting to Past Success	Assuming past strategies will work in future contexts	Scenario modeling with Bayesian priors to account for shifting variables
Data Confirmation Bias	Selectively using data that confirms existing beliefs	SHAP analysis to identify hidden variable influences and contradictory drivers
Linear Thinking in Nonlinear Systems	Oversimplifying causal chains in complex systems	System dynamics modeling (e.g., causal loops, feedback delays)
Anchoring on Early Information	Giving undue weight to the first data point or forecast	Monte Carlo simulations to test robustness across a range of input assumptions
Delegating Judgment to Algorithms	Blind trust in model output without interpretability or human validation	Interpretable ML tools (e.g., PDPs, LIME) paired with expert review
Escalation of Commitment	Persisting in failing strategies due to sunk cost logic	Real-time dashboard triggers tied to strategic kill-switch metrics

By embracing these countermeasures, leaders enhance not only their decision *outputs* but also the *integrity* of the strategic process itself. In the chapters to come, we'll explore how algorithm-supported leadership (12.4.2) builds on this calibration by embedding intelligence *into* leadership systems, not as a replacement for human wisdom, but as a complement to it.

13.4.2 Algorithm-Supported Leadership

In the STEM-enabled organization, algorithms are no longer confined to back-end analytics or isolated data science teams. They are increasingly part of the strategic conversation—informing decisions, shaping forecasts, and sometimes even recommending actions in real time. Yet, the presence of algorithms alone does not guarantee better leadership. The real question is: *How do leaders use algorithms as cognitive scaffolding rather than crutches?*

Algorithm-supported leadership is not about handing over decision-making to machines. It's about knowing when, where, and how to embed algorithmic tools to enhance insight, reduce bias, and strengthen the interpretability of strategic choices. This includes understanding both the *logic* behind a model and the *context* in which it operates.

Leaders must treat algorithms as *partners* in sense-making—not as infallible authorities. When leaders blindly trust model outputs without questioning the assumptions, design logic, or training data, they risk substituting one form of error (human bias) for another (model misalignment). Strategic integrity demands that leaders maintain *accountability* even in algorithm-supported environments. This demands three key capabilities:

1. **Model Literacy**: Understanding what the algorithm optimizes for, and why.
2. **Interpretability Fluency**: Being able to explain and critique outputs using tools like SHAP, LIME, or PDP.
3. **Contextual Judgment**: Recognizing when to override, modify, or delay action based on non-model factors such as ethics, legality, or system complexity.

The table below contrasts common misuses of algorithms with more effective, scaffolded approaches to leadership:

Table 13.12. Algorithm Use in Leadership — Scaffolding vs. Substitution

Leadership Behavior	Misuse (Substitution)	Effective Use (Scaffolding)
Delegating Strategy to AI	Allowing algorithms to define strategic goals	Using algorithms to simulate outcomes and stress-test options
Ignoring Model Limitations	Blindly trusting output without validation	Auditing training data, checking for bias, applying SHAP or LIME analysis
Misreading Probabilities as Certainties	Treating predicted outcomes as inevitable	Interpreting outputs as probabilistic guides, not absolute predictions
One-Size-Fits-All Tooling	Applying the same model across contexts	Matching tool to problem: e.g., PDP for transparency, Monte Carlo for risk
Avoiding Accountability	Blaming "the model" when strategy fails	Owning final decisions while citing algorithmic insights transparently

By supporting—not replacing—human reasoning, algorithms can act as a form of *augmented foresight*. They surface unseen relationships, project complex interdependencies, and simulate cascading effects that the human mind may overlook. But leaders must always remain in the loop: interpreting, questioning, and owning the final call.

13.4.3 STEM-Enhanced Leadership Diagnostics

As leadership becomes more interwoven with data systems and organizational analytics, the way we assess leaders must evolve. Traditional 360-degree reviews and personality-based instruments are no longer sufficient for evaluating effectiveness in complex, tech-enabled environments. Today's leaders need real-time, data-integrated feedback loops that align with the very systems they are leading.

STEM-enhanced leadership diagnostics refer to diagnostic tools and frameworks that connect behavioral data, strategic decision outcomes, and organizational signals to assess leadership performance dynamically. These systems allow leaders to track how their decisions ripple through teams, operations, and customer experiences—and to calibrate accordingly.

Rather than focusing on static personality traits, modern diagnostics emphasize **adaptive behavior**, **learning agility**, and **data-informed influence**. When integrated into enterprise dashboards, these tools transform leadership evaluation from a periodic exercise to a continuous improvement system. These diagnostics often incorporate:

- **Decision Traceability**: Logging decision paths and measuring impact over time.
- **Behavioral Analytics**: Using pattern recognition to detect overconfidence, delay bias, or herd behavior in leader choices.
- **Simulation-Based Feedback**: Running scenario-based labs that test ethical reasoning, ambiguity tolerance, and strategic consistency.
- **Team-Sourced Feedback Systems**: Linking anonymous team input to AI-generated behavioral summaries, providing honest reflection without fear of hierarchy.

The table below contrasts traditional leadership evaluation methods with STEM-enhanced approaches, highlighting how diagnostics are evolving to meet the demands of modern leadership:

Table 13.13 Evolution of Leadership Diagnostics in the STEM Era

Dimension	Traditional Approach	STEM-Enhanced Approach
Data Source	Self-report, peer evaluation	Behavioral logs, decision impact metrics, and AI-assisted summaries
Frequency	Annual or semi-annual	Continuous, real-time dashboards
Feedback Type	Subjective interpretation	Quantified behavioral analytics with interpretability layers
Evaluation Focus	Traits and perceptions	Patterns, adaptability, and ethics under pressure
Integration with Strategy	Low – separated from business outcomes	High – directly linked to organizational KPIs and system modeling
Development Utility	General self-awareness	Targeted feedback for strategic behavior and growth tracking

By embedding diagnostics within enterprise systems—just as we embed analytics into strategy—organizations create a **leadership intelligence layer**. This enables not only better decisions but also better decision-makers. Leaders can monitor their blind spots, detect overreliance on gut instincts, and model data-responsible behavior across the firm.

The strategic leader in a STEM world does not just rely on intuition and charisma—they grow through feedback that is *structured, ethical, system-aware,* and *behaviorally grounded.* These diagnostics do not replace judgment—they sharpen it.

13.5 Building the AI-Augmented Coaching Ecosystem

As leadership roles evolve in the STEM era, the demand for scalable, personalized, and ethically sound coaching grows. Traditional one-on-one mentorship, while impactful, lacks the adaptability and reach needed in fast-moving, data-intensive organizations. To address this gap, forward-looking firms are developing **AI-augmented coaching ecosystems**—hybrid platforms that blend behavioral analytics, AI nudging, real-time feedback, and human insight to support leadership development continuously and intelligently.

These systems are not about replacing human mentorship. Instead, they reframe coaching as a **cyber-physical ecosystem**—an integrated feedback loop that brings together AI tools, team dynamics, and human coaches to foster insight, accountability, and growth across all leadership levels.

This emerging model uses intelligent agents, behavioral feedback loops, simulation environments, and real-time performance data to offer personalized, ongoing leadership development at scale. AI tools can track decision patterns, detect ethical blind spots, and nudge leaders toward growth—while human coaches step in for high-touch interventions when needed. The result is a hybrid system: **precision coaching** that is both systemically intelligent and emotionally attuned.

Key Components of an AI-Augmented Coaching Ecosystem

1. **Leadership Behavior Trackers**

   Integrated tools log decision speed, diversity of inputs, consultation behavior, and risk-taking patterns—providing a digital signature of leadership style.

2. **AI Nudges and Reflection Prompts**

   Context-aware systems trigger behavioral nudges: "Have you consulted your data scientist?" or "Consider checking for bias in your assumption." These function as lightweight, real-time coaching cues.

3. **Digital Simulation Environments**

   Interactive labs powered by AI allow leaders to practice strategic scenarios—testing ethical reasoning, crisis responses, or stakeholder trade-offs in safe, simulated conditions.

4. **Feedback Loops from Stakeholders and Teams**

   Anonymized, continuous feedback is analyzed by AI to surface themes, emotional tone shifts, or collaboration bottlenecks—providing social mirrors beyond what human perception captures.

5. **Human Coach Integration**

   Trained coaches review AI-generated leadership diagnostics and simulate "critical incident" debriefs—helping leaders develop self-awareness and meta-cognitive judgment.

6. **Ethical Governance Layer**

   The coaching system is embedded with privacy protections, opt-in transparency, and accountability protocols, ensuring that growth does not come at the cost of surveillance or misuse.

13.5.1 Behavior Tracking and Feedback Loops

At the heart of an AI-augmented coaching ecosystem is the ability to capture leadership behaviors through digital traces—decision speed, stakeholder inclusion, data consultation patterns, and even tone of communication. These behavior trackers act as diagnostic mirrors, enabling real-time assessment of leadership dynamics.

AI systems aggregate inputs from dashboards, communication platforms, and collaborative tools to deliver **insightful feedback loops**. This feedback is shared not annually but continuously—guiding self-regulation and situational awareness. Patterns in decision-making (e.g., overconfidence, hesitancy under ambiguity) become teachable moments.

13.5.2 AI Nudges and Simulation Tools

Using behavioral data, AI agents can issue timely nudges and reflective prompts, much like a digital coach whispering, "Have you considered alternative scenarios?" or "Does this decision reflect stakeholder equity?" These nudges reinforce ethical vigilance and decision rigor.

In parallel, simulation labs powered by AI let leaders test their responses to volatile or ethically sensitive situations. Leaders can role-play crises, stakeholder dilemmas, or ambiguous regulatory scenarios, building **bias resilience and judgment calibration** in high-stakes contexts. These labs also support gamified feedback, improving engagement in learning.

13.5.3 Human-AI Coaching Collaboration

Rather than displacing coaches, the AI system enhances their effectiveness. Human mentors receive synthesized behavioral diagnostics, ethical risk alerts, and engagement maps, enabling **targeted coaching interventions**.

This co-active system also supports **leadership calibration at scale**. For example, AI might flag a regional manager's declining team engagement score, while a coach facilitates a reflection session focused on empathy and digital leadership gaps. The loop between **real-time data and human empathy** becomes a source of strategic growth.

13.5.4 Ethical Governance in Coaching Systems

With great insight comes great responsibility. AI-augmented coaching must embed transparent ethical frameworks. This includes:

- Opt-in data use with informed consent
- Explainable nudging algorithms
- Data minimization and anonymization
- Governance boards for digital coaching systems

As coaching ecosystems grow more predictive, the **ethical use of behavioral data** becomes central to maintaining trust and legitimacy, especially in multicultural, global teams. To understand how AI is reshaping coaching, the following table compares traditional coaching models with AI-augmented ecosystems across key dimensions:

By integrating machine learning, behavioral science, and human judgment, AI-augmented coaching ecosystems represent a **paradigm shift** in leadership development. No longer confined to a one-time training or feedback session, leadership becomes an always-on capability, guided by data, refined through reflection, and scaled ethically across the enterprise.

Table 13.14. Traditional vs. AI-Augmented Coaching Systems

Dimension	Traditional Coaching	AI-Augmented Coaching Ecosystem
Access	Exclusive, top-tier roles	Inclusive, role-based, and scalable
Frequency	Monthly or quarterly	Continuous and real-time
Feedback	Qualitative and episodic	Data-informed, visualized, adaptive
Customization	Coach-driven	Context-aware and behavior-informed
Learning Mode	Reflective	Reflective + predictive + scenario-tested
Ethical Risk	Low	Managed via embedded governance protocols
Measurability	Subjective outcomes	Linked to performance, decision logs, stakeholder sentiment

13.6. Chapter 13 Wrap-Up: Strategic Leadership and STEM Capabilities

In a world shaped by complexity, uncertainty, and intelligent machines, strategic leadership is no longer defined by charisma or hierarchy, but by systems thinking, data fluency, and adaptive learning. This chapter reframes leadership as a distributed capability empowered by diagnostics, AI augmentation, and ethical judgment—not merely an individual trait.

We began by redefining what leadership looks like in analytics-intensive organizations. The classic command-and-control model is giving way to collaborative, evidence-based ecosystems, where the leader is not a lone visionary but a systems integrator who bridges AI models, human dynamics, and policy impacts.

Section 13.1 introduced the new leadership profile: transparent, humble, ethically alert, and digitally fluent. We highlighted how high-impact STEM-era leaders combine narrative intelligence with behavioral adaptability and how traditional pipelines often misalign with today's strategic demands.

Section 13.2 explored the emergence of **data-driven leadership development**. We shifted from gut-driven evaluation to insight-driven diagnostics, using behavioral analytics and simulation tools to build ethical resilience, detect bias, and create continuous self-calibration. This section laid the groundwork for AI-enabled feedback loops and performance-linked reflection.

In Section 13.3, we identified the core **STEM-aligned leadership competencies**, including data literacy, modeling fluency, decision integrity, and systems navigation. We discussed how these skills are cultivated through simulation labs, multidisciplinary sandboxes, and cross-functional mentorship—not just classroom learning. We also emphasized why developing these competencies matters now: because leadership inertia in the face of AI and complexity is a silent risk multiplier.

Section 13.4 addressed **strategic integration,** how leaders operate across disciplines, navigate bounded rationality, and use algorithmic tools as cognitive scaffolding. We introduced the ethics of AI delegation and emphasized the importance of diagnostic ecosystems that feed leaders with transparent, action-ready insights.

Finally, in Section 13.5, we introduced the **AI-Augmented Coaching Ecosystem,** a visionary framework that fuses real-time behavioral feedback, AI nudges, and ethical safeguards to democratize leadership development at scale. Leaders are no longer shaped in corner offices; they are refined through transparent data, cross-disciplinary dialogue, and feedback-rich cultures.

As you move forward, remember: The most effective STEM strategist is not only a technical expert, but also a leadership architect, capable of building people systems that learn, adapt, and reflect our highest values.

13.6.1. Discussion Questions

1. How does the leadership profile required in a STEM-driven organization differ from that of traditional corporate leadership?
2. What are the benefits and risks of using behavioral analytics to assess leadership effectiveness?
3. How can AI-augmented coaching systems preserve trust while offering real-time performance feedback?
4. In what ways can bounded rationality lead even experienced leaders astray, and how can diagnostics help mitigate this?
5. Why is it important to integrate ethical safeguards in AI-assisted leadership decisions?
6. What challenges might arise in cultivating STEM-fluent leaders in legacy organizations?
7. How can firms align strategic planning with leadership development to ensure long-term adaptability?

13.6.2. Micro-Case: Zara's Decision Integrity Sprint

Beyond Performance—Building Trust Through Leadership Diagnostics

Zara, recently promoted to VP of Strategy Enablement at a global energy firm, is tasked with selecting a new leader for its AI Ethics & Innovation Lab. The stakes are high: the role requires technical literacy, emotional intelligence, and ethical maturity. Traditionally, candidates would be evaluated through panel interviews and resumes.

Instead, Zara deploys a **leadership diagnostic protocol**: behavioral analytics to assess prior decisions under uncertainty, AI simulations testing response to bias-prone scenarios, and a reflection-based storytelling platform. Finalist candidates are run through a "judgment lab" that includes ambiguous strategy simulations and digital twin role-play scenarios.

One finalist, highly praised on paper, shows erratic judgment in feedback loops. Another, previously overlooked due to a lack of flashy credentials, demonstrates data fluency and ethical resilience. Zara presents a data-backed recommendation—and the board unanimously supports the lesser-known candidate.

Zara's sprint shows how **strategic leadership can be selected, not just sensed, measured, or assumed.**

Discussion Questions

1. **How did Zara's use of diagnostics challenge traditional leadership selection methods?**
   What advantages did this data-driven approach provide compared to resume reviews or panel interviews?
2. **What risks and biases might still exist in leadership diagnostics—even when using AI-enhanced tools?**
   How can organizations balance algorithmic insights with human judgment?
3. **How does Zara's process illustrate the importance of ethical foresight in leadership selection?**
   What would have been the consequence of prioritizing performance over integrity?
4. **What lessons can be drawn from Zara's approach for organizations developing leadership pipelines in data-intensive industries?**
   How can this diagnostic model be scaled to support a broader talent strategy?

13.6.3. Interactive Exercise: Leadership Calibration Challenge

Scenario:
You are the Chief Talent Strategist at **Infogenia**, a fast-growing healthcare analytics firm that's struggling with inconsistent leadership across its regional offices. Some leaders rely heavily on instinct and intuition, while others are paralyzed by over-analysis and risk aversion. The CEO has asked you to design a leadership development pilot using diagnostic and AI-augmented tools.

Your Task:
Using principles from this chapter, design a **leadership calibration blueprint**. Your deliverable should be a one-page concept note addressing the following:

1. **Diagnostic Toolkit**:
   - Choose 2–3 diagnostic tools (e.g., behavioral analytics, cognitive bias tests, AI nudging) and explain how they will reveal leadership gaps.
2. **AI-Augmented Coaching Plan**:
   - Propose how real-time nudges and feedback can be ethically integrated into daily leadership routines.
3. **Competency Framework**:
   - List 3 critical competencies you want to develop and how they align with strategic goals.
4. **Feedback Governance**:
   - Outline how you'll ensure privacy, transparency, and opt-in trust in your data-driven approach.

This exercise simulates how strategic leaders design intelligent people systems, not just for performance, but for long-term ethical and adaptive growth.

13.6.4. Recommended HBR Cases

❖ **Change and Leadership at Yashashvi Rasayan Pvt. Ltd.**

* **Overview:** Published on **August 8, 2024**, and authored by Sunil Budhiraja and Koel Nath (both affiliated with the Tata Institute of Social Sciences), this 14-page case explores the organizational transformation journey of Yashashvi Rasayan Pvt. Ltd. (YRPL), a family-owned chemical manufacturing company in India. Despite achieving record productivity in the 2022–23 financial year, the company's director remained concerned about safety culture and people-related challenges. Two years earlier, YRPL had launched a business transformation initiative, and while the results were positive, leadership questioned whether the improvements were due to incremental operational changes or the deeper impact of organizational development (OD) principles. The case focuses on the director's decision to convene a leadership meeting to assess the transformation's effectiveness and plan the next phase. It raises key questions about change management, leadership alignment, and the sustainability of cultural transformation in a traditional industrial setting.

* Citation: Budhiraja, S., & Nath, K. (2024). *Change and Leadership at Yashashvi Rasayan Pvt. Ltd.* (Case No. W36143-PDF-ENG). Ivey Publishing.

- ❖ **Leadership Reflections on Stepping Down**

  - * **Overview:** Published on April 18, 2025, and authored by Steven John DeKrey and Ramee Liu from The Hong Kong University of Science and Technology, this 13-page case explores the leadership journey of Jacinda Ardern, former Prime Minister of New Zealand, with a focus on her decision to step down in January 2023. The case highlights Ardern's globally admired leadership style, marked by compassion, decisiveness, and authenticity, as she navigated crises such as the Christchurch mosque attacks, the COVID-19 pandemic, and economic recovery efforts. Despite her success, Ardern chose to resign, citing a lack of energy to continue leading effectively.
    - The case challenges readers to reflect on:
      - Leadership transitions
      - Emotional intelligence and self-awareness
      - Timing and responsibility in stepping down
      - Sustainable leadership and legacy

    It is ideal for discussions on adaptive leadership, succession planning, and values-based decision-making.
  - * **Citation:** DeKrey, S. J., & Liu, R. (2025). *Jacinda Ardern's Farewell: A Leadership Reflections on Stepping Down* (Case No. ST156). The Hong Kong University of Science and Technology.

- ❖ **Puthran and Associates: Enhancing Authentic Leadership**

  - * **Overview:** Published on July 3, 2024, and authored by Deepa R. and Aastha Tripathi, this 13-page case focuses on Puthran & Associates, an all-women intellectual property law firm based in India. In 2023, the firm was facing a significant challenge: high employee turnover, particularly among women who were resigning or relocating due to family responsibilities. The case centers on Rukmini Puthran, the firm's founder, who is deeply committed to creating a supportive and empowering work environment. She must now reflect on how to evolve her leadership style and organizational practices to retain talent and foster a culture of authentic leadership. The case explores how leaders can balance empathy, flexibility, and performance expectations in a professional services setting. This case is ideal for discussions on leadership development, employee engagement, gender and work-life integration, and organizational culture in small and mid-sized firms.
  - * **Citation:** Deepa, R., & Tripathi, A. (2024). *Puthran and Associates: Enhancing Authentic Leadership* (Case No. W37027-PDF-ENG). Ivey Publishing.

13.6.5. Strategy-in-Film: Learn by Watching

Leadership in a data-driven world demands more than traditional decision-making, it requires systems thinking, technological fluency, and ethical judgment. The films below explore how leaders guide organizations through complexity, uncertainty, and transformation. As you reflect on strategic leadership in this chapter, use these narratives to examine how modern leaders harness data, technology, and adaptive thinking to shape organizational success.

Film / Documentary	Strategic Leadership Theme
The Post	Ethical leadership under institutional pressure
Oppenheimer	Systems thinking, moral ambiguity, and leadership in scientific disruption
Apollo 13	Collaborative decision-making under high-stakes uncertainty
Moneyball	Data-driven leadership vs. intuition
Coded Bias	Leadership accountability in AI systems
The Imitation Game	Leadership in algorithmic innovation
Hidden Figures	Unrecognized leadership and data literacy
Her	Ethical dilemmas in human-AI interaction

References

Alon, I., & Higgins, J. M. (2005). Global leadership success through emotional and cultural intelligences. *Business Horizons*, *48*(6), 501-512.

Antonakis, J., & Day, D. V. (Eds.). (2017). *The nature of leadership*. Sage Publications.

Aziz, M. F., Rajesh, J. I., Jahan, F., McMurrray, A., Ahmed, N., Narendran, R., & Harrison, C. (2024). AI-powered leadership: a systematic literature review. *Journal of Managerial Psychology*.

Bachkirova, T., & Kemp, R. (2025). 'AI coaching': democratising coaching service or offering an ersatz?. *Coaching: An International Journal of Theory, Research and Practice*, *18*(1), 27-45.

Bornay-Barrachina, M., López-Cabrales, Á., & Salas-Vallina, A. (2025). Sensing, seizing, and reconfiguring dynamic capabilities in innovative firms: Why does strategic leadership make a difference?. *BRQ Business Research Quarterly*, *28*(2), 399-420.

Davies, B., & Brundrett, M. (Eds.). (2010). *Developing successful leadership* (Vol. 11). Springer Science & Business Media.

DeRue, D. S. (2011). Adaptive leadership theory: Leading and following as a complex adaptive process. *Research in Organizational Behavior*, *31*, 125-150.

Graßmann, C., & Schermuly, C. C. (2021). Coaching with artificial intelligence: Concepts and capabilities. *Human Resource Development Review*, *20*(1), 106-126.

Haslam, S. A., Alvesson, M., & Reicher, S. D. (2024). Zombie leadership: Dead ideas that still walk among us. *The Leadership Quarterly*, *35*(3), 101770.

Heifetz, R. A., Grashow, A., & Linsky, M. (2009). *The practice of adaptive leadership: Tools and tactics for changing your organization and the world*. Harvard Business Press.

Hitt, M. A., Haynes, K. T., & Serpa, R. (2010). Strategic leadership for the 21st century. *Business Horizons*, *53*, 437-444.

Ho, G. K., Lam, C., & Law, R. (2023). Conceptual framework of strategic leadership and organizational resilience for the hospitality and tourism industry for coping with environmental uncertainty. *Journal of Hospitality and Tourism Insights*, *6*(2), 835-852.

Jarrett, M., & Vince, R. (2024). Mitigating anxiety: The role of strategic leadership groups during radical organisational change. *Human Relations*, *77*(8), 1178-1208.

Johns, G. (2024). The context deficit in leadership research. *The Leadership Quarterly*, *35*(1), 101755.

Luo, X., Qin, M. S., Fang, Z., & Qu, Z. (2021). Artificial intelligence coaches for sales agents: Caveats and solutions. *Journal of Marketing*, *85*(2), 14-32.

Khan, E. A., Hossain, M. A., Jahed, M. A., Akter, R., & Pappas, I. O. (2024). Green strategic leadership capability: Construct development and measurement validation. *Journal of Cleaner Production*, *450*, 141575.

Khandelwal, K., & Upadhyay, A. K. (2021). The advent of artificial intelligence-based coaching. *Strategic HR Review*, *20*(4), 137-140.

Mahidhar, V., & Davenport, T. H. (2018). Why companies that wait to adopt AI may never catch up. *Harvard Business Review*, *6*(2018), 8-13.

Malafronte, O., & Loufrani, S. Artificial Intelligence (AI) as a Coach for Developing Leader Competence: Exploring Schema Changes in AI Coaching Conversations. *Available at SSRN 5004776*.

Messick, D. M., & Kramer, R. M. (Eds.). (2004). *The psychology of leadership: New perspectives and research*. Psychology Press.

Obolensky, N. (2017). *Complex adaptive leadership: Embracing paradox and uncertainty*. Routledge.

Peifer, Y., Jeske, T., & Hille, S. (2022). Artificial intelligence and its impact on leaders and leadership. *Procedia Computer Science*, *200*, 1024-1030.

Randall, L. M., & Coakley, L. A. (2007). Applying adaptive leadership to successful change initiatives in academia. *Leadership & Organization Development Journal*, *28*(4), 325-335.

Samimi, M., Cortes, A. F., Anderson, M. H., & Herrmann, P. (2022). What is strategic leadership? Developing a framework for future research. *The Leadership Quarterly*, *33*(3), 101353.

Schneider, J. (2020). Human-to-ai coach: Improving human inputs to AI systems. In *Advances in Intelligent Data Analysis XVIII: 18th International Symposium on Intelligent Data Analysis, IDA 2020, Konstanz, Germany, April 27–29, 2020, Proceedings 18* (pp. 431-443). Springer International Publishing.

Shrivastava, P., & Nachman, S. A. (1989). Strategic leadership patterns. *Strategic Management Journal*, *10*(S1), 51-66.

Singh, A., Lim, W. M., Jha, S., Kumar, S., & Ciasullo, M. V. (2023). The state of the art of strategic leadership. *Journal of Business Research*, *158*, 113676.

Sposato, M. (2024). Leadership training and development in the age of artificial intelligence. *Development and Learning in Organizations: An International Journal*, *38*(4), 4-7.

Useem, M. (2010). Four lessons in adaptive leadership. *Harvard business review*, *88*(11), 86-90.

Wu, Y. L., Shao, B., Newman, A., & Schwarz, G. (2021). Crisis leadership: A review and future research agenda. *The Leadership Quarterly*, *32*(6), 101518.

CHAPTER 14

Strategic Futures – Where Strategy Goes Next

"The best way to predict the future is to model it simulate it and shape it before it arrives"

by MAK

Learning Objectives

After completing this chapter, readers will be able to:

The strategist of the future is not a commander, but a conductor—fluid, sensing, and system-aware, like the jellyfish navigating the currents of a changing world.

- Reframe strategic foresight as a STEM-powered discipline for navigating uncertainty and systemic change.
- Evaluate how quantum computing, generative AI, and synthetic data transform risk, optimization, and competitive strategy.
- Analyze the concept of planetary strategy and its implications for circular systems, sustainability, and ethical governance.
- Assess the challenges and opportunities of post-human leadership in algorithmic and bio-augmented decision environments.
- Apply tools of strategic anticipation—including scenario simulation, futures wheels, and backcasting- to design resilient strategies.
- Articulate a forward-looking leadership philosophy grounded in intelligence, ethics, and adaptability.

Strategic management has always dealt with uncertainty. But the scale, speed, and complexity of today's uncertainty demand more than adaptive plans or responsive analytics; they require the ability to anticipate, simulate, and design for futures that have not yet unfolded. This is the terrain of strategic foresight, and it is no longer a luxury. In an era defined by technological acceleration, ecological tipping points, and the rise of post-human intelligence, the strategist must evolve into something more: a futures architect.

Traditional strategic planning focused on managing what is known, analyzing competitors, tracking market trends, and allocating resources within familiar boundaries. But as those boundaries dissolve under the weight of quantum computation, synthetic data, AI governance, and planetary constraints, the strategist's role expands. The strategist of tomorrow must think across decades, model systems that do not yet exist, and grapple with choices that carry ethical weight across generations.

This chapter explores what it means to lead strategy in a world that is constantly shifting beneath our feet. We move beyond quarterly KPIs and five-year plans to examine the deep structures of long-term advantage in a STEM-integrated future. You'll encounter quantum computing not as a theoretical breakthrough, but as a force for real-time strategic optimization. You'll engage with generative AI not as a tool for automation, but as a partner in designing alternative realities. You'll confront climate models, circular systems, and AI ethics—not as compliance challenges, but as foundations for regenerative, resilient strategy. And you'll ask what leadership looks like when cognition, decision-making, and even ethical reasoning are distributed between humans and machines.

Strategic foresight is no longer about predicting a single future; it's about preparing for many plausible ones, and shaping the conditions under which the most desirable ones can emerge. In that context, STEM is more than a set of tools. It is the language of complexity, the logic of modeling, and the infrastructure of anticipatory decision-making.

This final chapter challenges you to go beyond strategy as a plan and embrace strategy as a design science. As you step into the future of strategic management, you are not just solving problems. You are architecting the systems that will define what problems and what possibilities even exist.

Opening Case: Ping An and the Architecture of Futures

Imagine a company that began as a traditional insurer, but became one of the world's most advanced strategic laboratories.

Headquartered in Shenzhen, Ping An was once seen as just another financial services firm in China's rapidly growing economy. But by the early 2020s, its leaders realized that conventional strategy, based on past performance, market share, and operational improvements, was not enough. They understood that the future was becoming less predictable and more programmable. So, Ping An didn't just pivot; they rearchitected their entire strategic model around intelligent foresight.

At the core of Ping An's transformation was a commitment to modeling the future, not once, but continuously. They didn't just use AI to optimize claims or reduce fraud. They trained large language models to detect weak signals of customer dissatisfaction. They simulated the impacts of policy shifts, pandemics, and climate events on health insurance portfolios using digital twins and stochastic models. Their foresight infrastructure became a strategic asset, allowing the company to test scenarios before they unfolded and adapt faster than competitors who were still reacting.

Ping An also recognized the strategic power of synthetic data. Realizing that privacy regulations and data sparsity could limit innovation, the company developed generative models to simulate new customer segments and stress-test product-market fit under hypothetical but plausible futures. These synthetic simulations helped the firm explore new pricing strategies, delivery channels, and even partnerships—before committing capital.

But perhaps most transformative was their approach to leadership. Ping An redefined the strategist's role from planner to orchestrator—from controller of resources to designer of adaptive systems. Strategy was not a one-off event but a living model: monitored, recalibrated, and evolved in real time by AI systems, executive insight, and environmental feedback.

By 2030, Ping An had become more than a financial company. It was a foresight engine, one that made strategy not about reacting to what's coming next, but actively shaping it.

Play the Strategist

You've just been appointed Chief Strategy Officer at a multinational firm inspired by Ping An's model.

- What foresight tools or systems would you implement first?
- How would you help your leadership team shift from static planning to continuous simulation?
- What trade-offs might you face when balancing prediction, ethics, and execution?

14.1 Quantum Strategy and the Future of Strategic Logic

In the classical world of strategy, decisions are modeled as trade-offs, plans are built on linear forecasts, and optimization unfolds within fixed constraints. But what happens when those constraints no longer apply, when computing power explodes, uncertainty becomes mappable, and risk transforms from limitation into leverage?

Welcome to the quantum age of strategy.

Quantum computing represents more than a technological breakthrough; it is a paradigm shift in how organizations can perceive, model, and act upon complexity. With the rise of qubits, entanglement, and superposition, strategy no longer needs to be confined to "either-or" decisions. It can instead explore vast spaces of "what-if" possibilities, simultaneously.

This section explores how quantum computing challenges foundational assumptions about strategic logic. It introduces the concept of quantum advantage, explains how it redefines optimization, and examines the implications for risk, logistics, and competitive positioning. We shift from deterministic pathways to probabilistic landscapes, from fixed choices to continuously evolving state spaces.

For strategists in a STEM-powered world, quantum isn't science fiction. It's a new cognitive framework. One that demands a rethinking of how we define value, model choices, and structure decision-making under uncertainty.

> The next frontier of strategy is not about thinking faster—it's about thinking in more dimensions.

14.1.1 Quantum Advantage as Strategic Advantage

For decades, strategic modeling has relied on the limits of classical computing—sequential logic, binary outcomes, and brute-force scenario testing. Even our most advanced simulations, from Monte Carlo forecasts to dynamic optimization, have been bound by computational trade-offs: the more variables introduced, the more time and power required to solve them. But quantum computing challenges this paradigm at its core.

Quantum strategy begins with a simple but radical premise: what if we could compute not one scenario at a time, but all plausible scenarios in parallel? What if uncertainty were no longer a constraint, but a structure we could navigate probabilistically?

This is the promise of **quantum advantage,** not just faster computing, but *fundamentally different logic.* Unlike classical systems that process information in bits (0s and 1s), quantum systems operate using **qubits**, which can exist in multiple states at once (superposition) and interact through **entanglement**, allowing for exponentially greater complexity to be explored simultaneously. The implications for strategy are profound.

Imagine a pharmaceutical company evaluating 10,000 possible molecular pathways for a new drug. In a classical world, this takes weeks or months of simulations. In a quantum world, it could be reduced to hours, or less, while also uncovering non-obvious paths that would have been discarded by conventional models. Now apply that logic to your firm's strategic options: pricing structures, investment portfolios, supply chain routes, or geopolitical risk mitigation.

Strategically, this means **quantum advantage becomes strategic advantage**. Firms that understand how to leverage quantum-powered insights will outperform rivals not just by reacting faster, but by designing superior options that others never compute in the first place.

Yet this future is not automatic. To benefit from quantum systems, strategists must first rethink their models. Traditional scenario planning often assumes linearity and bounded rationality. Quantum-informed strategy requires embracing **probabilistic logic**, **entangled variables**, and **non-intuitive correlations**. It's not about predicting a single outcome; it's about understanding the full distribution of possibilities and optimizing across them.

Classical View: Simulate three supply chain routes based on cost and reliability scores.
Quantum View: Simulate *all* possible routes, factoring in weather, tariffs, labor unrest, and geopolitical volatility, *simultaneously*, and identify emergent optimal pathways across scenarios.

Firms like IBM, Google, and D-Wave are already deploying quantum prototypes to explore finance, logistics, and AI training. In the next decade, organizations that fail to engage with these technologies risk falling behind, not because they lack speed, but because they lack **strategic dimensionality**.

To illustrate the shift, the following table compares classical and quantum approaches to strategic modeling:

Table 14.1. Classical vs. Quantum Simulation in Strategic Modeling

Feature	Classical Strategy Modeling	Quantum Strategy Modeling
Decision Logic	Linear, sequential	Probabilistic, entangled
Speed	Limited by binary computation	Exponential parallelism
Scenarios Handled	Dozens to thousands	Millions to infinite pathways
Strategy Insight	Point-based optimization	Distributional foresight
Competitive Edge	Efficiency-driven	Dimensional foresight-driven

14.1.2 Rewriting Optimization: From Linear Models to Quantum Superposition

Optimization has always been a core pillar of strategy, from maximizing profit margins to allocating resources, from pricing structures to supply chain flows. Traditionally, strategists have relied on linear models, constraint-based decision trees, and sensitivity analysis to evaluate trade-offs. These tools have served well in a world of moderate complexity and bounded rationality. But in today's strategic environment, shaped by high-dimensional data, multiple interacting uncertainties, and exponential change, linear logic often collapses under its own weight.

Quantum computing introduces a new class of optimization: one rooted in **superposition** (the ability of a qubit to exist in multiple states simultaneously) and **quantum tunneling** (the capacity to bypass local optima in search of global ones). This allows quantum algorithms to explore and evaluate **millions of combinations at once**, not sequentially, but in parallel. It's not just faster, it's fundamentally more expansive.

In strategic terms, this means the firm is no longer constrained to "either/or" models. A company choosing between launching a product in Market A or Market B under specific assumptions can instead simulate **all permutations of launch conditions** across timing, pricing, regulation, competitor response, and consumer sentiment. This changes the question from "What is the best path?" to "What is the full geometry of the decision space, and where do emergent patterns reveal optimality?"

Quantum optimization also enables new forms of **dynamic equilibrium modeling**. For example, a logistics firm can use quantum annealing to continuously adjust shipping routes in response to changing weather, fuel prices, and geopolitical disruptions, *in near real time*. This adaptive modeling approach turns strategy into a fluid process, constantly responding to live variables without collapsing under computational load.

Perhaps most importantly, quantum superposition invites **a new way of thinking about uncertainty**. In classical optimization, risk is often modeled as noise to be minimized. In quantum models, risk becomes structure, an input to be navigated, not avoided. This unlocks new possibilities in portfolio design, risk management, and scenario exploration, particularly in volatile sectors such as energy, pharmaceuticals, defense, and fintech.

The strategist's role, then, shifts again: from planner and selector to **curator of quantum simulations,** someone who understands not just how to run models, but how to frame the right dimensions, interpret the entangled variables, and design meaningful output from probabilistic possibility spaces.

From Decision Trees to Quantum Graphs

Classical models often rely on decision trees or linear regressions that map inputs to outcomes in predictable paths. Quantum systems, by contrast, operate in **Hilbert space,** a multidimensional environment where outcomes are not fixed until observed. This requires strategists to develop fluency not in deterministic logic, but in **probabilistic interpretation**, **non-linearity**, and **quantum-enhanced heuristics**.

Old World: Optimize advertising spend across three channels using a linear budget allocation model.
Quantum World: Simulate all combinations of timing, message format, market volatility, and competitor noise to discover **non-obvious ad sequencing** with the highest emergent influence on behavior—*under multiple future conditions.*

14.1.3 Strategic Applications: Logistics, Finance, and Cybersecurity

Quantum strategy is not a distant frontier; it's already unfolding in high-stakes industries where complexity, uncertainty, and velocity converge. The most immediate and compelling applications are emerging in **logistics**, **financial modeling**, and **cybersecurity** domains where traditional optimization hits its limits and where quantum methods reveal powerful strategic leverage.

Logistics: Real-Time Routing and Dynamic Coordination

Global logistics networks face continuous disruption—from fuel price volatility and weather patterns to customs delays and political unrest. Classical optimization models struggle with such high-dimensional, stochastic variables. Quantum-inspired algorithms, by contrast, can evaluate millions of routes and constraint configurations simultaneously. Firms like **DHL** and **Maersk** are already experimenting with quantum annealing to improve delivery scheduling, warehouse optimization, and fleet routing under live, shifting constraints.

Instead of planning routes the night before, future logistics platforms will *co-model multiple disruptions in real time*, adjusting operations as new variables emerge. This is not just efficiency, it's **resilience by design**.

Financial Strategy: Portfolio Construction in Probabilistic Environments

Modern portfolio theory has long relied on variance, covariance, and risk-adjusted returns as the basis for strategic asset allocation. But these models are computationally expensive and quickly become unstable when extended to complex portfolios or nonlinear constraints. Quantum systems can optimize over massive, entangled investment spaces, balancing not just returns and volatility, but ESG factors, liquidity constraints, and macroeconomic scenarios *all at once.*

Leading firms like **Goldman Sachs**, **JPMorgan**, and **BlackRock** are now exploring quantum-enhanced optimization for credit risk modeling, derivative pricing, and scenario stress-testing. The future of finance will not belong to those who guess best; it will belong to those who simulate most broadly and reason most deeply across entangled data structures.

Cybersecurity: Strategic Defense Through Quantum Simulation

As organizations digitize everything, from financial systems to healthcare infrastructure, their exposure to cyber threats grows exponentially. Classical security protocols are increasingly vulnerable to both brute-force attacks and subtle, AI-driven infiltration. Quantum computing presents a dual role here: **threat and solution**.

On the one hand, quantum computers may eventually break today's encryption methods; on the other hand, they also enable **quantum key distribution** (QKD), a technique that leverages quantum entanglement to ensure secure communications. For strategic leaders, cybersecurity becomes not just an IT issue but a strategic frontier. Firms must now anticipate **quantum-class threats** before they materialize and invest in post-quantum defense architectures before competitors exploit the gap.

Logistics: UPS uses quantum-inspired routing to save fuel and reduce delays during climate disruptions.

Finance: A hedge fund simulates 10,000 market collapse paths and builds resilient portfolios against each.

Cybersecurity: A health tech firm uses QKD to ensure secure transmission of genomic data across global labs.

Quantum-Inspired Optimization vs. Full Quantum Computing

Even before full-scale quantum machines become commercially viable, many organizations are leveraging **quantum-inspired algorithms**—hybrid models that emulate quantum principles using classical hardware. These methods already offer superior performance in strategic modeling tasks.

The takeaway? You don't need a quantum computer to start practicing **quantum thinking**.

14.2 Generative Intelligence and Synthetic Strategy

Data is no longer discovered, it's invented. Strategy must now simulate futures, not just study the past.

If quantum computing redefines the logic of decision-making, generative intelligence transforms the inputs and materials of strategic design. In the traditional model, strategic analysis relies on historical data, customer surveys, and market reports, rich but static sources. But in a world where privacy concerns, data silos, and market volatility constrain access to real-time insight, a new question emerges: **What if we could generate the data we need?**

Generative AI, particularly through large language models (LLMs), diffusion models, and generative adversarial networks (GANs), is enabling a new class of **synthetic strategy tools**. These models don't just analyze existing datasets; they produce plausible, high-dimensional alternatives. They generate customer behaviors, simulate economic conditions, design products, test policies, and even explore organizational ethics, all before real-world deployment.

This is the rise of **synthetic foresight,** the ability to imagine, simulate, and rehearse strategic decisions using data that doesn't yet exist, but could.

14.2.1 Synthetic Data as Strategic Infrastructure

Synthetic data refers to artificially generated datasets that mimic the statistical properties of real data. Originally developed to protect privacy or augment machine learning pipelines, it is now being repurposed for **strategic exploration**.

Imagine a global retailer testing pricing strategies across 50 hypothetical economic shocks, without waiting for those shocks to occur. Or a healthcare startup modeling patient behavior in a market it hasn't entered yet. By using GANs to generate synthetic customer personas, firms can simulate strategic choices that would otherwise be too risky, too slow, or too ethically sensitive to test live.

Strategists now have access to **a simulation laboratory for the future**, no longer constrained by what has already happened.

14.2.2 Generative AI as a Strategic Co-Pilot

Large language models (LLMs) like GPT, Claude, and Gemini aren't just writing assistants. They are becoming **co-strategists,** able to synthesize scenarios, generate policy options, critique investment decisions, and frame strategic narratives in real time. More importantly, they can generate thousands of **"what-if" frames** across sectors, geographies, and stakeholder groups.

For example, a strategist might ask:

- "What would be the impact on consumer trust if we shifted from a subscription model to dynamic pricing?"
- "How would a generative AI system describe our value proposition to an emerging market segment?"
- "What are the likely societal objections to this strategy, and how could we preemptively respond?"

By embedding LLMs into the strategy development process, leaders gain **a real-time simulation partner,** one that reflects, tests, and expands their cognitive boundaries.

14.2.3 Simulating the Unseen: From Scenario Planning to Imagination Engineering

The fusion of synthetic data and generative AI gives rise to a new discipline: **imagination engineering**. Strategy is no longer limited to known risks and observable data. Instead, it becomes a process of constructing **hypothetical worlds**, testing strategic actions within them, and learning from those rehearsals.

Instead of building one strategic plan, future-ready firms will generate hundreds of "micro-futures," each shaped by different AI-driven assumptions, climate scenarios, regulatory shifts, consumer sentiment patterns, and select pathways that perform robustly across variance.

This is more than foresight. It is **strategic rehearsal at scale**—a shift from planning to **prototyping futures**.

Generative AI plays a central role in this transformation. It augments traditional strategic workflows by enabling the rapid creation of synthetic personas, simulated policy reactions, branching scenario libraries, and stakeholder-specific narratives. The table below illustrates how generative AI enhances key strategic tasks:

- A global telecom firm uses synthetic data to test 50 pricing strategies under 10 inflation paths.
- A fashion retailer uses GenAI to simulate cultural responses to sustainability messaging across regions.
- A healthcare NGO generates synthetic family case profiles to pre-model service allocation under budget constraints.

Table 14.2: Generative AI in the Strategic Workflow

Strategic Task	Classical Approach	GenAI-Augmented Approach
Customer Segmentation	Surveys and historical clustering	Synthetic personas generated via GANs
Policy Testing	Focus groups and pilots	LLM simulations of public reactions
Scenario Planning	Analyst-led narratives	AI-generated scenario libraries with dynamic branching
Strategic Communication	Executive memos and reports	LLM-crafted narratives tailored to stakeholder personas

14.3 Planetary Strategy and the Climate-Strategy Nexus

For much of modern strategy, the environment was treated as an external context—a set of risks, regulations, or cost factors to be managed. But this view is no longer viable. As climate thresholds tighten, biodiversity loss accelerates, and societal expectations shift, the Earth itself now enters the strategy room, not as scenery, but as an actor. This is the rise of **planetary**

The planet is no longer a backdrop to business—it is a stakeholder, a constraint, and a co-designer.

strategy: the deliberate integration of ecological systems, intergenerational ethics, and global limits into the strategic design of organizations.

Planetary strategy does not replace financial performance or technological innovation. Rather, it expands the definition of value, adds new variables to the strategy equation, and forces a shift from extractive logic to regenerative logic. In the same way STEM tools expanded decision modeling, climate constraints now reshape the boundaries of what counts as "strategic success."

14.3.1 Strategic Decision-Making Within Planetary Boundaries

Planetary boundaries are the ecological thresholds within which humanity can operate safely, covering climate stability, land use, ocean acidification, freshwater cycles, and more. Crossing these thresholds may trigger irreversible tipping points.

> **Interactive Exercise: Synthetic Strategy in Action**
> Using any generative AI tool (e.g., ChatGPT, Claude, Gemini), simulate a strategic dilemma:
>
> "You are a strategist at a fintech firm entering a new emerging market. Create three synthetic personas and simulate how each would respond to your product pricing, UX design, and regulatory terms."
> Debrief:
>
> - What insights did AI offer that human-only models may have missed?
> - What limitations did you observe in the synthetic logic?

Strategists can no longer ignore these limits. Firms must now **embed climate models, ecological risk analytics, and systems mapping** into core decision processes. This means:

- Using geospatial data and AI to assess environmental impact across supply chains
- Running lifecycle simulations to evaluate the carbon, water, and material footprint of strategic initiatives
- Designing KPIs that track resilience, circularity, and intergenerational outcomes, not just quarterly returns

Strategic advantage is now tied to how well a firm operates *within* these constraints, not in spite of them.

14.3.2 Circular Systems and Regenerative Value Loops

The traditional model of value creation, extract, produce, use, discard, is no longer sustainable. Leading organizations are embracing **circular strategies** that regenerate ecosystems, close resource loops, and internalize externalities.

Examples include:

- Product-as-a-service models (e.g., leasing instead of selling)
- Reverse logistics and material recapture
- Supply chain redesigns based on biomimicry and cradle-to-cradle principles

The strategist must now think like a system designer, crafting **circular ecosystems** that align business success with planetary health.

* A food conglomerate models water stress scenarios to shift crop sourcing.
* A city planner uses AI to redesign urban mobility within carbon-neutral boundaries.
* A clothing brand simulates supply chain emissions under various climate treaty scenarios.

Consider a company you admire.

* Are their long-term strategies aligned with planetary boundaries?
* How might their metrics change if climate limits were treated as non-negotiable constraints?

14.3.3 AI Governance, Ethics, and Climate Intersections

Planetary strategy is not just about carbon; it's also about the governance of the tools used to shape the future. As firms deploy AI to optimize sustainability, questions of transparency, fairness, and unintended bias become central.

Consider a machine learning model that optimizes energy usage but inadvertently deprioritizes low-income neighborhoods. Or an ESG scoring algorithm trained on biased data. Strategy in the planetary era must include **algorithmic accountability** as a pillar of ethical foresight.

To support this shift, organizations are increasingly integrating advanced climate models into their strategic decision-making. These models, when paired with AI, enable more nuanced, data-rich planning across environmental, social, and ethical dimensions. The table below outlines key model types and how they are being applied in practice:.

Table 14.3: Climate Models in Strategic Decision-Making

Model Type	Strategic Use	Example
Integrated Assessment Models (IAMs)	Link economic activity to carbon emissions and policy pathways	Used by Shell to evaluate decarbonization strategies
Life Cycle Assessment (LCA)	Simulate environmental impact across a product's life	Applied by IKEA to redesign furniture for reuse and disassembly
Geospatial Analytics	Assess location-based climate risk and resilience	Used by insurers to model wildfire and flood exposure
Climate-AI Hybrids	Forecast multi-variable ecological outcomes	Employed by Microsoft in their planetary computing initiative

14.4 Strategic Leadership in a Post-Human Era

Strategy has always been a human endeavor, crafted in boardrooms, debated by teams, driven by vision, intuition, and experience. But as artificial intelligence, robotics, neurotechnology, and synthetic biology advance, a foundational question emerges: **What does leadership mean when cognition is shared with machines?**

We are entering the **post-human era of strategy,** an age in which human decision-makers are no longer the sole originators of judgment, insight, or design. Intelligent systems now assist with investment decisions, optimize supply chains, write strategic narratives, and even participate in ethical deliberation. The strategist's role does not vanish, but it evolves dramatically.

This section explores the transformation of leadership under these conditions: from human exceptionalism to **human-machine collaboration**, from charisma to **computational reasoning**, and from reactive governance to **embedded ethics**.

14.4.1 Human-Machine Strategic Collaboration

In a post-human workplace, leadership involves **designing decision architectures**, not just making decisions. Strategic leaders must now coordinate hybrid teams, some composed of human experts, others of algorithmic agents and autonomous systems. The challenge is not only technical, it is relational, cognitive, and cultural.

Key implications include:

- Redesigning roles to distinguish between human creativity and machine efficiency
- Structuring oversight protocols for AI-influenced strategic choices
- Training leaders to read, question, and interpret algorithmic outputs, not passively accept them

> *When intelligence is no longer uniquely human, leadership must become uniquely ethical, systemic, and visionary.*

For example, an executive team may use AI to surface new markets or simulate pricing outcomes, but must still determine **which values guide those strategies**. Leadership becomes a system of judgment applied to systems of intelligence.

14.4.2 Trust, Explainability, and Algorithmic Integrity

As intelligent systems gain more autonomy, leaders must grapple with questions of **trust** and **transparency**. Will stakeholders accept decisions made, or even suggested, by algorithms? How do we ensure that these systems reflect organizational purpose, societal values, and ethical responsibility?

Enter the concept of **explainable AI (XAI),** designing models that can articulate the rationale behind their outputs. Strategic leaders must now ask:

- Can we trust this AI's recommendation under uncertainty?
- Does it reflect our goals, or merely optimize for incomplete metrics?
- What assumptions were embedded in its training data?

Leading in a post-human era requires not just using AI, but interrogating it.

14.4.3 Augmented Cognition and Bio-Enhanced Strategy

The frontier of post-human leadership extends beyond AI to **neurotechnology and cognitive augmentation**. Startups and research labs are developing brain-computer interfaces, attention-enhancement wearables, and decision-support implants. These tools raise profound questions: If a leader can process five times more information through augmentation, does that redefine competence? If some executives are bio-enhanced, how do we define fairness or access?

Strategists must begin to **think ethically and strategically** about bio-enhancement before it becomes a competitive differentiator.

In this context, leadership becomes:

- Part technologist: navigating frontier technologies
- Part ethicist: safeguarding human dignity and collective trust
- Part systems architect: designing hybrid workplaces where cognition is distributed intelligently and equitably

To illustrate how leadership is being redefined in this emerging landscape, the following table contrasts traditional leadership paradigms with those shaped by machine-augmented decision systems:

Table 14.4: Leading Through Machine-Augmented Decision Systems

Leadership Focus	Traditional Context	Post-Human Context
Decision Ownership	Human executives	Shared between human, AI, and data layers
Information Processing	Experience and intuition	AI-enhanced pattern recognition, neural augmentation
Trust and Accountability	Based on personal integrity	Based on model transparency and traceability
Team Composition	Human expertise only	Hybrid teams of analysts, bots, and systems advisors
Ethical Leadership	Moral judgment of humans	Ethical oversight of human-machine systems

14.5 Strategic Intelligence Tools for the Futures Architect

In a strategy landscape shaped by quantum computing, generative intelligence, planetary thresholds, and post-human decision-making, the strategist must evolve into a **futures architect,** a designer of resilient, adaptive, and ethically grounded systems. But architectural thinking requires more than vision. It requires tools, rigorous, structured, and dynamic instruments that help organizations explore uncertainty, simulate outcomes, and select robust paths forward.

This section introduces a new strategic toolkit, one designed not for linear plans or static reports, but for **modeling plausible futures**, navigating complexity, and embedding strategic foresight into daily practice.

Foresight is not about seeing the future—it's about building systems that are prepared for any version of it.

14.5.1 Futures Wheel, Backcasting, and Wind Tunneling

Traditional strategy often begins with trend extrapolation: projecting forward from the present. Futures architecture flips this logic by designing from the future backward or simulating multiple conditions around a fixed strategy.

- **Futures Wheel**: A structured brainstorming tool that maps first-, second-, and third-order consequences of a change or innovation. This helps visualize cascading effects of decisions across systems.
- **Backcasting**: Starts with a preferred future (e.g., "net-zero operations by 2040") and works backward to define necessary milestones, capabilities, and pivots.

- **Wind Tunneling**: A technique to stress-test a strategy against diverse future scenarios, like climate disruption, supply shocks, AI overreach, or geopolitical instability.

These tools don't predict the future; they **test the fitness of a strategy** under alternate futures.

14.5.2 Simulation Engines and Decision Sandboxes

With the rise of digital twins, generative AI, and quantum-inspired modeling, strategy can now be **rehearsed, iterated, and stress-tested** before resources are committed.

- A **decision sandbox** allows leaders to simulate actions (e.g., entering a market, shifting a pricing model) under different assumptions, observing unintended consequences before real-world launch.
- A **digital twin** replicates the structure and dynamics of a system, be it a supply chain, customer ecosystem, or regulatory regime, and allows experimentation in a risk-free environment.

Strategic planning becomes **less linear and more experimental**, guided by scenario probability, not just past performance.

14.5.3 Building an Anticipatory Culture

Even the best tools fail if the culture around them remains reactive. Futures architects must design **anticipatory cultures,** environments where sensing, scanning, and systemic exploration are embedded into workflows.

Key principles:

- Encourage cross-functional foresight teams
- Use **weak signals** (early, ambiguous indicators of change) to guide exploration
- Integrate strategic foresight into budgeting, innovation cycles, and risk management
- Reward scenario-building, not just forecasting accuracy

In these environments, foresight is not a static report; it becomes a **strategic habit**. It's about building the muscle to ask, *"What if?"* before disruption forces the question. To support this shift, organizations are adopting a range of foresight methods that align with STEM-era strategy. The following table outlines key tools and how they are being applied in practice:

Table 14.5: Strategic Foresight Methods for STEM-Era Strategy

Tool / Method	Primary Use	Example in Practice
Futures Wheel	Mapping consequences of innovation	Used by UNICEF to explore AI's long-term impact on education
Backcasting	Designing backward from preferred futures	IKEA’s net-zero strategy is anchored in 2040 goals
Wind Tunneling	Stress-testing a strategy across scenarios	Used by Shell to test the energy portfolio under geopolitical volatility
Digital Twins	Simulating operational systems in real-time	Siemens models smart factory ecosystems before launch
Decision Sandboxes	Safe experimentation with strategic moves	Startups simulate pricing before market entry with synthetic personas
Weak Signal Detection	Identifying emerging trends early	Amazon flags early shifts in consumer sentiment through sentiment AI

Which Tool Fits Your Strategy?
Select a current or past strategic decision you’ve observed (in a company, nonprofit, or public sector).

- Which foresight tool could have enhanced that decision?
- Would the outcome have been different under a futures-informed process?

Interactive Exercise: Timeline to 2040
Design a **strategic timeline** using backcasting.

- Step 1: Choose a preferred future outcome for a company, city, or industry.
- Step 2: Identify five critical shifts that must happen by 2040 to reach that future.
- Step 3: Build a timeline with milestones, technologies, policies, and strategic choices that backcast from that vision.
- Step 4: Present the risks and uncertainties at each inflection point.

14.6. Chapter 14 Wrap-Up: Strategic Futures – Where Strategy Goes Next

As the world accelerates into an era of radical uncertainty, the strategist’s role is being redefined. This chapter invited you to explore what lies ahead, not just in terms of emerging technologies, but in terms of evolving strategic logic. Strategy is no longer confined to annual plans or market analysis. It must now operate across dimensions of quantum logic, synthetic data, planetary boundaries, and post-human leadership.

We began by introducing the future strategist not as a planner, but as a **futures architect**—one who can anticipate systemic risks, design multiple scenarios, and guide organizations through ambiguity with both technical fluency and ethical grounding.

In Section 14.1, we explored the rise of **quantum strategy**, where traditional optimization models are replaced with parallel futures and quantum-enhanced decision structures. In this world, risk becomes probabilistic geometry, and leaders must prepare for disruption at computational speed.

Section 14.2 introduced the domain of **generative intelligence and synthetic strategy**. You saw how synthetic data, multimodal AI, and strategic simulations are transforming how decisions are modeled, rehearsed, and explained. This isn't about replacing human strategists—it's about amplifying imagination with data-rich foresight.

Section 14.3 pushed strategy into the domain of **planetary intelligence**, emphasizing ecological thresholds, circular systems, and AI governance. Here, strategy must engage with global systems—not just economic ecosystems, but Earth systems. The firm of the future must balance competitiveness with resilience.

Finally, in Section 14.4, we examined **strategic leadership in a post-human age**, where humans share cognitive authority with algorithms, robots, and intelligent agents. The strategist must now lead hybrid systems with transparency, interpretability, and moral foresight, balancing machine precision with human judgment.

Throughout the chapter, the tools and themes from earlier chapters, SHAP, digital twins, ethics, and simulation, reappear in future-forward form. The strategist of tomorrow isn't just a technologist or a decision-maker, but a designer of ethical systems and regenerative futures.

As you prepare to lead into this unknown terrain, remember: it's not just about predicting what comes next, but about preparing your organization to adapt, align, and act with purpose across multiple futures.

14.6.1. Discussion Questions

1. How does quantum strategy challenge the assumptions behind traditional forecasting models?
2. What strategic opportunities and risks are introduced by synthetic data and generative AI?
3. How can organizations design business models that respect planetary boundaries and social tipping points?
4. What new leadership challenges arise when AI agents participate in decision-making processes?

5. How would you build a governance framework to manage hybrid human–machine boardrooms?

14.6.2. Strategy Micro-Case: Ping An and the Predictive Planet

From AI to Planetary Intelligence – Ping An's Strategic Leap

In 2031, Ping An has expanded beyond finance into a fully integrated health-climate-technology ecosystem. Using its proprietary digital twin platform, it simulates not only consumer behavior but also carbon output, disease hotspots, and socio-political instability across Asia.

A new project—*PlanetRisk360*—promises predictive insurance and investment services based on AI-enhanced foresight of climate and geopolitical volatility. The platform recommends divesting from a series of lucrative agribusiness portfolios in regions facing water stress, but internal teams push back, citing short-term loss of shareholder value.

Ping An's strategy team is caught between two forces: the AGI-backed model warns of long-term systemic collapse, while human executives demand performance benchmarks by Q3.

Strategic Challenge:

As Chief Futures Strategist, you are tasked with crafting a five-year strategic response that:

- Reconciles planetary risk forecasts with financial performance expectations
- Designs a hybrid governance structure for machine-generated strategy
- Communicates explainable foresight to global stakeholders

Learning Outcome:
This case illustrates how strategic futures are no longer abstract—it is the lived terrain of cross-system foresight, ecological limits, and ethically bounded AI decisions.

Discussion Questions

1. How does Zara's use of strategic simulations illustrate the difference between traditional scenario planning and future-shaping foresight?
   (Explore the added value of AGI-backed forecasts, explainable simulations, and stress-tested planetary boundaries.)
2. What ethical dilemmas emerge when simulated strategies offer high returns but violate environmental or social limits? How should Zara weigh these factors?
   (Probe tensions between AGI-optimized outcomes and stakeholder values like biodiversity, labor equity, etc.)
3. How might Zara structure trust and accountability mechanisms between human decision-makers and AI-generated insights in a hybrid boardroom?

(Encourage debate on validation, explainability, and the strategist's role in mediating AI advice.)

4. What new skills or capabilities does Zara need to lead effectively in this emerging post-human strategy environment?
   (Push reflection on AI literacy, ethics, planetary systems thinking, and future-oriented leadership.)

14.6.3. Interactive Exercise: Designing a Strategy for the Post-Human Boardroom

Scenario:
It's 2033. You've just been appointed Chief Strategic Foresight Officer at *Veritas HealthTech*, a fast-growing global biotechnology firm known for its ethical AI and regenerative medicine products. The company has a hybrid board: 9 human directors and 2 autonomous General AI (AGI) systems with full voting rights. The board is debating a new expansion initiative that would increase profits by 30%, but potentially conflict with new climate and biodiversity regulations.

Your Task:
You must guide the board through a structured, future-oriented strategic simulation and present your recommendation within 72 hours.

Using this chapter's tools and principles, prepare a 2-page strategy dossier that includes:

1. **Scenario Modeling**
   - Use one future simulation method (e.g., quantum-informed tree of decisions, synthetic data modeling, or digital twin of planetary impact) to simulate two divergent paths.
2. **Ethical Audit**
   - Apply interpretability and AGI oversight principles to evaluate how the AGI systems arrived at their conclusions. Consider unintended impacts and ethical blind spots.
3. **Strategic Recommendation**
   - Present your decision (support, oppose, revise). Justify using principles of planetary strategy, synthetic insight, and stakeholder legitimacy.
4. **Leadership Framework**
   - Propose a new decision-governance structure that balances human and machine agency for future high-impact decisions.

This exercise helps you operationalize strategic futures thinking in complex, ethically ambiguous, and machine-augmented environments.

14.6.4. Recommended HBR Case Studies

❖ **Google Quantum AI**

- **Overview:** Published on September 12, 2024, and authored by David B. Yoffie, Michael A. Cusumano, and Matt Higgins, this 25-page case explores Google's strategic positioning in the emerging field of quantum computing.

  The case provides a deep dive into:

  - The technological foundations of quantum computing
  - Google's focus on superconducting qubits as its core hardware approach
  - The competitive landscape, including alternative technologies like neutral atoms
  - Strategic questions around business models, pricing, and resource allocation

  It poses four key strategic questions:

  - What business model(s) should Google adopt to deliver quantum computing?
  - Should Google hedge its bets on superconducting qubits or explore alternative technologies?
  - How should quantum computing services be priced?
  - Could Google accelerate its progress with more resources?

This case is ideal for discussions on technology strategy, platform innovation, emerging technologies, and competitive advantage in deep tech.

- **Citation:** Yoffie, D. B., Cusumano, M. A., & Higgins, M. (2024). *Google Quantum AI* (Case No. 725362-PDF-ENG). Harvard Business School.

❖ **Predicting the Future Impacts of AI: McLuhan's Tetrad Framework**

- **Overview:** Published in 2025, this case uses Marshall McLuhan's Tetrad of Media Effects as a conceptual framework to help students and professionals analyze the societal and organizational impacts of artificial intelligence.
  - The tetrad framework asks four key questions:
    - What does the technology enhance?
    - What does it make obsolete?
    - What does it retrieve from the past?
    - What does it reverse into when pushed to its limits?

In this case, students apply the tetrad to AI technologies, particularly in the context of algorithmic control in digital labor platforms. The exercise encourages critical thinking about how AI reshapes work, power dynamics, and human agency—making it ideal for courses in technology ethics, digital transformation, media studies, and organizational behavior.

- **Citation:** Rahrovani, Y.. (2025). *Predicting the Future Impacts of AI: McLuhan's Tetrad Framework* (Case No. W42595-PDF-ENG), Harvard Business Publishing.

❖ **Balancing Impact: Modeling the Future at British International Investment**

- **Overview:** Authored by Shawn Cole and Jonah Zahnd and published on January 21, 2025, this case is a supplement to case 225-047 and focuses on British International Investment (BII), the UK's development finance institution, as it seeks to model and measure the long-term impact of its investments. The case explores how BII balances its dual mandate of achieving financial returns while delivering positive social and environmental outcomes in emerging markets. It introduces students to the challenges of impact modeling, including:
    1. Forecasting long-term development outcomes
    2. Quantifying impact across diverse sectors and geographies
    3. Integrating impact metrics into investment decision-making

  This case is ideal for courses on impact investing, development finance, sustainability strategy, and financial modeling for social outcomes.

- **Citation:** Cole, S., & Zahnd, J. (2025). *Balancing Impact: Modeling the Future at British International Investment* (Case No. 225076-PDF-ENG). Harvard Business School.

14.6.5. Strategy-in-Film: Learn by Watching

The future of strategy will be shaped by forces that are still emerging, from advanced technologies to shifting global systems. The films below offer glimpses into future-facing challenges, disruptive innovations, and the evolving role of leadership in uncertain environments. As you explore strategic foresight in this chapter, use these narratives to reflect on how organizations anticipate change, prepare for multiple futures, and build resilience in an unpredictable world.

Movie / Documentary	Strategic Futures Theme Highlighted
Transcendence	Post-human intelligence, ethical dilemmas in human-AI evolution
The Creator	Governance of AI in global conflict zones, strategic simulations of risk and autonomy
The Social Dilemma	Strategic consequences of algorithmic decisions and loss of human oversight
Ex Machina	Emergent consciousness and moral ambiguity in AGI decisions
Her	Human-AI relational dynamics, synthetic environments, and emotion-aware strategy
Oppenheimer	Strategic ethics under technological acceleration and irreversible consequences
Don't Look Up	Strategic inaction in the face of planetary crisis and system-wide denial
The Imitation Game	Parallel logic, code-breaking, and the roots of AI-enabled strategy in existential situations

References

AbuGhanem, M. (2025). IBM quantum computers: Evolution, performance, and future directions. *The Journal of Supercomputing*, *81*(5), 687.

Batool, A., Zowghi, D., & Bano, M. (2025). AI governance: a systematic literature review. *AI and Ethics*, 1-15.

Bhandari, K., Xin, Y., & Ojanen, V. (2021, December). Understanding human-machine collaborative systems in industrial decision-making. In *2021 IEEE International Conference on Industrial Engineering and Engineering Management (IEEM)* (pp. 1402-1406). IEEE.

Borges, A. F., Laurindo, F. J., Spínola, M. M., Gonçalves, R. F., & Mattos, C. A. (2021). The strategic use of artificial intelligence in the digital era: Systematic literature review and future research directions. *International journal of information management*, *57*, 102225.

Bova, F., Goldfarb, A., & Melko, R. G. (2023). Quantum economic advantage. *Management science*, *69*(2), 1116-1126.

Burrus, D. (2017). *The anticipatory organization: Turn disruption and change into opportunity and advantage*. Greenleaf Book Group.

Camilleri, M. A. (2024). Artificial intelligence governance: Ethical considerations and implications for social responsibility. *Expert systems*, *41*(7), e13406.

Coccia, M., Roshani, S., & Mosleh, M. (2022). Evolution of quantum computing: Theoretical and innovation management implications for emerging quantum industry. *IEEE Transactions on Engineering Management*, *71*, 2270-2280.

Demneh, M. T., Zackery, A., & Nouraei, A. (2023). Using corporate foresight to enhance strategic management practices. *European Journal of Futures Research*, *11*(1), 5.

Duhaime, I. M., Hitt, M. A., & Lyles, M. A. (Eds.). (2021). Strategic management: State of the field and its future. Oxford Academic, DOI: 10.1093/oso/9780190090883.001.0001

Elgin, C. (2025). Anticipatory macroeconomic governance: exploring future-oriented strategies for economic resilience and sustainability. *European Journal of Futures Research*, *13*(1), 4.

Haesevoets, T., De Cremer, D., Dierckx, K., & Van Hiel, A. (2021). Human-machine collaboration in managerial decision making. *Computers in Human Behavior*, *119*, 106730.

Helbing, D. (Ed.). (2018). *Towards digital enlightenment: Essays on the dark and light sides of the digital revolution*. Springer.

Henry, A. (2021). *Understanding strategic management*. Oxford University Press.

Heracleous, L. (2013). Quantum strategy at Apple Inc. *Organizational Dynamics*, *42*(2), 92-99.

Keller, J., & von der Gracht, H. A. (2014). The influence of information and communication technology (ICT) on future foresight processes—Results from a Delphi survey. *Technological Forecasting and Social Change*, *85*, 81-92.

Maccoby, M. (2015). *Strategic intelligence: Conceptual tools for leading change*. OUP UK.

Madanchian, M., & Taherdoost, H. (2025). Ethical theories, governance models, and strategic frameworks for responsible AI adoption and organizational success. *Frontiers in Artificial Intelligence*, *8*, 1619029.

Mohebbi, G., Ezadikhah, J., & Esmaili, M. (2024). Simulation of Futures Scenarios in Strategic Management with Quantum Intelligence. *Journal of Intelligent Strategic Management*, *3*(4), 7-24.

Mujtaba, B. G. (2025). Human-AI Intersection: Understanding the Ethical Challenges, Opportunities, and Governance Protocols for a Changing Data-Driven Digital World. *Business Ethics and Leadership*, *9*(1), 109-126.

Nikolenko, S. I. (2021). *Synthetic data for deep learning* (Vol. 174). Cham: Springer.

Obrenovic, B., Gu, X., Wang, G., Godinic, D., & Jakhongirov, I. (2024). Generative AI and human–robot interaction: implications and future agenda for business, society, and ethics. *AI & Society*, 1-14.

Popescu, C. R. G., & Yu, P. (Eds.). (2024). *Intersecting Environmental Social Governance and AI for Business Sustainability*. IGI Global.

Preskill, J. (2018). Quantum computing in the NISQ era and beyond. *Quantum*, *2*, 79.

Semeraro, F., Griffiths, A., & Cangelosi, A. (2023). Human–robot collaboration and machine learning: A systematic review of recent research. *Robotics and Computer-Integrated Manufacturing*, *79*, 102432.

Smith, S., & Cox-Smith, S. (2023). *Future cultures: how to build a future-ready organization through leadership*. Kogan Page Publishers.

Raghunathan, T. E. (2021). Synthetic data. *Annual Review of Statistics and Its Application*, *8*(1), 129-140.

Rêgo, B. S., Jayantilal, S., Ferreira, J. J., & Carayannis, E. G. (2021). Digital transformation and strategic management: A systematic review of the literature. *Journal of the Knowledge Economy*, 1-28.

Ren, M., Chen, N., & Qiu, H. (2023). Human-machine collaborative decision-making: An evolutionary roadmap based on cognitive intelligence. *International Journal of Social Robotics*, *15*(7), 1101-1114.

Tani, M., Muto, V., Basile, G., & Nevi, G. (2025). A bibliometric analysis to study the evolution of artificial intelligence in business ethics. *Business Ethics, the Environment & Responsibility*.

Tran-Dang, H., Kim, J. W., Lee, J. M., & Kim, D. S. (2025). Shaping the future of logistics: Data-driven technology approaches and strategic management. *IETE Technical Review*, *42*(1), 44-79.

Woods, R. B. (2023). *Disruptive technologies in a world of chaos and opportunity: AI, quantum computing, and other advances in technology*. In Regent Research Roundtables Proceedings (pp. 325–333). Regent University School of Business & Leadership.

Xu, M., David, J. M., & Kim, S. H. (2018). The fourth industrial revolution: Opportunities and challenges. *International Journal of Financial Research*, *9*(2), 90-95.

Made in the USA
Las Vegas, NV
05 November 2026

57552598R00321